AGRO-FOOD MARKETING

AGRO-FOOD MARKETING

Edited by

D.I. Padberg

Department of Agricultural Economics
Texas A&M University
USA

C. Ritson

Department of Agricultural Economics and Food Marketing
University of Newcastle upon Tyne
UK

and

L.M. Albisu

Department of Agricultural Economics
Agricultural Research Service – DGA
Zaragoza
Spain

CAB INTERNATIONAL

in association with the

International Centre for Advanced
Mediterranean Agronomic Studies (CIHEAM)

CABI is a trading name of CAB International

CABI Head Office
Nosworthy Way
Wallingford
Oxfordshire OX10 8DE
UK

CABI North American Office
875 Massachusetts Avenue
7th Floor
Cambridge, MA 02139
USA

Tel: +44 (0)1491 832111
Fax: +44 (0)1491 833508
Email: cabi@cabi.org
Web site: www.cabi.org

Tel: +1 617 395 4056
Fax: +1 617 354 6875
Email: cabi-nao@cabi.org

A catalogue record for this book is available from the British Library, London, UK

Published in association with
Centro Internacional de Altos Estudios Agronomicos Mediterraneos
Instituto Agronomico Mediterraneo de Zaragoza
Apartado 202
50080 Zaragoza
Spain

Library of Congress Cataloging-in-Publication data
Agro-food marketing / edited by D.I. Padberg ... [et al.].
     p.   cm.
   Includes bibliographical references and index.
   ISBN 0-85199-144-0 (alk. paper)
    1. Farm produce-Marketing. 2. Food-Marketing. 3. Produce trade.
  4. Food industry and trade. I. Padberg, Daniel I.
  HD9000.5.A374 1997
  635'.068'8-dc20   96-27557

ISBN-13: 978-0-85199-144-3
ISBN-10: 0-85199-144-0

First published 1997
Reprinted 2002
Transferred to print on demand 2006

Printed and bound in the UK by CPI Antony Rowe, Eastbourne.

Contents

PART IV: RESEARCHING THE FOOD CONSUMER

Contributors

Heinz Ahrens, *Professur für Agrarpolitik und Agrarumweltpolitik, Landwirtschaftliche Fakultät, Martin-Luther-Universität Halle-Wittenberg, Emil-Abderhalden Str. 20, 06108 Halle (Saale), Germany*

Luis Miguel Albisu, *Department of Agricultural Economics, Agricultural Research Service – DGA, PO Box 727, 50080 Zaragoza, Spain*

Marianne Altmann, *67, rue de la Gare, 3377 Leudelange, Luxembourg*

Reimar von Alvensleben, *Lehrstuhl für Agrarmarketing, Institut für Agrarökonomie, Universität Kiel, Olshausenstraße 40, 24098 Kiel, Germany*

Julián Briz, *Unidad de Comercialización y Divulgación Agraria, Departamento de Economía y Ciencias Sociales Agrarias, ETS Ingenieros Agrónomos, Universidad Politécnica de Madrid, Ciudad Universitaria, 28040 Madrid, Spain*

Michael Burton, *School of Economic Studies, University of Manchester, Manchester M13 9PL, UK*

Hoy F. Carman, *Department of Agricultural Economics, University of California, Davis, California 95616–8512, USA*

Paul L. Farris, *Department of Agricultural Economics, 1145 Krannert Building, Purdue University, West Lafayette, Indiana 47907–1145, USA*

Isabel de Felipe, *Unidad de Comercialización y Divulgación Agraria, Departamento de Economía y Ciencias Sociales Agrarias, ETS Ingenieros Agrónomos, Universidad Politécnica de Madrid, Ciudad Universitaria, 28040 Madrid, Spain*

Philip Garcia, *Department of Agricultural and Consumer Economics, University of Illinois at Urbana-Champaign, 305 Mumford Hall, MC-710, 1301 West Gregory Drive, Urbana, Illinois 61801, USA*

Leslie Gofton, *Department of Social Policy, Faculty of Law, Environment and Social Sciences, University of Newcastle upon Tyne, Newcastle upon Tyne NE1 7RU, UK*

Raymond M. Leuthold, *Department of Agricultural and Consumer Economics, University of Illinois at Urbana-Champaign, 305 Mumford Hall, MC-710, 1301 West Gregory Drive, Urbana, Illinois 61801, USA*

Oswin Maurer, *Department of Agricultural and Consumer Economics, University of Kiel, Olshausenstrasse 40, 24118 Kiel, Germany*

Matthew T.G. Meulenberg, *Department of Marketing and Marketing Research, Wageningen Agricultural University, Hollandseweg 1, 6706 KN Wageningen, The Netherlands*

Mitchell Ness, *Department of Agricultural Economics and Food Marketing, Faculty of Agriculture and Biological Sciences, University of Newcastle upon Tyne, Newcastle upon Tyne NE1 7RU, UK*

Daniel I. Padberg, *Department of Agricultural Economics, Room 308G Anthropology Building, Texas A&M University, College Station, Texas 77843–2124, USA*

George G. Panigyrakis, *Department of Business Administration, The Athens University of Economics and Business, 76 Patission Street, Athens 104 34, Greece*

Christopher Ritson, *Department of Agricultural Economics and Food Marketing, Faculty of Agriculture and Biological Sciences, University of Newcastle upon Tyne, Newcastle upon Tyne NE1 7RU, UK*

Gerhard Schiefer, *Department of Agricultural Economics, University of Bonn, Meckenheimer Allee 174, 53115 Bonn, Germany*

Ronald W. Ward, *Food and Resource Economics Department, Institute of Food and Agricultural Sciences, University of Florida, 1125 McCarty Hall, PO Box 110240, Gainesville, Florida 32611–0240, USA*

Trevor Young, *School of Economic Studies, University of Manchester, Manchester M13 9PL, UK*

Foreword

The International Centre for Advanced Mediterranean Agronomic Studies (CIHEAM), established in 1962, is an intergovernmental organization of 14 countries: Albania, Algeria, Egypt, France, Greece, Italy, Lebanon, Malta, Morocco, Portugal, Spain, Tunisia, Turkey and the former Yugoslavia.

Four institutes (Bari, Italy; Chania, Greece; Montpellier, France; and Zaragoza, Spain) provide postgraduate education at the Master of Science level. CIHEAM promotes research networks on Mediterranean agricultural priorities, supports the organization of specialized education in member countries, holds seminars and workshops bringing together technologists and scientists involved in Mediterranean agriculture and regularly produces diverse publications including the series *Options Méditerranéennes*. Through these activities, CIHEAM promotes North/South dialogue and international cooperation for agricultural development in the Mediterranean region.

An important part of CIHEAM activity is focused on the field of agricultural economics. Within this area, and for more than ten years, the Mediterranean Agronomic Institute of Zaragoza has been organizing a training and research promotion programme on agro-food marketing. This programme includes a one-year advanced course of specialization, two-week courses addressed to professionals, organization of seminars and preparation of collaborative research projects involving teams from different countries.

This book stemmed from the programme of the 'Advanced Course on Marketing of Agricultural and Food Products' and the experience gathered during its different editions. The course, lasting one academic year, has been running since 1987 for postgraduate participants with different professional experience and of many different nationalities. The relevance of the lecturers, coming from prestigious academic and research institutions and from agro-food firms of Europe and America, and their interaction with the participants have resulted in an original approach to agricultural marketing.

The above-mentioned circumstances encouraged us to produce a book addressed to professionals and postgraduate students that would deal comprehensively with agricultural and food product marketing within an international scope. The scientific editors have done excellent work in planning and integrating the book content in order to achieve this goal. Both their efforts and the mutual collaboration with the authors of the different chapters have made this venture possible.

Because of its origin, objectives and preparation, this book contributes to the CIHEAM goals of cooperation and development. Besides, we are especially pleased because of the current importance of the topic dealt with. The improvement in efficiency and modernization of agro-food marketing processes is a key factor for the agricultural and economic development of any country, and thus, we believe the publication of this book is highly appropriate.

We would like to express our gratitude to everyone who has, directly or indirectly, contributed towards this endeavour, particularly the authors of the chapters for their valuable contributions and, very specially, to Prof. Ritson, Prof. Padberg and Dr Albisu, the scientific editors, who have made this book possible not only through their knowledge but also through their enthusiasm and dedication.

We would also like to express our satisfaction with the efficient collaboration initiated with CAB INTERNATIONAL whom we thank for their interest and attention to this project.

Miguel Valls
Director
Mediterranean Agronomic Institute of Zaragoza, Spain

Preface

Since 1987, the International Centre for Advanced Mediterranean Agronomic Studies has offered, through its Mediterranean Agronomic Institute of Zaragoza (IAMZ) (Spain), an 'Advanced Course on Marketing of Agricultural and Food Products' to postgraduate participants coming from many different countries. The course has been taught by well-known specialists from universities, research institutions and companies in Europe and North America. The international aspects of agricultural and food marketing have automatically been highlighted because of the diversity of the backgrounds of students and lecturers but, in addition, the international dimension has been purposefully emphasized.

Drawing on this experience, it was decided to develop a book based on the agro-food marketing course. The philosophy behind this project has been the integration of food product marketing topics with traditional agricultural commodity marketing. This approach fills a gap in the literature which is not covered elsewhere. The book is directed to postgraduate students and professionals.

The book is structured into six parts. Part I provides a background and context. Part II covers some fundamental building bricks for the analysis of agricultural commodity markets, and Part III extends this into various structural and organizational aspects. Parts IV and V are concerned with the application of business marketing principles to food products, and Part VI provides some strategic marketing extensions, particularly in an international context. The relationship between the content of specific chapters in the book, and the subjects of agricultural marketing and food marketing, is explored in more detail in Chapter 2.

In order to create a coherent textbook, the editors have taken a very proactive approach, not only to the structure of the book, but also to the content of the individual chapters. The editors would like to express their deep appreciation to the chapter authors for their contributions. Their perseverance through several drafts shows a serious commitment and willingness to share their special expertise. We have learnt much from this mutual effort.

With authors coming from such a variety of countries and academic backgrounds, there remains a considerable degree of variation in style and perspective with respect to the individual chapters, which the editors believe enriches the contribution of the book as a text directed at an international market. However, we have attempted, where possible, to use a consistent vocabulary. One important aspect of this has been to use the word *commodity* to refer to the largely undifferentiated outputs of farms; and *food products* to refer to manufactured consumer products. This has been to emphasize the distinction between the traditional subject matter of agricultural commodity marketing, and the more recently developed application of business marketing principles to food products. In some cases, however, the contexts in which particular terms are used lead to somewhat different meanings, particularly with respect to their use by North American and European authors respectively. This applies, for example, to the term *food chain* which can be used to refer to either vertical or horizontal relationships; and also to the term *strategic*. Clearly, we have not finished the task of integrating the many different parts of agro-food marketing. We hope, however, that this effort will come to be recognized as a substantial beginning.

D.I. Padberg
C. Ritson
L.M. Albisu

1

The Global Context of Agro-food Marketing

Daniel I. Padberg

Department of Agricultural Economics, Room 308G Anthropology Building, Texas A&M University, College Station, Texas 77843–2124, USA

INTRODUCTION

Agro-food marketing refers to buying and selling; the economic incentive structure; and the goods handling system for food, from the point of production through processing and distribution to the final sales to consumers. This economic sector is not always treated separately from other sectors. However, there are some unique economic and policy features of the agro-food sector which receive special attention by governments and business and therefore require special expertise. These special features include: (i) all people require a continuous supply of food; (ii) most food products and commodities are perishable; (iii) if improperly handled, food can cause human diseases; (iv) the food sector has a specialized and 'big business' distributor; and (v) in high-income countries, the food sector has a high level of product differentiation in relation to other consumer products industries.

Frequently, agro-food marketing has been a fundamental threshold in economic development. Developing countries must make major changes (involving significant improvements in the productivity of resources) in agro-food marketing to break away from a dependence upon low-income subsistence agriculture. The scope of agro-food marketing is greatly increased as economic development and consumer income rise. In developing countries, marketing activities may represent only 10–15% of the value of food to the consumer. In higher income countries, food marketing value added frequently accounts for 75% of the consumer's food expenditure.

Throughout the first three-quarters of this century, the food sectors of different economies were usually separate and autonomous: linkages were at the level of agricultural commodities. Major changes in this isolation have occurred since 1975. A much larger flow of international trade in food commodities has developed. In addition, there is a considerable industrial consolidation among the large manufacturing and distribution firms. The

© CAB INTERNATIONAL 1997. *Agro-food Marketing*
(eds D.I. Padberg, C. Ritson and L.M. Albisu)

flow of trade in manufactured food products is significant in relation to the flow of commodities. In addition, there is a substantial international transfer of product brands, processing technology and food marketing expertise. Both food manufacturing and distribution are becoming integrated across high-income economies.

Components of agro-food marketing

Agricultural commodities refer to unprocessed or minimally processed agricultural materials which are handled and marketed in bulk quantities. Grains are a good example, but there are many other agricultural commodities. Bulk milk, hay, onions, eggs, cotton, etc. are agricultural commodities. The orderly marketing of these commodities requires some public rules, such as grades and standards, food safety policies, market information, futures markets, etc. Experience throughout the last century gives a basis for gathering knowledge and expertise to guide the public rule setting and the functions of private business active in the food industries.

In developing countries, most of the economic activity in the food sector relates to transactions involving commodities. Typically, there is only the most modest structure of private firms and organizations. The commodities are frequently sold to households for processing in the home. Here there exists a minimum of rules concerning standards, or information. These local transactions are done by inspecting the commodity at each time of purchase. Where interregional movement of commodities is important, the network of rules, including standards, quality definitions, standard measures and handling methods, must be more developed. This burden is accomplished by public rules and by quality control measures within the larger firms dealing in the interregional product trade.

In more economically developed economies, only a small per cent of the population will be involved in agriculture. This allows agricultural production to undergo often extreme geographic specialization. Interregional commodity movement is necessarily much greater. To this point, the handling of agricultural commodities is important in both developing and more developed economies. In addition, another pattern of marketing is important mostly in high-income economies. This is that consumer food products are usually manufactured and then presented to the households. These products are much more prepared foods. A loaf of bread, breakfast cereals, prepared meals or entrées, canned or frozen vegetables or fruits are examples of food products.

The marketing of food products is very different from marketing of agricultural commodities. Vast private firms emerge to handle manufactured food products while agricultural commodities are frequently handled in an environment of primitive institutional arrangements. Science and technology are featured in the competition to introduce new food products. Product quality control programmes of large firms replace much of the network of public definitions. The nature of public regulation changes from deterministic requirements to disclosure requirements. Expertise for public rule setting and for the guidance of private business management draws little from the experience with agricultural commodities. It is a special pattern of economic activity conducted by a globally interactive set of many very large firms.

With global integration of the markets for food products and the more trade-oriented pattern of agricultural commodity flows, agro-food marketing takes on a more international character. The larger firms in particular deal in both agricultural commodities and food products. While it not usually attempted, it is appropriate to develop an integration of the separate disciplines involving agricultural commodities and food products. That is one of the major purposes of this book.

GLOBAL PATTERN OF FOOD MARKETING AND TRADE

The gains from trade are frequently studied and well understood. Essentially, trade allows each trading partner to use their resources in the most productive arrangement. The extent of potential gains from agricultural commodity trade are large and the certainty of obtaining them is not a matter of serious debate. At mid-century, only a very small amount (estimated at 2%) of the world's food supply crossed international boundaries. The 1970s saw an explosion of international agricultural commodity trade, bringing the portion of today's food supply crossing international boundaries to an estimated 15%.

Trade in agricultural commodities

Down through history, the pattern of agricultural commodity trade has been restricted by several factors. Agricultural commodities are bulky, perishable and expensive to transport. Low-income countries have much of the population in subsistence agriculture producing little exportable output. There is a great tendency to adopt national policies to protect the low-income rural sector. National policies are frequently motivated by concerns for food security or self-sufficiency. These factors led to the erection and maintenance of significant barriers to agricultural commodity trade.

Major changes in world politics are bringing changes to this historic impasse. The development of the European Union has opened borders of several high-income countries to agricultural commodity trade. The North American Free Trade Area (NAFTA) agreement is expected to have a similar effect. The Uruguay round of negotiation of the 130-nation General Agreement on Tariffs and Trade (GATT) included agricultural commodities in a major way for the first time. Many new independent nations have emerged following the decline of the British Empire and Soviet Union. In the new arrangements, almost all nations are more trade oriented.

Trade in manufactured food products

Trade in manufactured consumer food products has also increased significantly since the 1970s. Sales of manufactured consumer food products across international boundaries usually lead to the establishment of manufacturing facilities in the destination country or continent. Sometimes, the manufacturer builds subsidiary facilities but, most often, brands and product designs are licensed or franchised in a wide array of joint venture arrangements. These joint ventures seem to be an alternative to trade, but they may have the effect of increasing the flow of food across international boundaries. This will be likely to continue in the future with the tendency for new independent nations to emerge with the ethnic disintegration of traditional nations.

Trade in manufactured consumer food products is quite different from the classic agricultural commodity trade. Agricultural commodity trade statistics do not capture trade in manufactured consumer products, although the value recently has come to exceed that of commodity trade. While trade in agricultural commodities displays a responsiveness to comparative advantage in production, trade in manufactured food products seems to be less orderly. Intra-industry trade is a term relating to observed trade of the same product which may go in both directions. The same country that is exporting a famous brand of beer is also importing another one. We understand commodity trade as a device for rationalizing the effective use of a trader's basic productive resources. It may be that we will come to understand trade in manufactured food products as catering to the affluent consumer's need for variety and stimulation, just as these products are positioned to do in domestic trade.

THE MARKETING AND DISTRIBUTION SYSTEM

The food marketing and distribution system has perhaps the most developed industrial complex among consumer goods. It has a very high continuous volume. It is always established very early in the development of an economy so that it has more maturity and diversity. As a result, specialized institutions fill niches where their counterpart would not exist in other industries. In addition, there is great diversity in the structure of the distribution channel according to the level of consumer income.

Agricultural commodity distribution

Where low consumer income and subsistence agricultural conditions prevail, the marketing channel is short and direct. Households produce a significant part of their food and what is purchased comes from community shops or pushcarts, usually emphasizing local production. Low-income households process and prepare the family's food and typically have little access to a wide variety of food products or products coming from a distance. The logistics system for supplying the few manufactured products in distribution (salt, sugar, oil and other 'recipe items') is typically poorly developed and inefficient. These marketing channels attract little investment in modern facilities or business practices. Firms are usually very competitive household enterprises with relatively low volume and profits. This pattern of economic activity is the channel of food distribution for more than half of the human race.

Consumer food products

At the other end of the consumer income spectrum are the post-industrial economies. Here, as little as 1% of the population is involved in farm production, which is concentrated in the locations of highest resource productivity, frequently a great distance from the consuming population. Elaborate and efficient logistics systems evolve to handle commodity marketing including mass transportation, commodity futures markets, grades and standards, market information, etc. A great deal of food preparation is handled within the food marketing and distribution channel. Consumers place value on food product variety, convenience and 'status'. In response to this market emphasis on variety, convenience and status, the distribution channel has evolved to provide enormous product variation and experimentation. In this arrangement, food marketing activities can account for 75% of the consumer value of food. This ratio reflects not only added marketing activity but increases in farm production efficiency.

It is not surprising that the major food processors have become massive marketing firms. They may have a history of commodity handling, such as milling or slaughtering, but more recently they have acquired a focus on new product development and introduction. In the commodity handling activities, the extent of scale economies was frequently studied and generally well understood. Economies of scale or scope in relation to new product development and introduction are less well understood. There is evidence, however, that economies of large size are substantial and that they continue to massive size. The manufacturing and sale of branded food products is usually the province of multinational conglomerates which are large, not only in total, but in several countries and in many product lines.

Food retailing and wholesaling

Food distribution provides a connection between the food manufacturer level and the final consumer. Distribution of food is somewhat more specialized than distribution of most consumer goods. Many consumer goods are distributed in large units which last a long time, such as cars, televisions and microwave ovens, usually called consumer durables. By contrast, food is distributed in units having an average value of only a few dollars, but distribution must be repeated at frequent intervals. For these reasons, food has a specialized institution for distribution which must be located conveniently and must accommodate the wide variety of experimental products. In the USA, that means a supermarket. In many other high-income countries, a dense settlement of population was in place before the birth of modern food distribution, which precludes the allocation of space for these large stores. In this setting, food distribution facilities are provided by a combination of small and medium sized food stores within metropolitan areas and (less specialized) hypermarkets at a city's edge.

Efficiency in goods handling within the distribution level of the market channel requires sophisticated pre-retailing organization. It is typical for food stores to be organized as chains or 'multiples' served by an integrated wholesale operation. In addition to providing delivery of goods from the wholesale level, this structure allows the centralization of accounting and other managerial functions. The multinational nature of food distribution firms is less well developed than is the case in food manufacturing, but it has increased significantly during the past decade. Scale economies apply to some retail functions more than others. In general, scale economies seem to be less of a driving force in distribution than in manufacturing.

A special phenomenon of the large food distribution firm is the development and offering of 'private label' food products. These are products bearing the brand of the retail distributor. In the early stages of industry development, private label products may be competitively positioned at any quality level. As the multinational food manufacturer has come into the picture, they dominate the 'high-quality–higher-price' activity. In response, the retail brands tend more toward a 'standard quality–economy price' image. The extensive and superior logistics capability of the large distribution firms contributes to an economy image better than to competing in variety, convenience and status development. Typically, private label products are packed by smaller processors, unable to compete in the vast advertised new product business. Distributors are able to place their economy products (usually discounted 25% or more) beside the more expensive advertised brands to emphasize the price comparison.

This description of the food marketing channel is more fully developed for each end of the consumer income continuum. In between, there are many different patterns. Rather than an orderly transition, food industry development is affected by history and tradition as well as economics and marketing. A pattern of vertical coordination is frequently seen in agro-food industries. While vertical integration is seen in other industries as well, there seem to be special advantages flowing from the vertical integration of some food industries.

Vertical coordination

In addition to the large multinational manufacturers and distributors, it is not uncommon to find large vertical organizations. Poultry is produced in this type of organization. A half century ago, it would have been normal to find separate industries performing hatchery activities, poultry feed manufacturing, growing of broilers, egg production, veterinary and pharmaceutical services, slaughtering, and rendering of slaughtering wastes. Today, all of these activities are integrated into a single firm. Most of these components are within assets owned by the integrator. A major one, however, growing broilers, is largely done by small and separate firms connected to the larger system through contracts. In addition to these activities, the modern firm develops markets in other countries for chicken parts not having a good market at home. Pigs seem to offer a potential for the same type of organization. These large organizational patterns seem to be driven by economies of integration. It is likely that significant quality management, risk reduction and operating cost reduction are achieved by this organizational arrangement.

FOOD MARKETING PROBLEMS

Several problems challenge the food marketing industries. The way problems are dealt with is changing with increases in the level of international interactiveness within several parts of the food industry.

Efficiency

In low-income countries, efficiency is one of the greatest concerns. Food marketing institutions struggle with low-tech methods and equipment. The industry is often overcrowded earning low returns on investments. The presence of cheap and often redundant labour seems to discourage serious investment in new technology. Many of the items handled in the small firm structure are perishable and risky. Refrigeration and other capital-intensive equipment is scarce. Handling costs seem too high and waste seems too great.

Orderly commodity markets

Many new emerging nations and former Communist nations with a desire to move their food marketing system closer to free market incentives seek to set up orderly commodity markets. It is easy to underestimate the difficulty of a transition to a new set of public rules and private incentives. Incentives are related to some operational aspects of each culture. The set of public rules which provide constructive and functional incentives in one culture may not be the most effective set for another culture.

Market performance

The presence of very large firms in the food industry in daily contact with consumers and farm producers leads to questions of monopoly or excessive market power. While large firms naturally evoke analogy to the classic monopoly, the presence of competitive features almost unique to the food industry substantially alters the competitive balance of power. For example, the tendency for large distributors to offer price-competitive products in competition with the products of large manufacturers mitigates the market power that would otherwise accrue to manufacturers. In addition, the determined focus of 'large firm' food manufacturing competition upon new product development and introduction reduces the likelihood of oligopolistic collusion. With the complex structure in the food industry, the assessment of market performance is more difficult and multidimensional.

Environmental problems

Several aspects of the modern food system have the potential for disturbing environmental ecosystems. Large organizations, including the large vertical organizations, tend to concentrate the production of food products into more geographically specialized areas. This regional concentration also concentrates animal waste and the flow of waste products from food processing. This may have a greater tendency to inject residues from production or processing into the ground water or surface water of the area. In addition, a greater emphasis on selling ready-to-eat foods creates a need for 'preservatives' which did not exist in earlier times. The ingredient commodities for the bakery industry are quite stable in the marketing channel. When these products are processed into pastries, they are very perishable – inviting the use of a mould suppressant, for example. It is argued that the publicly determined grades and standards for fresh fruit and vegetables put emphasis on blemish-free fruit and, therefore, provide an incentive to use more pesticide than is necessary for the effective production of wholesome food.

The transition from commodities to products

As income rises around the world, patterns within households change to purchasing more developed and specialized food products rather than the commodities more directly from farm production. This transition encourages the development of large marketing firms and reduces the emphasis on public commodity markets. Policy towards food products changes as well. Consumer food products have evolving definitions where commodities' definitions or standards are more stable. This complicates food marketing, price reporting and especially consumer information. While changes in food products may be responsive to consumers' preferences, they may not always be improvements in either nutrition or food safety. Concerns about nutrition and food safety seem to grow with consumer income and product variety. Governments have become heavily involved with legislation relating to food additives, labelling, etc.

FOOD MARKETING INDUSTRY ANALYSIS

Agro-food marketing as it functions around the world has many dimensions. The analysis of agricultural *commodity* production, distribution and trade is a direct application of classic economic theory. Agricultural commodity marketing is especially important in low-income countries, but is important in all countries. Food *product* marketing is much more informal, drawing on many disciplines, because it departs significantly from classic economic models. The high rate of evolution of products is a major food industry departure from classic microtheory. The presence of a 'big business' or oligopoly distribution sector is a factor essentially undeveloped in economic theory. Food industry questions and the answers are often different from those in classic economic analysis.

In this situation, agro-food marketing analysis has rarely developed a successful integration across the whole industry. Some analyses concentrate on commodities and deal with economic efficiency and trade. Others deal with the emerging large firms in manufacturing and distribution, concentrating on product progress, distribution logistics, monopolizing questions and/or strategic management. In a global context, it may seem that the large firm dimensions of the food marketing system are not so important because they relate only to a small (high consumer income) portion of world population. At the same time, it is within these industry structures that new products and methods are developed and introduced. It is very important for higher-income countries, but it sets a sense of direction for developing countries as well. Thus, one purpose of this book is to integrate these components and to seek to understand synergism in developing a more complete and comprehensive analysis.

2 Food Marketing and Agricultural Marketing: The Scope of the Subject of Agro-food Marketing

Christopher Ritson

Department of Agricultural Economics and Food Marketing, Faculty of Agriculture and Biological Sciences, University of Newcastle upon Tyne, Newcastle upon Tyne NE1 7RU, UK

INTRODUCTION

On the face of it, one would expect 'agro-food marketing' to be a specialization of a subject called 'marketing' – applying marketing theory and practice to agricultural commodities and food products – comparable to, for example, services marketing, transport marketing, industrial marketing, or whatever. One would further expect such a specialization to follow the development of the core subject – to grow out of it, not to precede it. The truth is somewhat different.

Writing recently in the journal of the American Marketing Association on 'The changing role of marketing in the corporation', Webster (1992) comments:

> It is sobering to recall that the study of marketing did not always have a managerial focus. The early roots of marketing as an area of academic study can be found, beginning around 1910, in midwestern American land-grant universities, where a strong involvement with the farm sector created a concern for agricultural markets and the processes by which products were brought to market and prices determined.

Similarly, Jones and Monieson (1990), writing on the early development of marketing philosophy, refer to a number of articles and books published in the early 1920s on various aspects of the marketing of farm products. Thus the subject of agricultural marketing has a much longer pedigree than mainstream business marketing. It developed during the first half of the twentieth century, as a branch of applied agricultural economics, concerned with understanding the behaviour and functions of agricultural commodity markets, and of government policies to control the marketing of agricultural commodities – as epitomized, for example, by 'marketing orders' in the USA, and 'marketing boards' in the UK.

In contrast, 'marketing' only took off in the second half of the century, as part of business management education in business schools, first in North America and subsequently in Europe, and 'respectability' as an academic subject took some time to arrive. Indeed, the Chair of Agricultural Marketing at the University of Newcastle, established in 1963, is reputed to be the first in Europe with the word 'marketing' in its title and, at that time, agricultural marketing seemed to have little in common with the subject of marketing (or 'marketing management') then developing in university business schools.

The point was most cogently made by Bateman (1976) in his classic review article on agricultural marketing for the *Journal of Agricultural Economics*:

> Agricultural marketing policy has been the traditional subject matter of agricultural marketing. … The subject 'agricultural marketing' has a longer history as a recognized area of study … than does 'marketing'. … Marketing has become an academic subject in Britain only since about 1960. Its points of contact with agricultural marketing … are not conspicuous. … Marketing has developed with a business orientation, agricultural marketing with a policy one, and this accounts for the fact that the two approaches sometimes appear to have the same language but to be unable to communicate.[1]

It is only over the past 15 or so years that a degree of fusion between marketing and agricultural marketing has developed. This book can be seen as a product of that fusion. The present chapter sets out to introduce the subjects of marketing and agricultural marketing; to explain why they developed with different orientations and why increasingly 'points of contact' have emerged. The result might reasonably be described as 'agro-food marketing'. Reference is made throughout to subsequent chapters, placing them in the overall context of the scope of the subject, as conceived by the editors of this book.

AGRICULTURAL MARKETING

In what is perhaps the best-known textbook on the subject, Kohls and Uhl (1990) define agricultural marketing as:

> The performance of all business activities involved in the flow of food products and services from the point of initial agricultural production until they are in the hands of consumers.

Thus, in agricultural marketing, the word 'marketing' is usually used descriptively, to refer to a part of the economy – what one might call the agricultural (or food) marketing sector. For much of the history of mankind, this was relatively insignificant, with most of what was eaten produced in the locality: arguably, storage was the main marketing activity. However, since the industrial revolution there has been a progressive increase in the distance between the typical location of agricultural production and that of food consumption, and in the value added to produce as it moves from farmer to consumer. Eventually this process results in the contemporary 'globalization' outlined in the previous chapter; but earlier, the growth in the importance of the agricultural marketing sector, relative to that of farm production, brought increasing concern that many of the problems confronting farmers originate in the marketing sector. Agricultural marketing therefore developed out of a need to balance the previous concentration of agriculture faculties on farm production economics and management with corresponding work relating to post-farm-gate activity. Indeed it was precisely this view which led the British Ministry of Agriculture to finance the Chair (and Department) of Agricultural Marketing at the University of Newcastle:

> Experience has convinced both the Ministry and King's College (Newcastle University) that there is an extreme shortage of economists adequately trained to study distribution economics, and that the changes in the agricultural policy of the UK that may be expected will make it desirable to build up studies in marketing to the point at which they at least equal the national effort in production economics. ... The study of agricultural marketing is concerned with the efficiency of the use of resources in processing, handling and distributing food, fibres and other agricultural products. It applies economic principles to the problems of identifying and satisfying the needs and preferences of consumers, by the most effective use of markets, processing plants, transport, advertising and retail outlets .
>
> (Bosanquet, 1965)

The view that the subject of agricultural marketing developed both as a consequence of the increasing significance of the food marketing sector, and the view that many of the problems confronting farmers originated in that sector, provides a convenient framework for summarizing the subject, and a context for the chapters in Parts II and III of this book. Figure 2.1 provides a categorization based on the identification of perceived problems in agricultural marketing, the analysis of these problems and the potential policy solutions.

Briefly, three kinds of problems are identified. These are, first, the view that growing concentration in food manufacturing and distribution allows the marketing sector to exploit market power to the detriment of the farm sector and perhaps also consumers. This power might be expressed in the form of excess profits or efficiency losses due to the lack of competitive pressure, but in either case would be viewed in agricultural marketing as 'excessive margins'. But in addition to the impact of market power, margins might be excessive, so it is often believed, because of inefficiency in the structure and organization of the marketing sector. Third, much attention has been devoted to the efficiency of the agricultural market price mechanism in communicating information between farmer and consumer – in particular, the problems of price instability obscuring useful price messages between consumers and producers, and price cycles delivering false messages to expand and contract production. In addition, there is the question of whether price formation reflects efficient relations between markets over space and time, and the contribution of futures markets to pricing efficiency.

Agricultural market analysis is also divided into three categories: (i) the

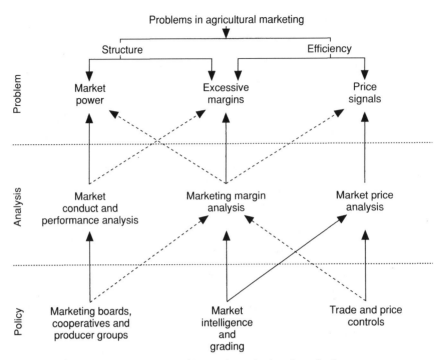

Fig. 2.1. Schematic categorization of issues in agricultural marketing.

application of structure/conduct/performance analysis (and the idea of 'effective competition'); (ii) the analysis of marketing margins; and (iii) the analysis of supply and demand relations and the explanation for price movements over time and space. Finally, a series of marketing policy initiatives are listed: (i) price controls (for example, intervention buying, margin controls, import taxes, etc.); (ii) the formation of producer marketing groups and boards, partly to exercise 'countervailing power'; and (iii) various initiatives aimed at improving marketing efficiency, such as market intelligence agencies and the establishment of quality standards and grades.

Although Fig. 2.1 lists the problems, analysis and policies in respective groups of three, there is not a single, and separable, relationship in which each set of policy solutions flows out of a particular category of problem and analysis – and the various arrows attempt to indicate the main interrelationships. Similarly, although the diagram embraces the content of Part II and Part III of this book, the chapters do not always fit neatly into single components of the diagram. It is, nevertheless, helpful to relate the chapters which follow to this framework.

Chapter 3 provides the fundamental underlying theory of supply and demand response, underpinning the agricultural price analysis and discussion of government commodity programmes in Chapter 4. Chapter 5 extends the price analysis into the international arena, and Chapter 6 introduces the techniques for modelling as an aid to price and structure analysis. Chapter 7 covers various aspects of market structure and performance analysis, and

Chapter 8 focuses on margins. Chapter 9 covers modern systems for improving market intelligence.

To conclude this section, then, the subject of agricultural marketing developed as the study of the economic structure and efficiency of the agricultural marketing sector, and the government's role in intervening to improve the performance of agricultural markets and increasing the share of expenditure on food received by farming. McLeay and Zwart (1993) analysed the content of six agricultural marketing textbooks. In decreasing order of importance, the subject areas addressed were government policy, pricing behaviour and analysis, futures and hedging, voluntary cooperatives, competition, commodity markets, grading, demand and supply analysis, marketing information, marketing efficiency and distribution channels.

MARKETING

The term 'marketing' is widely used, and most marketing specialists would regard it as frequently being widely misused. A common view of the subject is that it is concerned with the way an organization 'markets' its products – by advertising and other methods of communicating with its customers. Marketing specialists take a much broader view and usually regard the term as describing a philosophy in business in which the needs and wants of the customer are central to the approach adopted by the organization. For example:

> Marketing is not only much broader than selling. It is not a specialized activity at all. It encompasses the entire business. It is the whole business seen from the point of view of its final result, that is, from the customer's point of view. Concern and responsibility for marketing must therefore permeate all areas of the enterprise.
>
> (Drucker, 1958)

A neater version is: 'Marketing is the way in which any organization or individual matches its own capabilities to the need of its customers' (Christopher *et al.*, 1980). Or: 'Marketing is the process whereby, in order to fulfil its objectives, an organization accurately identifies and meets its customers' wants and needs' (Ritson, 1986).

Another way of characterizing the subject is by reference to the 'marketing concept'. For example:

> The marketing concept is a philosophy, not a system of marketing or an organization structure. It is founded on the belief that profitable sales and satisfactory terms of investment can only be achieved by identifying, anticipating and satisfying customer needs and desires – in that order. It is an attitude of mind which places the customer in the very centre of the business activity and automatically orientates a company towards its markets rather than towards its factories. It is a philosophy which rejects the proposition of production as an end in itself, and that the products manufactured to the satisfaction of the manufacturer merely remain to be sold.
>
> (Barwell, 1965)

The marketing concept, then, has been used to try to explain what marketing is; similarly, three alternative concepts, or 'orientations', have been used to indicate what it is *not*. The first is the *production concept*, in which market demand is assumed to be plentiful and the successful company will be the one which produces and distributes its products efficiently at low cost.

In contrast, the *product oriented company* is one which is obsessed by its product or products and believes that it will be successful because consumers will prefer products that are of the highest quality and possess the best range of features and performance. The dangers of focusing a business on its products rather than on its customers was most famously described by Levitt (1960) as 'marketing myopia'.

Third, under the *selling concept*, a firm is still production oriented, but it no longer regards the market as insatiable for a low-cost product. Competition is strong and although consumers may be interested in the organization's products, they will not actively seek to make a purchase. Thus, business success depends upon aggressive and effective selling. The selling concept is neatly described by Kotler (1994) as: 'Selling what you make rather than making what you sell'.

Production and sales orientation have also sometimes been used to suggest what marketing *was*, with low-cost production and distribution dominating successful business in the nineteenth century, and aggressive selling being linked to the inter-war depression. However, the heyday of the marketing concept may now be over, with increasing attempts to redefine it. This is an issue to which we refer later in the chapter.

Any student picking up a marketing textbook for the first time will find some of its contents familiar. In one sense the subject is very parasitic, applying fundamental disciplines to marketing issues and problems as appropriate. Because of this, most writers would agree that marketing does not really possess a core of theory which it can call its own. It is a cocktail of applications of other social sciences. In particular, it involves:

1. *Economics*. Successful marketing requires an understanding of the behaviour of the economic system, in particular the behaviour of markets, the economic behaviour of consumers and the relationship between price, revenue, costs and profit within a firm.

2. *Behavioural science*. This refers to a mixture of psychology, sociology and anthropology directed, in the main, to understanding the behaviour of consumers, but also the buying behaviour of organizations.

3. *Quantitative techniques*. A range of sophisticated quantitative techniques have now been incorporated into marketing, used mainly to analyse and interpret data about the market, but also formal methods for developing marketing strategies.

The outcome is that different aspects of the subject of marketing will be dominated by different disciplines as appropriate, and this is evident in this book with, for example, Chapters 3–5 being predominantly economics-based, Chapters 10 and 11 incorporating the behavioural science approach, and Chapters 6 and 12 concerned with the application of quantitative techniques. Other chapters involve a more balanced mix.

One broad division of the subject matter of marketing, which the author has found helpful, is between those aspects concerned with understanding the marketing environment – what might be described as marketing analysis, research and forecasting – and that which is concerned with an organization's own attempt to influence the market for its product. The latter is often referred to as 'marketing decisions' or 'marketing communication'. These can be defined as the set of variables available to an organization with which it may influence its market, and this is conventionally known as the 'marketing mix'. The term was originated by Borden (1964) who quoted a colleague describing a business executive as a 'decider', an 'artist' – a 'mixer of ingredients'. It is usual to divide the marketing mix into four main categories (McCarthy and Perreault, 1984), namely the nature of the product itself, the price, the way it is advertised and promoted, and distribution – though various other disaggregations of the marketing mix variables have been attempted.

This broad division of the subject of marketing corresponds to the division made in this book between Part IV and Part V, and Part V is built round the four elements of the marketing mix, as applied to agricultural and food products.

MARKETING AND AGRICULTURE

Identifying successful marketing as the effective deployment of the marketing mix variables provides one way of explaining why the subject of agricultural marketing took a different path from that of marketing. It also allows us to establish the 'points of contact' between the subjects and to explain why a degree of fusion is emerging.

There are three features of agriculture which have led to the somewhat detached and individual nature of the subject. First is the structure of farming with, typically, many thousands of small businesses supplying an agricultural commodity market. There are, of course, many other sectors dominated by small businesses, but in most cases they are different from farming in the sense that the structure is market-related; they serve local markets. Many are service industries, which *have* to be located at the point of consumption. Farming is unusual in that the structure is production, not market, related; that is, the small business pattern emerges because of the land-based nature of production rather than because of a requirement to be located near to the customer.

This leads on to the second important feature of farming. What is produced is sometimes referred to as 'undifferentiated' (or 'a commodity'). As produce leaves the farm, in most cases the output of one farm is much the same as that of others. Milk is milk; corn is corn; and, despite widespread attempts by individuals and groups to introduce differentiation, much the same applies with, for example, the main meats.

It is these two features together which have made effective manipulation of the marketing mix variables of limited relevance to individual farmers. This can be explained further by introducing another way of describing the manipulation of the marketing mix, as:

> The object of commercial marketing is profit maximization through the creation and exploitation of monopoly advantages. The monopoly advantage can be anything from a product-idea to favourable access to consumers.
>
> (Lesser, quoted in Ritson, 1986)

This is a disturbingly honest definition – and normally we would euphemistically refer to a *'marketing advantage'*. Monopoly means the ability to supply something that others cannot. New products can of course provide the seller with only a 'temporary monopoly advantage'. The monopoly can range from complete control of a market for a product, to some slight product difference – if only in the minds of the consumers – which allows a premium price to be obtained. Most industries dominated, because of location reasons, by small businesses have the opportunity to 'improve their marketing' – there are opportunities for creating and exploiting monopoly advantages in local markets – be it hairdressers or taxi firms or restauranteurs or whatever. With some notable and interesting exceptions, farmers do not.

The third important distinguishing characteristic of agriculture is the remoteness of the farmer from the final consumer. The value of food typically more than doubles between farm gate and retail sale, and during this process is usually controlled by businesses under ownership independent of farming. Thus marketing, in the sense of efficiency of distribution, was, and still is, of great importance to the prosperity of farming. Further, most of the opportunities for profitably matching business activity to consumer requirements occur in the food marketing sector.

There are two important conclusions from this. First, food marketing *is* different from agricultural marketing. As we move from farmer to consumer through marketing channels, so opportunities for businesses to exploit marketing advantages by manipulation of the marketing mix progressively increase; and in the case of branded food products, marketing techniques are as sophisticated, and consistent with the 'marketing concept', as in any sector of the economy. Thus, any modern marketing textbook is likely to be sprinkled with examples of food products (and is unlikely to include any relating to agricultural commodities).

The second point is somewhat more novel – it is that many of the government marketing policy initiatives identified in Fig. 2.1, and outlined in subsequent chapters of this book, can be interpreted also as manipulation of marketing mix variables – but to the collective benefit of farming. For agriculture as a whole, improvements in marketing in order either to obtain a bigger share of consumer food expenditure (efficiency in food distribution, or market power *vis-à-vis* the marketing sector) or to share in enhanced consumer food expenditure (by creating monopoly advantages in retail markets) require government intervention, government control or government support.

Many of these marketing policies have of course revolved round price and quantity control, but another area in which government policy has been very active is in advertising and promotion. When it comes to advertising manufactured branded food products, the techniques used are as sophisticated and subtle as for any sector (see for example Lannon, 1986), and mainstream marketing textbooks invariably illustrate discussion of advertising and promotion in the marketing mix with food product examples. However, mobilizing the weapon of effective advertising to the collective benefit of farmers poses special problems and these are both much less familiar and often peculiar to agricultural marketing. For this reason our chapter on this subject in this book (Chapter 15) concentrates on generic advertising.

Thus, it is perfectly possible to interpret agricultural marketing policies as government control of the marketing mix for farm products. Viewed this way, there is a much greater degree of consistency between the traditional subject matter of agricultural marketing, with its policy orientation, and modern business marketing; there always has been a significant point of contact.

MARKETING AND THE FARMER

Throughout much of the world, governments are attempting to loosen their grip on agricultural markets with the consequence of exposing farmers to the exigencies of market forces. At the same time, farmers are increasingly being encouraged to 'improve their marketing', and some commentators appear to view the application of modern business marketing techniques at the farm level as a substitute for government agricultural marketing policies. Traditional agricultural marketing theory suggests that, for the reasons explained above, the opportunities for farmers to do this are limited. There are, however, two main ways in which this interpretation is being modified.

The first is to see the application of modern marketing techniques at the farm level as not so much the adoption by farmers of the 'marketing concept', but as an aspect of contemporary strategic marketing management theory, in which marketing has incorporated concepts from strategic management and industrial economics, in particular the well-known work of Porter (1980). For example, McLeay and Zwart (1993) list 24 empirical studies of farm business marketing strategies, all published between 1985 and 1990. Many of these are concerned with choice of sales outlet, method or time. They argue that farmers typically utilize much more sophisticated marketing strategies than is recognized in the agricultural marketing literature. Another paper (McLeay *et al.*, 1996) applies factor and cluster analysis (as explained in Chapter 12) to data collected from intensive crop farms in New Zealand, identifying a variety of strategic marketing farmer groups.

The second way in which business marketing can be seen to be of increasing relevance to at least some successful farming is by recognizing that contemporary developments in food marketing are now providing opportunities for farmers to adopt the modern marketing concept. This is by identifying some product characteristics which can be created at farm level to provide differentiation. We have had the growth in farm-gate sales, fuelled by increasing car ownership and the desire for quality and variety in food purchases. Here there are opportunities for developing local monopoly advantages. Another good example of product differentiation which can *only* be introduced on-farm is organically produced food. The same applies to various welfare-oriented branding of livestock.

In addition, certain sectors are increasingly dominated by a small number of firms. It is no accident that farmer marketing seems to have been more successful in the fruit and vegetable sector. Generally, there are fewer suppliers, and the product is less processed and is intrinsically more varied, so there are more opportunities for product identity to be imparted at grower stage. One way of doing this is branding by variety. For example, in the UK we have pre-packed 'Cox's Orange Pippins' and 'Maris Piper', with no mention of 'apples' or 'potatoes'. But even in the higher value-added product areas, such as meat, concentration at the retail end has also had the effect that the provision of certain consumer product characteristics may require product differentiation at the raw material stage, giving producers some opportunities to establish (admittedly often weak) marketing advantages, by attention to product and, sometimes, promotion.

It is here that the relationship between 'improved marketing' and 'cooperation' appears. By grouping together in the disposal of their produce, farmers are more likely to be able to supply the required quantity of differentiated produce; more likely to be able to be involved in promotion; more likely to be able to exercise a local monopoly advantage; and may wish to become involved in activity down the food marketing channel, taking their business interests nearer to the final consumer. Farmer cooperation allows for the possibility of individual farmers sharing in the ownership of brands and thus their involvement in marketing mix variables extending to include, in particular, advertising and promotion. More often, however, this is concentrated in cooperative generic advertising for the product category (see Chapter 15).

Recently, another interesting example of this kind of application of marketing principles at the farm level has appeared in the European Union. This is to extend the idea of a controlled designated name of origin (Appellation Contrôllée) into new product areas. In 1992 a European Union Council regulation came into force 'on the protection of geographical indications and designations of origin for agricultural products and foodstuffs' in an attempt to harmonize legislation of this kind, and extend the protection involved in countries which had adopted this approach (for example France) throughout the European Union:

> ... it has been observed in recent years that consumers are tending to attach greater importance to the quality of foodstuffs rather than to quantity; ... this quest for specific products generates a growing demand for agricultural products or foodstuffs with an identifiable geographical origin; ... the desire to protect agricultural products or foodstuffs which have an identifiable geographical origin has led certain Member States to introduce 'registered designations of origin'; ... these have proved successful with producers, who have secured higher incomes in return for a genuine effort to improve quality,

and with consumers, who can purchase high quality products with guarantees as to the method of production and origin.[2]

Thus the introduction of a controlled designated name for a food from a geographical region: (i) has the potential to enhance the value of raw material from a region by ensuring that it is used for differentiated, premium products; (ii) may ensure that processing activity occurs within the region; and (iii) grants a monopoly advantage to producers and processors within a specific area by restricting the use of area names·to products genuinely supplied from the region. At its extreme, product differentiation by area of production becomes a proxy for national branding – with governments seeking to win a greater share of the world food market for produce from their nation's farmers by investing in schemes concentrating on national product characteristics and promotion.

THE SCOPE OF MARKETING

This chapter has attempted to explain why the traditional subject matter of agricultural marketing has differed from that of business marketing (and thus also food marketing); but how, by recognizing the role of government in the marketing mix for farm products, the traditional subject matter of agricultural marketing can be seen to be quite consistent with that of business marketing. It has also suggested reasons why marketing management is increasingly seen as relevant to farming.

The conclusion would be that it is perfectly proper for 'agro-food marketing' to encompass both the traditional subject matter of agricultural marketing *and* be viewed as an application of mainstream marketing management to both agriculture and food businesses. As such, however, the scope of the subject might appear much broader than that of modern business marketing management. Meanwhile, however, there has developed an interesting debate over whether the subject of marketing is adequately characterized by the 'marketing concept' as outlined earlier in the chapter; or whether a much broader view of the scope of marketing as a subject of study is now appropriate.

A very narrow view would be that marketing is a branch of management, restricted to profit-seeking firms. Broader views extend marketing to non-profit making organizations and include the study of the marketing behaviour of firms and the explanation of the function of markets. More fundamental is the question of the incorporation of the study of the extent to which the marketing system meets the public interest.

There are two interrelated issues which underlie these questions. The first is whether the managerial approach to marketing replaces or merely adds to the earlier view of marketing, as a study of markets, processes and institutions. The second concerns whether the marketing concept (previously known as the 'modern' marketing concept) remains valid in the modern world. To return to the article quoted at the beginning of this chapter, Webster (1992) continues:

These early approaches to the study of marketing are interesting because of the relative absence of a *managerial* orientation. Marketing was seen as a set of social and economic processes rather than as a set of managerial activities and responsibilities. The institutional and functional emphasis began to change in 1948, when the American Marketing Association defined marketing as: 'The performance of business activities directed toward, and incident to, the flow of goods and services from producer to consumer or user.'

He then goes on to point out that, nevertheless:

In academia, the functionalists and institutionalists held their ground well into the 1960s, stressing the value of understanding marketing institutions and functions and viewing marketing from a broader economic and social perspective. … The argument against the managerial point of view centred on its inability to consider the broader social and economic factors and issues associated with marketing, beyond the level of the firm.

The re-emergence of the view that the subject of marketing extends into broader social and economic factors beyond the narrow interest of the firm itself can be traced to the development of environmentalism and consumerism during the late 1970s, and following from this, whether the marketing concept remains valid in the wider interests of society in the modern world. Here, Kotler (1980) seems to have led the way, when the prevailing approach of marketing specialists (if they thought about it at all) was that if consumer wants are accurately identified and fulfilled, this *must* meet the consumer and public interest.

Kotler (1994) argued as follows:

Some marketers have raised the question of whether the marketing concept is an appropriate organizational goal in an age of environmental deterioration, resource shortages, explosive population growth, world-wide inflation, and neglected social services. The question is whether the firm that does an excellent job of sensing, servicing, and satisfying individual consumer wants is necessarily acting in the best long-run interests of consumers and society. The marketing concept side-steps the conflict between consumer wants, consumer interest, and long-run societal welfare.

The concern over the relationship between successful marketing and the public interest can be separated into three issues. First there is what economists refer to as a divergence between private and social costs and benefits. The most prominent example here is the environmental impact, i.e. pollution, unsightly factories, litter, depletion of scarce resources. Other examples might be rural depopulation caused by industrial concentration, and even unemployment. (In many labour-surplus low-income countries, businesses accept a 'social responsibility' for employing more labour than business economics would dictate.)

Second is the question of whether the individual is the best judge of his own welfare. This raises complex philosophical issues over 'freedom of choice', but most people accept that there is an area of personal consumption where, in the interests of the consumer himself, government intervention is required (for example, certain narcotics, seat-belt legislation, and of course products for children).

Kotler (1994) used the example of the US fast-food industry to illustrate both points:

> The fast-food hamburger industry offers tasty but not nutritious food. The hamburgers have a high fat content, and the restaurants promote fries and pies, two products high in starch and fat. The products are wrapped in convenient packaging, but this leads to much packaging waste material. In satisfying consumer wants, these restaurants may be hurting consumer health and causing environmental problems.

The third issue is the claim that consumer wants may be 'created' by successful advertising and promotion. Most marketing specialists would deny that consumers' wants are 'created' by business – only 'accurately identified'. Others are concerned that, in some sense, people are made less satisfied (and therefore less 'happy') as a consequence of becoming aware of the possibilities for enlarged consumption (see for example Hirsch, 1977, and Mishan, 1984).

From the point of view of the 'marketing concept', the question revolves around whether government legislation can be expected to deal adequately with the public interest aspects of successful marketing (e.g. by taxing pollution, subsidizing employment, controlling advertising and so on).

A growing belief that it cannot led to the call for a new concept to revise or replace the marketing concept. Kotler (1980, 1994) proposed 'the societal marketing concept' (more recent versions are 'ethical marketing' and 'green marketing'):

> The societal marketing concept is a management orientation that holds that the key task of the organization is to determine the needs and wants of target markets and to adapt the organization to delivering the desired satisfactions more effectively and efficiently than its competitors in a way that preserves or enhances the consumers' and society's well being. ... The underlying premises of the societal marketing concept are:
>
> 1. Consumers' wants do not always coincide with their long-run interests or society's long-run interests.
> 2. Consumers will increasingly favour organizations which show a concern with meeting their wants, long-run interests, and society's long-run interests.
> 3. The organization's task is to serve target markets in a way that produces not only want satisfaction but long-run individual and social benefit as the key to attracting and holding customers.

Agro-food marketing abounds with examples of where the successful adoption of the marketing concept may not be serving the wider interests of society. Two will suffice. First, there is the potential impact of efficient production of raw materials for the food industry on the environment – broadly defined to include, for example, nitrogen accumulation in drinking water, scenic aspects of the countryside, and animal welfare issues. Second is the growing concern over the relationship between food consumption, diet and health.

In one sense, the reaction of food manufacturers and retailers to food and nutrition is quite consistent with the marketing concept. If changes in the pattern of demand for food products are occurring on account of people's concern over the relationship between diet and health, then anticipating these changes is a vital part of forecasting in marketing for the food and farming sector, and knowledge of consumer attitudes to food and health is an important area of research in order to devise a successful marketing mix for a food firm's products. The successful firm will identify and meet new market segments. There is, however, a second stage in this process – which is very apparent in the current reaction of the UK food manufacturers and retailers to the British Government's Nutrition Task Force initiative following publication of the 'Health of the Nation' White Paper. Quite genuine attempts are being made to provide better information for consumers over the relationship between food product purchases and nutritional health guidelines, and in reducing average levels of fat, sugar and salt content of processed products. This is clearly the 'societal marketing concept' at work. The intriguing question is, though, does 'social responsibility' mean that ethical marketing extends beyond the long-term commercial interests of the firm?

Ness (1994) argues not:

> It is evident … that there is no theoretical justification for firms and organizations to adopt social responsibility as formal policy … the justification for social responsibility is that firms perceive that the changing business environment, and pressure from stakeholder groups, point the way to this course of action and that it makes good business sense in the context of social investment to do so. In the new climate of societal marketing and with increasing consumer sensitivity to environmental issues there is likely to be an increasing number of opportunities for food firms and food industry organizations to consider the attractiveness of such a commitment.

There is, however, another way in which the validity of the marketing concept has been challenged which may lead to a different view. This has been termed 'product related goals'. According to Hirschman (1983): 'The marketing concept is not applicable to two broad classes of producers (artists and ideologists) because of personal values and social norms that characterize the production process.'

Extended to any producer, what in effect is being argued is that successful marketing *matches* customer wants *to* producer objectives, and to the extent that such objectives go beyond simple profitability into a desire to be consistent with public interest values – for example animal welfare or the amenity value of the countryside – then it is quite legitimate to include such values within the concept of successful marketing management. This does not, however, resolve the problem of conflict between, for example, owners and managers in the importance of public interest values.

A CLASSIFICATION OF THE SUBJECT AREA

Table 2.1, which is based on a development by Oliver (1980) of a classification suggested by Bartels and Jenkins (1977), helps to show the relationship between marketing management and the broader interpretations of the subject discussed in the previous section.

Table 2.1. A classification of marketing subject areas (source: Oliver, 1980).

	Positive	Normative
Micro	Micromarketing theory undertakes to explain how and why marketing processes are managed as they are within firms	Micromarketing models are constructs of how marketing should be conducted for the best achievement of the objective of the firm
Macro	Macromarketing theory undertakes to explain the functioning of the composite marketing mechanism both as a result of and as a determinant of the economic and social environment	Macromarketing models are constructs of how the general marketing process should be conducted in the best interests of society

In Table 2.1, 'micro' refers to the individual firm or household; 'macro' to aggregation at the level of the market. 'Positive' refers to an attempt to gain a better understanding of some aspect of behaviour; 'normative' means assessing to what extent that behaviour is achieving (or is likely to achieve) certain specified objectives. It will be clear that this classification is certainly broad enough to be consistent with our interpretation of agro-food marketing. A few examples might help.

Study of the marketing behaviour of firms (for example, the decision-making process in a marketing cooperative), or of the behaviour of food consumers, would be classified as micro/positive. The core area of agro-food business marketing management would fall into the micro/normative category. The traditional subject matter of agricultural marketing, in which we attempt to understand the behaviour of the food marketing sector (e.g. marketing margins, the behaviour of farm product prices, the impact of government marketing policies) is macro/positive. Finally, the various public interest and efficient aspects of the marketing of agricultural and food products would be described as macro/normative.

Using this approach, Table 2.2 attempts a broad classification of the various topics taught under the agricultural and food marketing banner – or 'agro-food marketing'.

It is worth noting that Bartels and Jenkins (1977) see 'management' as the implementation of normative models and thus restrict the use of the term 'marketing management' to the normative side. Interestingly, we often use the term 'management of the market' to refer to government agricultural marketing policies.

It should also be pointed out that a 'narrow' view of marketing would not restrict the subject solely to the micro/normative category. The successful

Table 2.2. Agro-food marketing – a classification of subject areas.

	Positive	Normative
Micro	The behaviour of food consumers. Study of the marketing behaviour of firms in the agro-food sector	Application of marketing principles to firms in the food marketing sector. Farmer marketing (including cooperative marketing). Government marketing initiatives on behalf of farmers (e.g. marketing boards)
Macro	The behaviour of agricultural and food markets (e.g. marketing margin analysis, price analysis, effect of agricultural policies)	Application of structure/conduct/performance approach to the agro-food sector. Public interest aspects of agricultural policies. 'Green Marketing'. Food and nutrition policy

implementation of a marketing plan in business may well require an understanding of the behaviour of the consumer, of the behaviour of markets, and an understanding of the wider social implications of the firm's policies; but this would be a means to an end. In a broader definition of marketing they are legitimate subjects of study in their own right.

CONCLUDING REMARKS

This chapter has introduced the subjects of marketing and agricultural marketing. It has explained why these subjects have developed along rather different lines and related this to the way this book has been structured. It has also argued, first, that once the role of government agricultural marketing policies is properly appreciated, the subjects have much more in common that is usually realized. Second, it has argued that, for a variety of reasons, the subjects are coming closer together. In the title of this book we have characterized the outcome as 'agro-food marketing'.

NOTES

1. Breimyer (1973) and Meulenberg (1986) also discuss how agricultural and business school marketing had developed down different paths during the postwar period.
2. Council Regulation (EEC) No. 2081/92.

FURTHER READING

Bateman, D. (1976) Agricultural marketing: A review of the literature of marketing theory and of selected applications. *Journal of Agricultural Economics* 27(2), 171–226.

Breimyer, H. (1973) The economics of agricultural marketing: A survey. *Review of Marketing and Agricultural Economics* 41, 115–165.

Jones, D. and Monieson, D. (1990) Early development of the philosophy of marketing thought. *Journal of Marketing* 54(1), 102–113.

McLeay, F. and Zwart, A.C. (1993) Agricultural marketing and farm marketing strategies. *Australasian Business Review* 1(1), 80–98.

Meulenberg, M. (1986) The evolution of agricultural marketing theory: Towards better coordination with general marketing theory. *Netherlands Journal of Agricultural Science* 32, 301–315.

Ritson, C. (1986) *Marketing and Agriculture: An Essay on the Scope of the Subject Matter of Agricultural Marketing.* Discussion Paper. University of Newcastle upon Tyne, Newcastle upon Tyne.

Webster, F.E. (1992) The changing role of marketing in the corporation. *Journal of Marketing* 56(4), 1–17.

REFERENCES

Bartels, R. and Jenkins, R.L. (1977) Macromarketing. *Journal of Marketing* 41(4), 17–20.

Barwell, C. (1965) The Marketing Concept. In: Wilson, A. (ed.), *The Marketing of Industrial Products.* Hutchinson, London.

Bateman, D. (1976) Agricultural marketing: A review of the literature of marketing theory and of selected applications. *Journal of Agricultural Economics* 27(2), 171–226.

Borden, W.H. (1964) The concept of the marketing mix. *Journal of Advertising Research* 4, 2–13.

Bosanquet, C. (1965) Quotation from British Ministry of Agriculture press release announcing the establishment of a Chair and Department of Agricultural Marketing at Newcastle.

Breimyer, H. (1973) The economics of agricultural marketing: A survey. *Review of Marketing and Agricultural Economics* 41, 115–165.

Christopher, M., McDonald, M. and Wills, G. (1980) *Introducing Marketing.* Pan, London.

Drucker, P.F. (1958) Marketing and economic development. *Journal of Marketing* 22(1), 252–259.

Hirsch, F. (1977) *Social Limits to Growth.* Routledge and Kegan Paul, London.

Hirschman, E.C. (1983) Aesthetics, ideologies and the limits of the marketing concept. *Journal of Marketing* 47, 45–55.

Jones, D. and Monieson, D. (1990) Early development of the philosophy of marketing thought. *Journal of Marketing* 54(1), 102–113.

Kohls, R.L. and Uhl, N.U. (1990) *Marketing of Agricultural Products,* 7th edn. Macmillan, New York.

Kotler, P. (1980) *Marketing Management: Analysis, Planning and Control,* 4th edn. Prentice-Hall, Englewood Cliffs, NJ.

Kotler, P. (1994) *Marketing Management: Analysis, Planning and Control*, 8th edn. Prentice-Hall, Englewood Cliffs, NJ.

Lannon, J. (1986) How people choose food – advertising and promotion. In: Ritson, C., Gofton, L. and Mckensie, J. (eds), *The Food Consumer*. Wiley, Chichester, pp. 241–256.

Levitt, T. (1960) Marketing myopia. *Harvard Business Review* 38(4). Reprinted in Enis, B.M. and Cox, K.K. (eds) (1981) *Marketing Classics*. Allyn Bacon, Boston, MA, pp. 1–23.

McCarthy, J. and Perreault, W. (1984) *Basic Marketing: A Managerial Approach*. R.D. Irwin, Homewood, IL.

McLeay, F. and Zwart, A.C. (1993) Agricultural marketing and farm marketing strategies. *Australasian Business Review* 1(1), 80–98.

McLeay, F., Martin, S. and Zwart, T. (1996) Farm business marketing behaviour and strategic groups in agriculture. *Agribusiness* 12, 339–351.

Meulenberg, M. (1986) The evolution of agricultural marketing theory: Towards better coordination with general marketing theory. *Netherlands Journal of Agricultural Science* 32, 301–315.

Mishan, E.J. (1984) G.N.P. – Measurement or mirage? *National Westminster Bank Quarterly Review* November, 2–13.

Ness, M.R (1994) Corporate social responsibility in the food sector. In: Henson, S. and Gregory, S. (eds), *The Politics of Food*. Occasional Paper No. 2. Department of Agricultural Economics & Management, University of Reading, Reading, pp. 35–52.

Oliver, G. (1980) *Marketing Today*. Prentice-Hall, Hemel Hempstead.

Porter, M. (1980) *Competitive Strategy*. The Free Press, New York.

Ritson, C. (1986) *Marketing and Agriculture: An Essay on the Scope of the Subject Matter of Agricultural Marketing*. Discussion Paper. University of Newcastle upon Tyne, Newcastle upon Tyne.

Webster, F.E. (1992) The changing role of marketing in the corporation. *Journal of Marketing* 56(4), 1–17.

3 Supply and Demand of Agricultural Products

Trevor Young and Michael Burton

School of Economic Studies, University of Manchester, Manchester M13 9PL, UK

INTRODUCTION

In this chapter we review that part of economic theory concerned with two main actors in the economic system: firms and consumers. The theory is developed to explain, in terms of a set of basic rules and assumptions, observed behaviour of these actors. In addition, the theory provides us with a basis for predicting the response to changes in market conditions.

It is important to remember that all theory is an abstraction from reality. It would not be possible to describe an economic system in all its complexity; some abstraction is inevitable. The economist has developed relatively simple theoretical models of economic behaviour that, it is hoped, capture the essentials of economic processes. In the construction and use of these models many 'outside' forces are held constant; this is the *ceteris paribus* (other things the same) assumption. Another key assumption is that economic agents behave rationally, choosing the most preferred option, when faced with a number of choices.

Whether the economist's approach provides a useful and significant contribution to our understanding of the way the economic system functions rests on the theory's power of explaining and predicting real-world events. Confronting an economic model with data from the economy itself provides the ultimate test of its validity.

SUPPLY OF FARM PRODUCTS

Foundations

Much of the analysis of agricultural producers' behaviour is based upon the neo-classical theory of the firm. As we shall see in later chapters, empirical

© CAB INTERNATIONAL 1997. *Agro-food Marketing*
(eds D.I. Padberg, C. Ritson and L.M. Albisu)

analysis does not always conform to this theory, and there are features peculiar to agricultural and food markets which mean that the theory has to be modified to suit the particular product, but here we start with a review of the theoretical foundations.

The neo-classical theory of the firm assumes that the producer's only objective is to maximize profits. It abstracts entirely from the relationship between the producer and the means of production, and requires only that the producer can coordinate the use of inputs. Thus, the fact that a producer may supply labour, or own some of the inputs, is not relevant to the allocation of inputs and the production of output as long as all quantities are valued at the prevailing market prices. The constraints that the producer faces in trying to achieve maximum profit are twofold: the market conditions (the prices of outputs and inputs), and the technical relationships governing production.

As we will show in the following section, the quantity of product a producer will supply will be determined by: (i) the price of the output; (ii) the price of the (variable) inputs used to produce the output; and (iii) the quantity of any input that is fixed and cannot be varied by the producer.

The last factor is of importance in defining the short-run and the long-run. In the short-run there are some fixed factors, due to institutional constraints or market rigidities: in the long-run, all factors are variable. One can represent this mathematically by a *supply function* of the form $Q = f(P, r_1, r_2, ... r_{n-1}, x_n)$ for the short-run case, where in this example P denotes the output price, there are $n-1$ variable input prices (r_i), and only one input quantity (x_n) which is fixed. In the long-run the supply function will be given by $Q = f(P, r_1, r_2, ... r_n)$. The exact functional form of this relationship will depend on the production technology, to which we now turn.

Derivation of the supply function

The *production function* is simply a conceptualization of the relationship between the inputs used and the outputs: it is a description of the physical or technical world within which the farmer operates. The descriptions used may be either narrative, graphical or mathematical but all share a common feature: they are highly simplified representations which have certain characteristics imposed upon them in order to aid the subsequent economic analysis. There are a number of ways of representing the production relationships graphically: Fig. 3.1a shows a 'typical' relationship for a single output, Q, while varying a single input, x_1. The shape of the curve indicates that there may be increasing returns to the input initially (increases in input lead to proportionally greater increases in output), but eventually decreasing returns set in. An alternative representation explicitly identifies the relationships between inputs, as revealed by the *isoquant map* (Fig. 3.1b). Line AA is an isoquant, showing the minimum combinations of the two inputs that produce a fixed level of output, Q_1. Higher levels of output are associated with isoquants further from the origin. The fact that the isoquants are convex to the origin implies diminishing marginal productivity, i.e. for any given level of x_2, increases in x_1 lead to ever smaller increases in output. The

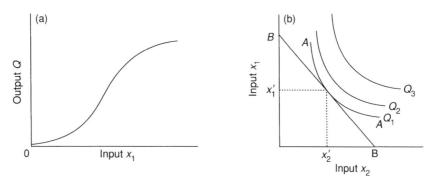

Fig. 3.1. (a) The output–input relationship. (b) The isoquant map.

curvature of the isoquant reflects the marginal rate of technical substitution between the two inputs in the production of the output. The relationship between isoquants indicates the returns to scale. The returns to scale will be reflected in the change in output as one moves along a ray drawn through the origin in Fig. 3.1b.

If one sets aside the question of how much to produce, a profit maximizing producer will minimize the cost of producing any given level of output. Assume we have the case where there are three inputs, one of which is fixed, and hence Fig. 3.1b has been drawn for the two variable inputs. The combination of variable inputs that minimize costs is shown on Fig. 3.1b as x_1' and x_2'. These are identified by the tangency of the isoquant and the iso-cost line BB. The latter line shows the combinations of inputs that can be purchased for a given expenditure level. The slope of the line is given by the ratio of the input prices $-r_2/r_1$. Formally the iso-cost line is given by:

$$x_1 = \frac{C}{r_1} - \frac{r_2 x_2}{r_1}$$

where C is the level of expenditure. Thus x_1' and x_2' are the input levels which are on the lowest possible iso-cost line, but still allow the production of Q_1. This gives us the first marginal condition for profit maximization: the marginal rate of technical substitution is equal to the ratio of input prices (MRTS$_{1,2} = -r_2/r_1$).

If one identifies the minimum cost of producing a range of output levels (holding input prices constant) then one can plot the total cost function (Fig. 3.2a). This includes the cost of both variable and fixed inputs. Simple manipulation yields the average total cost (ATC) curve, i.e. total costs divided by the output level. One can also identify the average variable cost (AVC) by excluding the cost of the fixed inputs. The marginal cost (MC) is defined as the cost of producing an additional unit of output, and is given by the slope of the total cost curve at each level of output. The shape of these functions will be determined by the nature of the production function: those in Fig. 3.2 illustrate the cost functions associated with a production function that has increasing, constant and decreasing returns to the variable inputs. One is

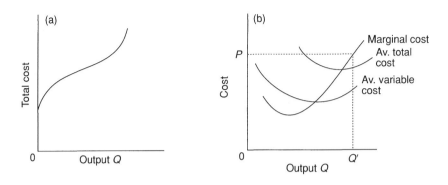

Fig. 3.2. (a) The total cost curve. (b) Average and marginal cost.

now in a position to identify the optimal level of output. A producer will continue to produce output if there is a marginal profit to be made: this will be the case if the marginal value of the output (i.e. the market price) exceeds the marginal cost. Thus, in Fig. 3.2b, if the market price is P then the profit maximizing output will be at Q', where price equals marginal cost. Thus, the supply curve for the firm is given by that (upward sloping) portion of the marginal cost curve above the average variable cost curve. The latter restriction is needed because if the firm operates at a point on the marginal cost curve below the average variable cost curve then it is not even recovering the cost of the variable inputs, and would be better off stopping production entirely.

In the long-run, all factors are variable. One can now repeat the same derivation as before, but with the producer choosing the level of all three inputs. This is difficult to draw, as the relevant isoquant line becomes a three-dimensional bowl, and the iso-cost line a plane, but the same principles apply. Strictly, the long-run total cost curve will never be greater than the short-run total cost curve, because the inability to vary x_1 cannot be to the advantage of the producer in minimizing costs. The long-run marginal cost curve (LRMC) (given by the slope of the long-run total cost curve) will normally have a shallower slope than the short-run cost function. Thus, the long-run supply will be more responsive to price changes than the supply in the short-run, and this reflects the greater flexibility available to the producer.

At this point it is important to note the role played by the returns to scale of the production function in determining the firm's supply function. If there are decreasing returns to scale, then the long-run marginal cost curve is upward sloping and it is possible to identify the supply curve. If there are constant returns to scale then the LRMC is constant. Any price above that level should induce unlimited output, any price below will result in no output, and a price equal to the LRMC will mean that output is indeterminate. If there are increasing returns to scale then the marginal cost is falling everywhere and any price will induce unlimited output by the firm. Thus a well-behaved firm-level supply function can only be identified in the long-run if the production function has decreasing returns to scale at some level of output. The problem need not arise in the short-run, as some input is fixed, and hence one might expect decreasing returns to the remaining inputs.

So far we have considered the firm-level supply only, and seen that the distinction between short- and long-run can be reduced to the appropriate derivation of the cost functions. At the level of the industry, this is not the case. The industry short-run supply function can be identified by the aggregation of the individual firms within the industry, but the long-run supply curve cannot. This is because, if all inputs are variable, then the number of firms in the industry is not pre-determined. Even with decreasing returns to scale at the firm level, if individual firms are making profits then one would expect to see an increase in the number of firms in the industry. However, one could not expect the price of the output to be invariant to the level of production in the long-run, and hence one would expect the price of the product to fall as output increases, until it reaches the point where the price equals the minimum long-run average cost. At this point there will be an unlimited number of producers willing to enter the industry and supply will be determined by aggregate demand conditions. A further caveat, however, is that one may not be able to assume that the input prices are fixed at the industry level. Thus, as aggregate output increases, input demands increase and input prices may be bid up. This would have the effect of shifting upwards all of the firm-level cost functions. The result is that the long-run industry supply function may be upward sloping, but this is determined by conditions in the input markets as much as by the production technology.

Producer surplus

When conducting an analysis of changes in the economic environment facing an agricultural sector, whether those changes are induced by policy or other factors, it is convenient to have some measure of the impact on the welfare of producers. In fact, one can utilize the supply curve for this purpose, as it can be used to identify what is known as 'producer surplus'. Recall that the supply curve for the individual producer is defined by the marginal cost of production: hence, in Fig. 3.3, the producer would be prepared to supply the marginal unit of output Q_1 for price of P_1. However, the market price is P_m, so the difference, $P_m - P_1$, represents a surplus on that unit of production. When the optimal level of output (Q') is produced, total producer surplus is represented by the area between the marginal cost curve and the market price. Note that in Fig. 3.3 at low levels of production the marginal cost exceeds the market price, and hence in this case producer surplus is given by area '*a* minus *b*'.

At the market level, the supply curve is derived by the aggregation of marginal cost curves over individual producers, and hence the area between the market supply curve and the market price represents the aggregate

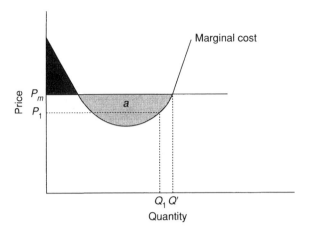

Fig. 3.3. Producer surplus for the individual producer.

producer surplus. It is important to note that if the supply curve has been estimated empirically, then there will often be few data available to identify the marginal cost at very low levels of output, making the estimate of producer surplus derived from the estimated supply curve an unreliable indicator of overall welfare levels. However, it will be able to give a more accurate estimate of the change in welfare following some intervention, such as the increase in market price shown in Fig. 3.4, from P_m to P'_m. Here one can identify the increase in producer welfare as equal to area 'c', the increase in producer surplus. This will be independent of the actual marginal cost at low levels of output.

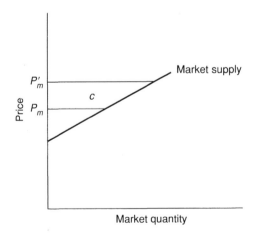

Fig. 3.4. Change in producer surplus for the industry.

Elasticities of supply and input demand

It is possible to generalize the model further, to the case where there are a number (k) of outputs which the firm may produce. In that case, the price of all other outputs, not just the price of the output under consideration, are included as arguments to the supply function. A convenient way of summarizing the relationships within the supply function is through elasticities. These are defined as the proportionate changes in output i as a result of changes in the prices, and can be classified into three types:

1. *The own price elasticity*
This elasticity measures the proportionate change in output that is induced by a proportionate change in the output price, *ceteris paribus*. Formally, it is defined as:

$$\varepsilon_{ii} = \frac{\Delta Q_i / Q_i}{\Delta P_i / P_i} = \frac{\Delta Q_i}{\Delta P_i} \cdot \frac{P_i}{Q_i} = \frac{\partial Q_i}{\partial P_i} \frac{P_i}{Q_i}$$

where Δ denotes a small change. For an infinitesimal change, the expression in partial derivatives applies.

2. *The cross price elasticity*

$$\varepsilon_{ij} = \frac{\partial Q_i}{\partial P_j} \frac{P_j}{Q_i}$$

which gives the proportionate change in output that is induced by a proportionate change in the price of another output, *ceteris paribus*.

3. *The input price elasticities*

$$\varepsilon_{iv} = \frac{\partial Q_i}{\partial r_v} \frac{r_v}{Q_i}$$

which gives the proportionate change in output that is induced by a proportionate change in the price of an input, *ceteris paribus*.

The elasticities are evaluated at specific values of the prices and quantities. One would always expect that an increase in own price will induce an increase in output for a profit-maximizing firm. With the cross price elasticity one might expect that this should be negative, i.e. increases in the price of another product will lead to reductions in the supply of the product under consideration, but it is possible that the products are in some sense complementary (i.e. the production of milk and beef) so that increases in the price of one may lead to increases in the supply of the other. With the input price elasticities the situation is more straightforward: increases in an input price will reduce supply unambiguously.

Relationships between supply elasticities

The most important of these relates to the lack of 'money illusion'. If all prices (both input and output) are changed by the same proportion, then there should be no change in output quantities or input use. Thus, in Fig. 3.1b above the use of inputs to produce a certain level of output is determined by relative prices of inputs, not their absolute level. The increase in all input prices will have the effect of increasing marginal cost by the same amount, and hence if output price is increased by the same amount there will be no change in the equilibrium output either. Formally, if there are k outputs and n inputs, this implies:

$$\sum_{j=1}^{k} \varepsilon_{ij} + \sum_{v=1}^{n} \varepsilon_{iv} = 0 \quad i = 1, \dots k$$

It is also possible to identify certain relationships between the marginal effects of price changes. Thus it is the case that:

$$\frac{\partial Q_i}{\partial r_v} = - \frac{\partial x_v}{\partial P_i}$$

i.e. the marginal change in the level of output Q_i following a change in the price of input v (r_v) is equal to the negative value of the marginal change in the use of input v following a marginal change in the output price P_i. This arises because of the smooth, convex nature of the isoquants, and the close relationship between output supplies and the demand for inputs. One can go further, and identify symmetry relationships between the input demands:

$$\frac{\partial x_u}{\partial r_v} = \frac{\partial x_v}{\partial r_u}$$

which suggests that the marginal change in the use of input u in response to a change in input price v is equal to the marginal change in the use of input v following a change in input price u. Because of the close interlinkages between output supply and input demands, empirical work which is based firmly upon the idea of profit maximization subject to a production function often estimates both supply and input demands within the same framework. Table 3.1 reports elasticities for USA agriculture that are generated from a study of this kind.

Note that the signs of the elasticities are as expected, and that the elasticities sum to zero across a row. Although not derived above, the own price elasticities of the inputs are negative, i.e. as the price of an input increases, the demand falls, which one would expect. The fact that the cross price effects are the same (e.g. the elasticities of input use with respect to the price of energy are all 0.0362) is not something one would expect *a priori* and is a result of the particular functional form of the production function used.

Extensions

A number of extensions are now briefly introduced. Some of these lead to major issues in agricultural marketing and are developed in subsequent chapters.

Table 3.1. Output supply and input demand elasticities for USA agriculture* (source: derived from Ozanne, 1992).

	Elasticity with respect to price of:					
	Aggregate output	Durable equipment	Farm produced durables	Hired labour	Energy	Other inputs
Aggregate output	0.3199	−0.0602	−0.0474	−0.0319	−0.0243	−0.1561
Durable equipment	0.5032	−0.9159	0.0413	0.0758	0.0362	0.2593
Farm produced durables	0.5032	0.0842	−0.9587	0.0758	0.0362	0.2593
Hired labour	0.5032	0.0842	0.0413	−0.9242	0.0362	0.2593
Energy	0.5032	0.0842	0.0413	0.0758	−0.9638	0.2593
Other inputs	0.5032	0.0842	0.0413	0.0758	0.0362	−0.7407

*Elasticities calculated at the sample means.

Institutional constraints

The models outlined above imply that the producer is operating in a free market, whereas for many commodities in most countries, there will be institutional or government influences that will also determine supply. The most obvious of these are fixed quotas on production or marketing quotas, which are imposed upon producers. However, there may be more subtle factors, such as restrictions on credit availability, which make it difficult to purchase inputs.

Biological constraints

The most obvious of these is that although producers may plan to produce certain outputs, the outcome may deviate from this due to weather variations, damage by pests, etc. Often, these may be more important in determining supply than any economic factors.

If one is to use a production function-based approach to the analysis of supply, then one loses the ability to include much of the detail associated with agricultural production. Thus, the production functions often use very aggregated definitions of outputs (sometimes only one), relatively few inputs and the nature of the representation of the technological relationship is constrained by the requirements of the formal analysis. However, in fact, agricultural products are diverse, with very differing production systems (ranging from intensive livestock to long-lived tree crops) with strong biological constraints on the responses to economic stimuli. For example, the biological nature of livestock and crops may mean that it is difficult to alter supply in the short-run. For beef production, one has to retain additional female cattle in order to increase the breeding herd, and only when they have reached maturity may one expect an increase in the supply of beef. Not only does this mean there are lags in response, but there may be constraints on the size of the response in the short-run, as there may be a limit on the number of cattle that can be retained at any point in time. One may want to conceptualize this as a case where there is a fixed input in the production system, which in the very short-run would be true, but the issue here is that the transition path to the long-run will be constrained by very specific, biological factors. In the case of milk production, Chavas and Kraus (1990) found that the milk supply in the USA Lake States responds very little to a change in milk price – in the year that the change occurs, the value of the supply elasticity is given as 0.055. However, after a period of time the response increases, so that the elasticity of supply for a change in milk price is greater than two for a change in milk price that is sustained for 25 years. This lag in supply response is attributed to the nature of the biological time lags involved. There have been some models of supply which have been entirely conditioned by the biological system (e.g. Rayner and Young, 1980) with no economics. This may be an extreme approach, but it does emphasize the biological constraints on the supply system.

It is also important to note that, in the short-run, these factors may lead to perverse signs in the supply elasticities, as compared to those identified in the section on elasticities of supply and input demand. For example, if the price of beef rises, and farmers want to increase their long-run supply of beef in response, in the short-run they have to retain more female cattle which otherwise they would have slaughtered, i.e. the short-run elasticity of beef supply may be negative with respect to own price.

Time and price expectations

The model presented in the Foundations Section has assumed that output is produced instantaneously, as soon as inputs are applied, and all prices are known with certainty. However, because of the lags in the production system it is unlikely that, when inputs are committed to production, the price of the output will be known (and neither may some input prices). Under these circumstances the producer has to make decisions on the basis of expected prices. There is a considerable literature on the appropriate method of modelling expectations, which is made difficult by their unobservability. These range from 'naive expectations', which are simple extrapolations from the previous period, through to 'rational expectations', which utilize all relevant information, or other instruments such as futures market prices. Conceptually, the introduction of expectations does not present any real difficulties for the theory of the firm; it does however raise difficulties for empirical analysis, and may generate interesting dynamic behaviour, such as cobwebs or chaos (see Burton, 1993), when the supply side of the food market is combined with price formation.

Storage

Once one introduces time into the analysis then one has to allow for the possibility that there is a divergence between production and supply, as a result of storage. It is conventional to distinguish between products that are governed by continuous production processes (such as milk), where a flow of inputs over time produces a flow of outputs, and point production, where there may be a single production event (such as wheat). Of course the division between the two is arbitrary and depends upon the relative size of the unit of time used in the analysis and the production period. For a product that is governed by point production, there is then a need for some mechanism (i.e. storage) to meet demand in those time periods where there is no production. Under these circumstances the analysis of supply has to include two factors: production and then allocation of production over time. The latter can be analysed in terms of the supply and demand for storage. Thus consumers require the producer (or some other intermediary further up the marketing chain) to store the commodity so that it is available for consumption at later dates. One can consider the supply of storage as any other activity, with the operators trying to generate profits subject to the costs of providing that storage. The interaction of these demands and supplies will result in a certain amount of the commodity being stored over time, and a steady supply entering the market between production periods. Storage as a factor influencing supply is of course not restricted to the case of point production; for example, beef is produced throughout the year but it is still possible to store it if prices in the future are expected to be relatively high. This converts current production into future supply.

Technical change

The standard theory of the firm suggests that changes in supply will be driven by changes in (expected) prices alone. However, simple observation suggests that for some products there have been significant increases in production that cannot be explained solely by prices. The source of these increases are, of course, changes in technology. In the theory of the firm outlined above, the technology is implicitly embedded in the production function, and although we have not stated this explicitly, the use of a production function that is invariant over time implies that technology is constant. This can of course be easily amended, by allowing the production function to vary over time, which would yield time-signatured supply equations of the form: $Q_t = f_t(P_t, r_{vt})$. This would not alter any of the formal analysis of producers' decisions, as at any point in time they simply take the current technology as given. However, it does then impose further requirements on the empirical analysis, as one now has to be careful in attributing changes in supply to changes in the technology, and changes in prices. The method of modelling the source of technical change then becomes important: at its simplest one can make the parameters of the production function a function of a time trend, implying some autonomous, external source for technology. Other methodologies include using the level of Research and Development (R&D) expenditures, as it is presumed that these are the source of technical change, and hence the rate of change will be captured by this variable.

Capalbo and Denny (1986) provide estimates of the rate of technological change in agriculture during the 1960s and 1970s in the USA and Canada of approximately 1.9% and 1.4% per annum respectively. This means that in the USA output would have increased by 1.9% per annum even if the level of inputs had been held constant. However, the estimates will be sensitive to the countries studied and methodologies employed (see Capalbo and Vo, 1988, for a review of the issues involved).

DEMAND FOR FOOD

Foundations

In the neo-classical theory of demand the decision-making unit is the individual consumer. A consumer's demand for a product is the amount of it which he/she is willing and able to purchase, per unit of time, in a specified market, and at specified prices. The focus of the analysis is on choice expressed in the market-place, not demand simply as desire or need.

In contrast to the approach in supply analysis, the economist here is not overly concerned with what the particular goals or motivations of the economic agent (the consumer) might be (though they may be central to food marketing). Usable predictions of consumer behaviour can be generated, provided it can be assumed that a *preference ordering* can be defined. This is a scheme that enables the consumer to rank different bundles of goods in terms of their desirability or order of preference. Given the choice between different consumption bundles, the consumer can say whether one bundle is preferred to the other or whether he/she is indifferent between them. It is important to stress that we do not require that the consumer can say by how much one bundle is preferred to another; a simple ranking is sufficient.

In the analysis of consumer choice, economists continue to make use of the term *utility*, first introduced by Bentham in the nineteenth century. However, although we might equate utility loosely with 'satisfaction' or 'well-being', the term is now taken to reflect nothing more than the rank order of preferences: in finding the most preferred position, the consumer maximizes utility. The consumer's choice is, however, constrained by his or her limited resources. The consumer's purchasing power or budget constraint, the ability to translate preferences into purchases, is governed by the consumer's income and prices in the market-place.

The problem then is one of constrained optimization: to maximize utility, subject to a budget constraint. The solution which emerges is that, given the consumer's tastes and preferences, the demand for a product will be determined by: (i) the price of the product; (ii) the prices of all other products; and (iii) the consumer's income or total expenditure.

This relationship is conveniently expressed in mathematical terms as the *demand function*:

$$Q_1 = f(P_1, P_2, \ldots P_n, M)$$

which expresses the demand for good 1 (Q_1) in a given time period. There are n goods in the system, P_i denotes the price of good i, and M denotes the consumer's income or total expenditure. This relationship (sometimes referred to as the Marshallian demand function) is specified given the consumer's tastes.

Graphical analysis often focuses on the relationship between the demand for a product and its own price, all other factors held constant. This is the demand curve, given in Fig. 3.5. Mathematically, the demand curve (or schedule) is represented as:

$$Q_1 = f(P_1 | P_2, \dots P_n, M)$$

We may also wish to concentrate on the relationship between the quantity of the good purchased and consumer income, *ceteris paribus*. This is the Engel curve, named after Ernst Engel, a German statistician working in the nineteenth century. Figure 3.6a depicts a 'normal' good (the relationship between demand for the product and changes in income is positive), while Fig. 3.6b illustrates an 'inferior' good (the relationship is negative). Mathematically, the Engel relation can be expressed as:

$$Q_1 = f(M | P_1, P_2, \dots P_n)$$

Engel's own research findings gave rise to *Engel's Law*, that as income rises, the proportion of income spent on food falls. This 'law' has been empirically verified for many countries and time periods.

Derivation of the demand function

For some forms of analysis it is useful to have a function which provides a numerical representation of the preference ordering and this is known as the *utility function*:

$$U = U(Q_1, Q_2, \dots Q_n)$$

This function assigns a number to each possible bundle of products such that, if bundle A is preferred to bundle B, the number associated with A is greater than that for B, and if the consumer is indifferent between the two bundles, the numbers assigned by the function are the same. The utility function is simply an analytical device to express a preference ordering. Utility is not being measured in an absolute sense; the numerical representation is arbitrary, only the ranking is important.

The consumer's preferences can be described graphically by a map of *indifference curves* where each curve indicates those combinations of

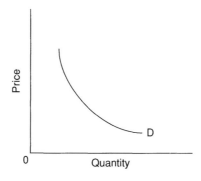

Fig. 3.5. The demand curve.

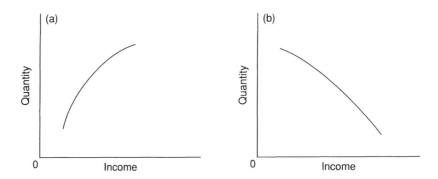

Fig. 3.6. (a) Engel curve for a normal good. (b) Engel curve for an inferior good.

products which yield the same level of utility[1]. Higher indifference curves represent more preferred combinations of the two goods. The slope of a given indifference curve indicates the marginal rate of substitution in consumption, i.e. how much of Q_1 would be required to compensate for a loss of Q_2 such that the same level of utility is maintained.

It is assumed that the underlying preference ordering, and hence the consumer's indifference map (Fig. 3.7), exhibits certain properties[2]. Specifically: (i) indifference curves are negatively sloped – the more of one good we consume the less of the other is required to maintain our level of satisfaction; (ii) indifference curves do not intersect; (iii) there is an indifference curve through each point in the graph; and (iv) indifference curves are 'convex' – they bulge towards the origin – reflecting the property that as we consume less and less of one good, we require greater and greater amounts of the other to maintain the same level of satisfaction. These properties ensure that the solution to the constrained optimization problem is unique and yields maximum attainable utility.

The budget constraint is illustrated in Fig. 3.8. Given the price of Q_1 and the consumer's income, the maximum amount of Q_1 which could be purchased is represented by point A ($=M/P_1$) on the horizontal axis. Similarly, if the consumer spent all of his or her income on Q_2, the maximum quantity obtainable, given the price of Q_2, is B ($=M/P_2$). Various combinations of the two goods are also affordable – illustrated by the area OAB, the *market opportunity set*. The consumer's problem then is to find the most preferred affordable bundle of products – the highest indifference curve consistent with the market opportunity set. In Fig. 3.9, the consumer's optimal choice is denoted as A' of Q_1 and B' of Q_2. Clearly, the solution is governed by the level of prices and income, given preferences. The analysis can be generalized to the multi-product case, yielding the demand functions described earlier.

The budget constraint is the constraint which receives most attention in the economic analysis of choice. However, it may be appropriate to adapt the analysis in order to incorporate other constraints. These might include:

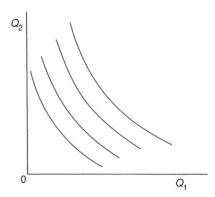

Fig. 3.7. Indifference curves.

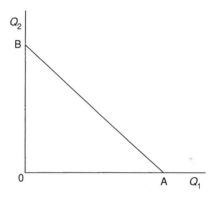

Fig. 3.8. The budget constraint.

1. Time. Becker (1965) has stressed that time can be seen as an important input in the consumption process and as a scarce resource limiting consumption choices. The enjoyment of food, for example, requires expenditure of time, in purchasing, preparation and cooking, as well as money expenditure on the raw material. As time is limited, the consumer must decide on its allocation between labour market and consumption activities. The consumer problem in this approach is thus specified as one of utility maximization given the budget and time constraints.

2. Legal or institutional constraints. A range of constraints arise as a result of macroeconomic policy and the need to protect the consumer or the environment. For example, consumer purchasing decisions may be constrained by controls on credit or foreign exchange availability, by import bans, by controls on the use of food additives or drugs, by age limits on alcohol and cigarette purchase, etc. These all restrict the consumer's choice

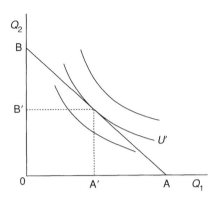

Fig. 3.9. The consumer's optimal choice.

in the market-place, irrespective of underlying preferences or purchasing power.

3. Socio-cultural constraints. To some extent, consumption patterns may be governed by social norms and customs. For example, the consumption of horsemeat is not socially acceptable in the UK; orthodox Jews and Muslims do not eat pork; and so on. Social norms may also be enshrined in the legal and institutional framework of the country.

The analysis can be modified to take account of these constraints, but most empirical work is founded on the basic model of consumer choice with which we started: taking prices and incomes as given, the consumers' task is to allocate their budgets to best serve these well-defined preferences.

As already noted, the neo-classical theory of demand focuses on the individual consumer. However, in applied economics we are rarely concerned with the actions of the individual. Rather we are interested in the aggregate demand of a number of consumers in a specified market: viz. the market demand. It is usually taken for granted that the market demand relationships conform to those specified for the individual consumer[3].

Consumer surplus

It is often useful to have a monetary measure of the extent to which the consumer benefits from a transaction. A commonly used measure is 'consumer surplus' (analogous to producer surplus discussed earlier), which is defined as the difference between the maximum outlay the consumer would be willing to pay for some quantity of the product and the actual total cost of purchasing that quantity in the market. The easiest way to proceed is to take the ordinary demand curve as the basis of the calculation[4]. Consider Fig. 3.10a. The height of the demand curve at any quantity indicates the most the consumer would be willing to pay for an extra unit of the product: P_1 is the maximum price the consumer would pay for the first unit, P_2 the maximum for the second unit and so on. If the market price is in fact P_m, the consumer purchasing Q_m would benefit by paying less than the maximum he or she would be willing to spend for that amount. By making the increments on the horizontal axis infinitesmally small, the smooth demand curve of Fig. 3.10b is obtained and the consumer surplus is measured as the area ('a') under the demand curve and above the market price. When dealing with aggregate market demand, consumer surplus is measured simply by the area under the market demand curve and above the price line. Finally, if the market price were subsequently to rise, there would be a loss of consumer surplus (area 'b' in Fig. 3.11).

Demand elasticities

It is often useful to have measures of the responsiveness of consumers to market stimuli. For this purpose, as with the analysis of supply, we define elasticities as follows.

The *own price elasticity of demand* is defined as the proportionate change in quantity demanded of good i relative to the proportionate change in its price, *ceteris paribus*:

$$\varepsilon_{ii} = \frac{\Delta Q_i / Q_i}{\Delta P_i / P_i} = \frac{\Delta Q_i}{\Delta P_i} \cdot \frac{P_i}{Q_i} = \frac{\partial Q_i}{\partial P_i} \cdot \frac{P_i}{Q_i}$$

The *cross price elasticity of demand* measures the proportionate change in the quantity of good i relative to the proportionate change in the price of good j, *ceteris paribus*:

$$\varepsilon_{ij} = \frac{\partial Q_i}{\partial P_j} \cdot \frac{P_j}{Q_i}$$

The *income elasticity of demand* measures the proportionate change in the

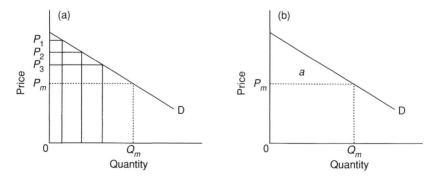

Fig. 3.10. (a) The demand curve measure of consumer surplus. (b) Consumer surplus generalized.

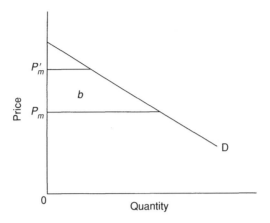

Fig. 3.11. Loss of consumer surplus.

quantity of good i relative to the proportionate change in consumer income, *ceteris paribus*.

$$\eta_i = \frac{\partial Q_i}{\partial M} \cdot \frac{M}{Q_i}$$

The own price elasticity of demand is expected to be negative. The cross price elasticity may be positive, where product i is a substitute for j, or negative where i is complementary. The income elasticity will be positive for normal goods and negative for inferior goods. Engel's Law is often interpreted as meaning the income elasticity of food (in the aggregate) is less than unity (inelastic).

Relationships among demand elasticities

Although not proved here, the following interrelationships, sometimes termed 'the general restrictions', should hold:

1. *Homogeneity*: each demand function is homogeneous of degree zero, i.e. if all prices and income are increased by the same proportion, demand for a given product will remain unchanged. As in the analysis of supply, there is no 'money illusion'. From this proposition the following condition can be derived:

$$\varepsilon_{i1} + \varepsilon_{i2} + \ldots + \varepsilon_{in} + \eta_i = 0 \quad i = 1, \ldots n$$

i.e. for a given product (i), the own price, all cross price and the income elasticities must sum to zero.

2. *Slutsky symmetry*: cross price elasticities are not symmetric but the following relationship can be derived:

$$\varepsilon_{ij} = \frac{w_j}{w_i} \varepsilon_{ji} + w_j(\eta_j - \eta_i) \quad i \neq j$$

If the income elasticities of the two goods are of similar magnitude, then:

$$\varepsilon_{ij} \approx \frac{w_j}{w_i} \varepsilon_{ji}$$

3. *Engel aggregation*: following from the budget constraint:

$$w_1\eta_1 + w_2\eta_2 + \ldots + w_n\eta_n = 1$$

i.e. the weighted sum of all income elasticities is unity. The weights are the budget shares of the respective goods. This implies that any increase in the consumer budget must be fully allocated across the range of market goods.

These interrelationships are particularly pertinent in the analysis of 'complete demand systems' in which the researcher attempts to model the demand for all goods or a self-contained sub-set of products. If investigating the demand for a single product only homogeneity is directly relevant.

Table 3.2 presents a set of empirically estimated elasticities which illustrate these relationships. Here meat and fish are treated as a self-contained sub-set of food products, so that 'total expenditure' refers to the total outlay on the four meats and fish. In particular, note that: (i) all the general restrictions hold; (ii) the own price elasticities are all negative; (iii) beef, lamb and pork appear to be substitutes for each other, but some of the cross price elasticities related to chicken and fish are less easy to rationalize.

Examples of similar studies of food demand systems include Moschini *et al.* (1994) for the USA, Mergos and Donatos (1989) for Greece, and Molina (1994) for Spain.

Extensions

The analysis can be extended in a number of ways. Again, as with supply,

Table 3.2. Price and total expenditure elasticities for meat and fish, Great Britain 1961–1987* (source: Burton and Young, 1992).

	With respect to the price of:					With respect to:
	Beef	Lamb	Pork	Chicken	Fish	Total expenditure
Beef	−1.66	0.24	0.00	0.09	0.02	1.32
Lamb	0.58	−1.42	0.21	−0.13	−0.32	1.07
Pork	0.15	0.14	−1.03	−0.03	−0.01	0.79
Chicken	0.34	−0.15	−0.15	−0.90	−0.11	0.98
Fish	0.19	−0.22	−0.01	−0.05	−0.67	0.76

*Mean budget shares are: 0.292 (beef); 0.139 (lamb); 0.293 (pork); 0.11 (chicken); 0.166 (fish).

these extensions reflect marketing issues – such as marketing margins, storage, logistics, and changing consumer tastes and preferences – which are covered in the rest of this book.

Demand at different market stages

In moving the product from the farm-gate to the retail consumer, there are a number of changes of ownership and economic processes (assembly, processing, storage, transportation and retailing). The theory of demand outlined above is concerned primarily with aggregate consumer response; demand at other market levels (e.g. wholesale, farm-gate) can be viewed as being derived from the consumer-level function.

Dynamics

The theory presumes instantaneous and full adjustment to changes in market stimuli. There are, however, a number of factors which may cause lags in response to price or income changes. These include the influence of habits, the storability of some goods (which allows consumers to consume out of stocks as well as current purchases), expectations (which may be based on the past history of price and income changes), and the institutional and legal constraints referred to earlier. A dynamic specification allows the researcher to investigate both short-run and long-run responses to changes in market conditions.

Taste changes

In the orthodox theory tastes are assumed to be constant and the economist's job is simply to predict the outcome of a given set of tastes. Recent empirical work suggests that better predictions may be obtained if the possibility of systematic changes in tastes is permitted. Some of this work makes explicit use of advertising expenditures as an explanatory variable in the empirical analysis of demand.

'New' theories of demand

In the approaches of Lancaster (1991) and Becker (1965) the consumer does not derive utility from market goods *per se* but from another set of entities derived from market goods. In Lancaster's model, the consumer is interested in the characteristics of goods (for example, fuel consumption and engine capacity of the motor car, the nutritional content of food products, etc.). In Becker's model the basic variables of consumer interest are more nebulous, e.g. nourishment, comfort, family life and social distinction. These 'commodities' are produced in the household by combining market goods with time inputs (that is to say, a household production function is specified). Both approaches can provide insights in the analysis of some economic problems for which the orthodox theory is relatively barren.

CONCLUDING REMARKS

The analysis of supply and demand offers an important tool that can be applied to a wide variety of economic problems. Indeed it underpins much of the subject matter of this book. When we try to understand the workings of a product market we need to have some framework within which to order and analyse the large amount of information available. Although the neo-classical models outlined here are based on assumptions that may appear to be overly restrictive, their merit is that they provide a mechanism for imposing structure on the data, and perhaps most importantly, imposing a consistency of interpretation. Thus the models give information on what factors we would expect to be important in determining supply and demand, and perhaps more significantly, exclude some outcomes. As noted earlier, the value of the models is finally judged by their ability to explain market behaviour. If they fail to do so, then they still have a useful role, in providing a springboard for further theoretical development.

NOTES

1. This is analogous to the isoquant in production, referred to earlier. However, whereas in production each isoquant has an unambiguous meaning (it is specified for a given level of physical output), the number assigned to an indifference curve, denoting a level of utility, is quite arbitrary.
2. More formally, these properties correspond to a set of axioms concerning the preference ordering: Completeness; Transitivity; Non-satiation; Continuity; Strict Convexity.
3. In fact the formal theoretical requirements for what is known as 'consistent aggregation' are quite strict. See Deaton and Muellbauer (1980).
4. By taking this approach, we will, strictly, only obtain an approximation of the true welfare benefits to the consumer. If the real income effect of a price change is small, however, the approximation will be acceptable. For more details, see, for example, Varian (1993).

FURTHER READING

Colman and Young (1989) and Tomek and Robinson (1991) cover this material in more detail. Goodwin (1994) presents, in a non-technical way, a number of applications for agricultural markets. To explore the theoretical foundations in more detail, see Beattie and Taylor (1985) and Chambers (1988) for supply, and Thomas (1987) and Deaton and Muellbauer (1980) for demand.

REFERENCES

Beattie, B.R. and Taylor, C.R. (1985) *The Economics of Production*. John Wiley and Sons, New York.
Becker, G.S. (1965) A theory of the allocation of time. *Economic Journal* 75, 493–516.

Burton, M. (1993) Some implications of chaos for models in agricultural economics. *Journal of Agricultural Economics* 44, 38–50.

Burton, M. and Young, T. (1992) The structure of changing tastes for meat and fish in Great Britain. *European Review of Agricultural Economics* 19, 165–180.

Capalbo, S.M. and Denny, M. (1986) Testing long-run productivity models for the Canadian and US agricultural sectors. *American Journal of Agricultural Economics* 66, 615–625.

Capalbo, S.M. and Vo, T. (1988) A review of the evidence on agricultural productivity and aggregate technology. In: Capalbo, S. and Antle, J. (eds), *Agricultural Productivity: Measurement and Explanation*. Resources for the Future Press, Washington, DC, pp. 96–137.

Chambers, R.G. (1988) *Applied Production Analysis. A Dual Approach*. Cambridge University Press, Cambridge.

Chavas, J.-P. and Kraus, A.F. (1990) Population dynamics and milk supply response in the US Lake States. *Journal of Agricultural Economics* 40, 75–84.

Colman, D. and Young, T. (1989) *Principles of Agricultural Economics. Markets and Prices in Less Developed Countries*. Cambridge University Press, Cambridge.

Deaton, A. and Muellbauer, J. (1980) *Economics and Consumer Behavior*. Cambridge University Press, Cambridge.

Goodwin, J.W. (1994) *Agricultural Price Analysis and Forecasting*. John Wiley and Sons, New York.

Lancaster, K. (1991) *Modern Consumer Theory*. Edward Elgar, Aldershot.

Mergos, G. and Donatos, G. (1989) Demand for food in Greece: An almost ideal demand system analysis. *Journal of Agricultural Economics* 40, 178–184.

Molina, J. (1994) Food demand in Spain: an application of the almost ideal demand system. *Journal of Agricultural Economics* 45, 252–258.

Moschini, G., Moro, D. and Green, R. (1994) Maintaining and testing separability in demand systems. *American Journal of Agricultural Economics* 76, 61–73.

Ozanne, A. (1992) *Perverse Supply Response in Agriculture. The Importance of Produced Means of Production and Uncertainty*. Avebury, Aldershot.

Rayner, A.J. and Young, R.J. (1980) Information, hierarchical model structures and forecasting. *European Review of Agricultural Economics* 7, 289–313.

Thomas, R.L. (1987) *Applied Demand Analysis*. Longman, London.

Tomek, W.G. and Robinson, K.L. (1991) *Agricultural Product Prices*, 3rd edn. Cornell University Press, Ithaca, NY.

Varian, H.R. (1993) *Intermediate Microeconomics. A Modern Approach*, 3rd edn. W.W. Norton and Co., New York.

4 Agricultural Price Analysis

Hoy F. Carman

Department of Agricultural Economics, University of California, Davis, California 95616–8512, USA

INTRODUCTION

Agricultural commodity prices, particularly their level, variability and determinants, are of central importance in the study of agricultural marketing. Prices for agricultural commodities are an important determinant of the level of farm income, the cost of food to consumers, export income for countries engaged in commodity trade, the profits for agricultural marketing firms and returns to commodity traders and speculators. Because of the importance of agricultural commodity prices to economic growth and development, most national governments have extensive and often complex policies and programmes to deal with commodity prices in both domestic and export markets. The nature of these policies and programmes is politically important because of their impact on food prices, farmers' incomes, and relations with trading partners.

This chapter is devoted to description and analysis of the basis for price variations through time and space, price differences related to quality and the impacts of government programmes on prices. It begins by defining a market, followed by a short discussion of the concepts of supply, demand and equilibrium price within the market. Changing market prices through time, as related to seasonality of supply and demand, costs of storage and price cycles are examined. Geographic price relationships are presented within the framework of a simple two-region trade model using excess supply and demand functions. Relationships between grades, product quality and prices are outlined in terms of buyer benefits, the contribution of grades to efficiency and hedonic prices. The chapter concludes with an examination of the impacts of government programmes designed to deal with perceived commodity pricing problems. The programmes outlined include government purchases, limits on production, tariffs, export subsidies, deficiency payments and price controls. Subsequent chapters will discuss price variations by level in the marketing system (marketing margins) and pricing related to the competitive structure (market structure).

SUPPLY, DEMAND AND MARKET PRICE

The concepts of individual consumer demand and firm supply were presented in the previous chapter. We will now move from the individual to the market and combine the aggregate supply and demand functions to determine market price. There are many definitions of a market, ranging from brief to detailed. To help students, Houck (1984) attempts to clarify the concept of a market. He states that:

> A market is a collection of actual or potential buyers and sellers of a specific good or service. This collection has two characteristics: (1) none of the buyers has the option to purchase the item from sellers outside this collection, and (2) none of the sellers has the option to sell the item to buyers outside this collection. The interaction of these buyers and sellers generates a set of interrelated prices and conditions of sale or use. The principles or facts determining which buyers and sellers are in this collection identify the market spatially, temporally and politically.

Houck's definition provides a framework for moving from the individual to the market. The market demand curve, which is the sum of individual demands, shows the total quantities of a good that will be purchased by all consumers in the market at all possible prices, other things equal. The horizontal summation of individual demands for a market with three consumers is illustrated in Fig. 4.1. At a price of p_1 consumers will purchase $q_1 + q_2 + q_3 = Q$ of the product. Other points on the market demand curve are obtained similarly. This reveals an important characteristic of market demand – that the position of market demand is based on the number of consumers in the market, and that the market demand will shift to the right (increase) as the number of consumers in the market increases. While each of the demand curves is drawn in the price-dependent form, horizontal summation of the functions requires that they be expressed in the quantity-dependent form, i.e. $q = f(p)$.

The market supply curve is, likewise, the horizontal summation of the supply curves of each of the firms in the market. This market supply curve shows the various quantities of the good that all sellers will place on the market at all possible prices, other things equal. The summation of firm supply curves to obtain market supply, which uses the same procedure as market demand, is illustrated in Fig. 4.2. As shown in Chapter 3, each firm's marginal costs above average variable costs provides the quantitative expression for its supply curve.

Market equilibrium

The market demand and supply functions are combined to form the model of competitive price determination in the short-run. This can be illustrated

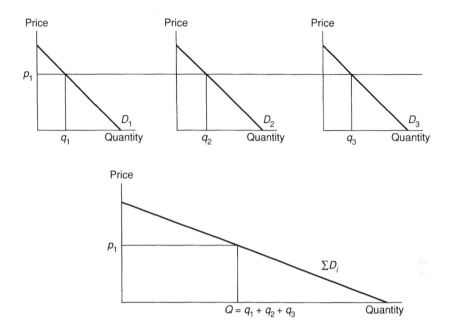

Fig. 4.1. Summation of individual demands to obtain market demand.

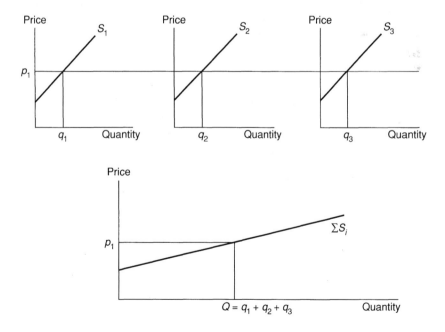

Fig. 4.2. Summation of firm supplies to obtain market supply.

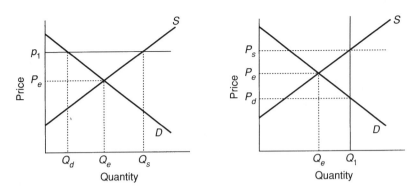

Fig. 4.3. Market supply, demand and equilibrium.

by the familiar Marshallian supply–demand cross, probably the single most useful tool for economic analysis. There is just one price where the quantity supplied by firms will equal the quantity demanded by consumers. As shown in Fig. 4.3, the intersection of the market demand and supply curves defines the market equilibrium price (P_e) and quantity (Q_e). This is an equilibrium because, once established, it tends to remain and, if disturbed, price and quantity will tend to return to the equilibrium. This can be easily illustrated by the case where price is temporarily established above the equilibrium price (P_1) or quantity is temporarily established above the equilibrium quantity (Q_1). Either event will set forces in motion which will tend to return price and quantity to the equilibrium. In the case of an above equilibrium price, shown in the left panel of Fig. 4.3, producers are willing to supply (Q_s) units of product but consumers will demand only (Q_d) units. The resulting surplus will encourage firms to cut prices to move the product. If the price was temporarily below equilibrium, the excess demand would cause consumers to bid the price up to acquire the product. In the case of an above-equilibrium quantity (Q_1), shown in the right panel of Fig. 4.3, marginal costs exceed what buyers are willing to pay and this provides incentives for firms to reduce output. In reality, movements to equilibrium are likely to include both price and quantity adjustments.

It is important to note that markets do not adjust immediately to equilibrium because of transactions costs. These include the costs of obtaining information about the market and the costs of finding a place in which to transact business. It takes time to learn prices, determine product availability, establish contacts between buyers and sellers, ascertain terms of trade, and to analyse quality.

Pricing with inelastic supply and demand

The aggregate demand for food and for groups of food products (i.e. meat, dairy products, etc.) tends to be very inelastic at the farm level. For the developed countries, this demand tends to increase with increased income and population, but the increase will not be as great as for many other non-farm products. The growth in food demand due to population is usually proportional (in the range of 1% for each 1% increase in population), but the impact of increased incomes is much smaller, with income elasticities of demand in the range of 0.2–0.3. While the total demand for food may change very little in response to increased income (because of the limited capacity of the stomach), there may be important changes for given commodities as consumers attempt to upgrade their diets or purchase more built-in services. The demand for particular products or groups of products may shift rapidly in response to changes in the prices of related products.

The supply of individual food products as well as the aggregate supply of food tend to be quite inelastic in the short-run. Because of the biological production processes found in agriculture, there are usually significant lags between the time a production decision is made and the time that production is available for consumption. Producers have a 'planned' level of production that is reflected by planted acreage and average or expected yields. As harvest approaches, planned production may shift due to factors affecting yields and, after harvest, actual supply will be fixed until the next harvest. Farmers may require months or years to respond to increased or decreased prices, so that relatively high or low prices may persist for considerable time periods. For example, to increase production in response to increased prices may require a year for annual field crops, more than a year for pork, three years for beef, and six or more years for tree crops such as apples, oranges or almonds. Yields may vary considerably from region to region and from year to year because of favourable or unfavourable weather or because of pests and disease.

The inelastic nature of short-run supply and demand can be used to illustrate the relatively large impact on the level of prices of small changes in either supply or demand. As shown in the left panel of Fig. 4.4, a small difference between expected and actual supply, as caused by weather, can result in a large change in price. Suppose that the actual supply, shown by the completely inelastic supply curve S_2, is larger than expected supply S_1. This will result in a proportionally greater change in price because demand is relatively inelastic. Ideal weather conditions can result in a bumper crop. Likewise, abrupt and unpredictable reductions in supply caused by a frost, too much rain, or a drought occur all too frequently around the world. A sudden change in demand is not as common but does occur. This is illustrated in the right panel of Fig. 4.4 by the shift in the short-run demand curve from D_1 to D_2, resulting in a price decrease from P_1 to P_2. A recent example in the USA was the impact of adverse publicity concerning the use of the pharmaceutical Alar on apples and its possible adverse impact on consumers who ate large amounts of apples or processed products such as juice. A decrease in demand severely depressed the prices of apples and led the government to buy apples as a form of relief to producers. Publicity can also increase demand, as occurred for red wines in the USA. A television news programme titled 'The French Paradox' attributed the French

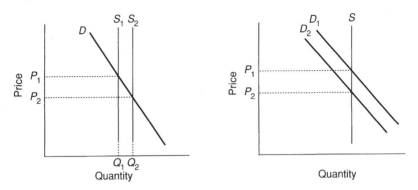

Fig. 4.4. The effects of inelastic supply and demand on relative price changes.

population's lower incidence of heart disease, in the presence of relatively high consumption of saturated fats, to moderate consumption of red wine. A sharp increase in demand and sales of red wine soon led to a price increase.

With the internationalization of agriculture, price-making forces are no longer confined to national boundaries. A freeze in Florida affects the price of oranges in Brazil; a reduced hazelnut crop in Turkey affects the price of California almonds; a drought in Australia affects Canadian and USA wheat prices; and a freeze in Brazil affects the price of a cup of coffee around the world. Likewise, government policies in one country can have significant impacts on commodity demand and prices in other countries.

Price relationships

Markets may exist in time, form or space, and price relationships are maintained through storage, processing and transportation. These marketing functions add utility to products, with time utility added by storage, form utility added by processing and place utility added by transportation. The pricing relationships between these markets under competitive conditions is expressed by the law of one price (LOOP) which states that, under competition, all prices within a market tend to equality when allowances are made for the costs of storage, processing and transportation. Most empirical applications of the LOOP, however, have dealt with transportation and geographic markets.

The prices of agricultural products may be observed at different levels in the marketing system (i.e. farm, processor, wholesale, retail) and they may be recorded and reported on a daily, weekly, monthly, crop year or annual basis. Because of systematic variability in prices, one must be very careful to define the product (quantity, quality) and other relevant characteristics such as time period, level in the marketing system and physical location when quoting or using price data.

COMMODITY PRICES THROUGH TIME

Many firms and individuals with interests or involvement in agricultural marketing are concerned with commodity price movements over time. Producers and middlemen, for example, must make decisions concerning production, storage, timing of purchases, and timing of marketing. Those businesses that carry large inventories must evaluate the risks of price changes over time and pursue strategies for hedging. Commodity speculators continually attempt to forecast prices in order to profit from changing price relationships on futures and cash markets. Price changes over time may be classified as: (i) long-term price trends due to changing supply and demand or macroeconomic factors such as inflation; (ii) seasonal price movements due to regular seasonal changes in supply and demand that are repeated each 12 months; (iii) cyclical price movements that occur over several years; and (iv) year-to-year and irregular price changes. A commodity price observed at any point in time may include a combination of trend, seasonal, cyclical and irregular effects.

Seasonal price changes

Agricultural commodity prices often tend to vary predictably through time as a result of seasonal variations in supply and demand. The harvest period often extends over only one, two or three months, with consumption over the remainder of the year provided from storage stocks. This pattern of production and storage, characterized by the grains and oilseeds, tends to result in prices that are lowest at harvest time, with prices increasing in line with charges for storage over the year. Animal product prices may vary seasonally as a result of biological characteristics, the availability of feed, or management practices. Beef cow prices, for example, are often low in the autumn due to ranchers culling cows after calves have been weaned; egg production tends to be highest, and prices lowest, in the spring; manufacturing milk production is also highest in the spring due to birth of calves and favourable pasture conditions. Seasonal variation in demand can also have an impact on prices. Changing temperatures are often associated with changes in the demand for cold beverages, ice cream, roast beef, bacon and lemons, while holidays are responsible for 'spikes' in the demand for products such as turkey (Thanksgiving and Christmas), cranberries (Thanksgiving), ham (Easter) and marzipan (Christmas). Seasonal price characteristics may change over time as a result of changing production practices, such as occurred with milk shifting from seasonal to year-around production, or with new product development, as has occurred with turkeys and cranberries. Seasonal price changes for some products have been reduced by the development of large-scale specialized production facilities and vertical integration through ownership or contractual arrangements.

Seasonal price patterns are typically illustrated by graphing commodity prices over time. The theoretical pattern of prices for a storable commodity

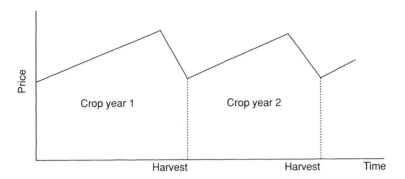

Fig. 4.5. Seasonal price relationships for a storable crop.

over a two-year period is shown in Fig. 4.5. Prices are lowest at harvest, then increase over time in line with the costs of storage, and finally begin to decrease as the next harvest period nears and begins. It is important to note that actual seasonal price patterns are seldom as exact as shown in Fig. 4.5, even though they may be similar from year to year. The magnitude of high and low points will change because of factors such as weather or because of shifts in other determinants of the position of supply and demand curves. Changes in storage technology or the costs of storage will also alter seasonal price changes, as will the actions of buyers and sellers as they attempt to take advantage of price movements.

The analysis of seasonal price changes may be based on simple graphics, such as illustrated in Fig. 4.5, on price indexes, or on regression analysis. An index of seasonal price changes is probably the easiest way in which to view average prices over the year. Using this approach, an average over 12 months or over several periods of 12 months is the base value (index value of 100). Dividing each weekly or monthly price by the base value provides a measure of the seasonal price variability. For example, if the overall average (base value) for the year is US$80 per tonne and the price for March is US$100 per tonne, the March index will be US$100/US$80 × 100 = 125. This index indicates that March prices were 25% higher than the annual average. Seasonal retail price relationships for Red Delicious apples in the USA are illustrated in Fig. 4.6. This ten-year average shows that prices were lowest after the harvest was completed in November at 92% of the crop year average, increasing to an annual peak of 115% in August, and then decreasing as harvest got underway in major producing areas in late September.

Commodity storage

Commodity storage and product inventories link markets together through time. Through this linkage, storage creates time utility by bridging the gap between discrete production and consumption. As a productive activity, storage uses resources and involves costs. The nature of these storage–cost

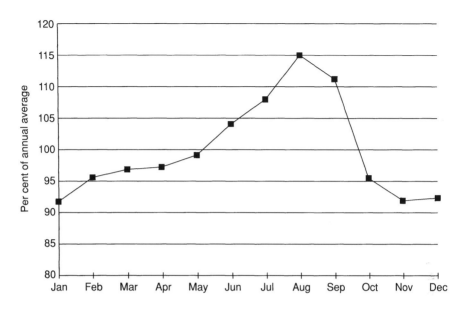

Fig. 4.6. Seasonal index of monthly average retail prices for Red Delicious apples, 1984–1993.

relationships is important in storage decisions and in the price relationships for a product through time. Going back to Fig. 4.5, the seasonal increase in prices, represented by the slope of the price line, is expected to cover the costs of storage and risks associated with price changes over a number of periods. However, during the short-run in a competitive market, the increase in product prices may be more or less than the cost of storage. With that in mind, it is useful to consider the costs involved with storing a product and the nature of storage–cost relationships.

Storage involves physical facilities such as warehouses, grain elevators, refrigerated rooms, or storage tanks. Thus, storage costs include fixed costs consisting of the usual overhead associated with facilities as well as handling costs for placing and removing products that do not vary with the length of the storage period. Variable costs, which are related to time in storage rather than output, consist of charges for fuel and utilities, protection (monitoring, inspection, chemical treatment, etc.), labour costs related to time in storage, insurance, and interest. The nature of fixed and variable costs related to time in storage, which are likely to be linear with time, are illustrated by the lower line in Fig. 4.7. The fixed costs are shown by the intercept value 'a' at time zero when the product is placed in storage and the variable costs are given by the slope of the storage cost function. Since most agricultural products are perishable and subject to changing quality when stored, it may be necessary to consider product deterioration and shrinkage as a cost of storage. Loss of product value may be due to pest and insect damage, loss of weight or volume, spoilage, or decreased quality. The top line in Fig. 4.7 provides a simple illustration of storage costs over time when there is a loss in value. The vertical (shaded)

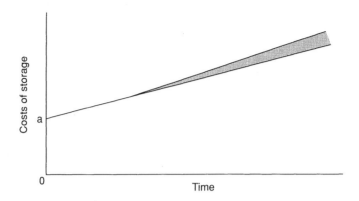

Fig. 4.7. Fixed and variable direct costs of storage and indirect costs associated with product deterioration.

difference between linear storage costs and net costs of storage is the loss in product value due to storage. There may be an increase in market value during storage for a few commodities such as wine and cheese, but even these products will suffer quality loss at some point in time.

While total production of some crops, such as the grains and oilseeds, are placed in storage as a regular part of distribution and marketing, only a portion of other crops, such as fresh fruits and vegetables, may be stored. For these latter crops, the storage–cost relationship and expected seasonal price relationships are very important in the storage decision. Apples, for example, can be marketed at the time they are picked, they may be placed in regular refrigerated storage for up to four months, or they may be placed in controlled atmosphere (CA) storage for up to one year. Quality will vary with each marketing option for a given time period and both fixed and variable costs will vary by option. Because of comparatively high costs associated with construction and maintenance of air-tight rooms, CA storage is typically used only for longer storage periods.

The effect of storage costs on monthly sales and seasonal prices can be illustrated by a simple diagram. Assume that perfect market conditions prevail so that the price in any month must be the price at harvest plus storage costs and the sum of the monthly quantities sold must equal the harvested supply. The demand curve (DD') in Fig. 4.8 represents the monthly demand, which is uniform and stable across months (Bressler and King, 1970). In month 1 (the harvest period), the monthly quantity sold (Q_1) will be the highest, the monthly price (P_1) will be the lowest, and the quantity stored will be greatest, immediately following harvest. Note that prices increase abruptly from the harvest period to the first month of storage, as a result of the fixed costs associated with facilities (the constant a in addition to the monthly price increase of e), and then increase regularly as a result of linear variable storage costs (equal to e).

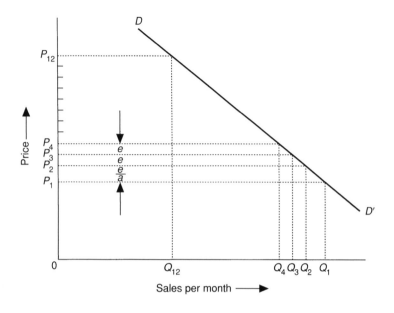

Fig. 4.8. Seasonal prices and quantities illustrating storage with uniform monthly demand (Bressler and King, 1970).

Price cycles

A price cycle is a pattern of prices that repeats itself regularly over time, usually in response to a production cycle. Because of the inverse relationship between price and quantity expressed by the law of demand, high prices are associated with low production and low prices with high production. In the USA, production and price cycles have been observed for livestock and perennial crops. The price and production cycles for hogs are fairly well defined, as illustrated by Fig. 4.9. The inverse relationship between prices and production is evident, with prices peaking every three to five years. The individual cycles observed for cattle are longer, with peaks occurring every seven to twelve years. Analysis of the California lemon industry indicates that there is a cycle which takes about 20 years from peak to peak. It should be noted that, while the pattern of prices and production is similar for each cycle for each product, each is clearly different because of the existence of trends, seasonal components and random factors. The length of a cycle tends to vary directly with the time required for the quantity supplied to respond to changing prices.

The mechanics of production and price cycles are often explained as follows: high prices lead producers to expand production, but this expansion takes time because of the lags involved in producing a new generation of breeding livestock or in establishing a new orchard. High prices continue until new production begins from the expansion of capital stock; then prices

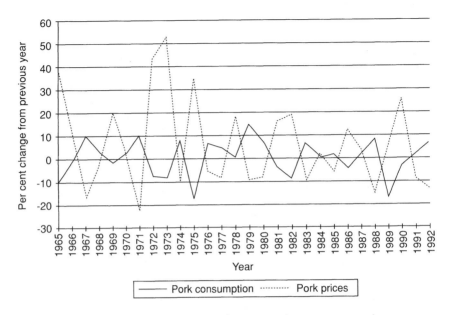

Fig. 4.9. USA hog prices and per capita pork consumption, percentage changes, 1965–1992.

decrease until it is no longer profitable to invest in new capacity. The length of the cycle is related to the time that it takes to produce a new generation, with lagged responses to changes in prices and other factors being the key to cyclical behaviour.

The cobweb model

The 'cobweb theorem' provides the basis for the economic model most often used to explain cyclical price movements, as just described[1]. The cobweb model makes three assumptions which are necessary for cyclical behaviour of prices and quantities. They are: (i) production decisions are based on current or recent prices; (ii) there is a lag of at least one full period between the decision to produce and the resulting production; (iii) current price is a function of current supply, which is determined by current production.

Given the time lags expressed by these assumptions, pricing becomes a recursive process that can be illustrated by a simple causal chain. The current quantity supplied is a function of past prices; the quantity produced in the current period is sold in the current period; and, current price is determined by current sales. The functions are:

Supply: $Q_t = f(P_{t-1})$, or in linear form, $Q_t = a + b P_{t-1}$
Demand: $P_t = f(Q_t)$, or in linear form, $P_t = c - d Q_t$

The causal chain can be illustrated as:

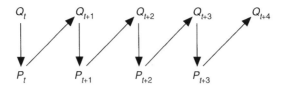

Using the linear functions just specified, we can illustrate the way the cobweb model works, and see that the pattern of price–quantity movements traced out on the supply–demand diagram is the reason it is called a cobweb model. In Fig. 4.10, the price in period 1 (P_1) leads producers to produce quantity Q_2 in period 2, which results in a price of P_2 in period 2. The path of price and quantity adjustments over time, illustrated in Fig. 4.10, converges to an equilibrium price and quantity. This is because the supply function has a steeper slope than does the demand function. If the demand function has a steeper slope than does the supply function, then the cycle would diverge, while if the slopes are equal, the cycle will be continuous and prices will vary by the same amplitude each cycle.

Note that the length and severity of cycles depend on two factors: the manner in which producers form their price expectations, and the length of the adjustment process. The cobweb model is clearly too basic for direct application to current agricultural pricing problems but it does highlight the relationships found in a recursive model. Applied econometric models have to be modified to allow for factors which shift the functions and a complete model will typically require more than two equations. The conceptual

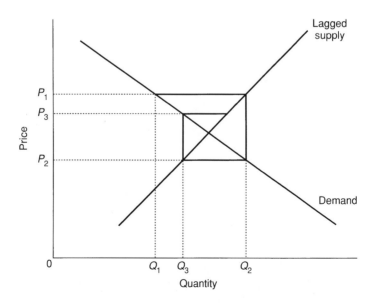

Fig. 4.10. An illustration of price and quantity adjustments in the cobweb model.

specification of the model is relatively simple; definition of appropriate variables to estimate the model is not.

SPATIAL PRICES

The dealing of individual buyers and sellers in a geographic area will result in a market price, but that price typically varies from point to point in the market because of transportation costs. However, since there are similar supply and demand factors affecting prices and arbitrage is possible, the prices at different points in the market are expected to move together.

Geographic price relationships

The commonly accepted definition of a competitive market area that can be traced back to Marshall[2], states that 'A market for a good is the area within which the price of the good tends to uniformity, allowance being made for transportation costs.' In terms of observable data, if a good is produced at point A and supplied to point B with transportation costs between A and B of T_{ab}, then we would expect the equation $P_b = P_a + T_{ab}$ to hold with perfect knowledge and no impediments to exchange. This equation states the law of one price in geographic terms. In real world commodity markets, however, prices seldom differ by exactly the transportation costs for several reasons. There may not be a unique transportation cost function and costs of movement may be less for some buyers than for others, because of size of shipments, back hauls, modes available, etc. In addition, perfect information and foresight do not exist. Thus, there can be divergent price movements that persist for some time. With this warning in mind we now turn to an examination of spatial price relationships.

Tomek and Robinson (1990) summarize the principles that underlie price differences between regions in a competitive market with a homogeneous commodity as follows: (i) price differences between any two regions (or markets) that trade with each other will be exactly equal to transfer costs; and (ii) price differences between any two regions (or markets) that do not trade with each other will be less than or equal to transfer costs.

Price differences cannot exceed transfer costs in a perfect market, for the simple reason that if they do, buyers will purchase goods in the lower price market and ship to the higher price market. This arbitrage will tend to bid up prices in the low price area and reduce prices in the high price area until prices differ by the transfer cost and shipping between areas is no longer profitable. Given these principles, the theoretical structure of prices over space can be determined. This price structure will be a function of transfer costs per unit of product and the pattern of trade. One may even be able to determine a price structure in the absence of trade. For example, if all surplus-producing areas ship to one central market (but not to each other), the price in each area will be the central market price minus transfer costs. The difference in prices in the different surplus-producing areas will be only the difference in transfer costs to the central market. Since net prices to any producer in the market area are the central market price minus transfer costs, two observations are relevant. First, the size of the market area will tend to be a direct function of the cost of transportation; and, second, net returns (and the capitalized value of resources used in production) will decrease as the distance from the central market increases.

Determination of the price structure can be much more complex when there are several consuming centres to which several surplus-producing areas can ship. Given the level of demand in each centre, producers will ship to the centre that yields the highest net price and the size of supply areas will change until market prices less transportation costs are equalized. A given producer will be located on a boundary between two markets (or consuming centres) when the net prices of shipping to either centre are equal. The market boundary will shift if there is a change in the price in either centre, causing a change in relative prices, or a change in transportation costs. The determination of a boundary between two consuming centres is illustrated by the following simple example, as shown in Fig. 4.11. Suppose there are two consuming centres, A and B located 400 miles apart, that the per unit prices for the product are US$5.00 in A and US$4.00 in B, and that it costs US$0.01 per mile for a producer located between A and B to ship a unit of product to either centre. Thus, a producer located midway between the two markets (200 miles from either market) would face a cost of US$2.00 per unit to ship the product to either market and would net US$3.00 per unit in market A and US$2.00 per unit in market B. This producer would clearly supply market A, where net returns are highest. We can determine the boundary line for shipments to either centre by solving the following expression:

$$US\$5 - 0.01D_a = US\$4 - 0.01D_b$$

where D_a and D_b are the distances from consuming centres A and B respectively, and $D_a + D_b = 400$. To solve, substitute the constraint for either D_a or D_b in the original expression (i.e. $D_a = 400 - D_b$) and find the distance. The calculation is $US\$5 - 0.01(400 - D_b) = US\$4 - 0.01D_b$, or $D_b = 150$ and $D_a = 250$.

Thus, any producer located within 250 miles of consuming centre A will ship to A and any producer located within 150 miles of consuming centre B will ship to B. Producers located on the boundary ($D_b = 150$ and $D_a = 250$) would realize the same net price (US$2.50 per unit) from shipping to either A or B. The interested reader can investigate the impact of changes in transportation costs or market prices on market boundaries.

Transfer costs

Location differentials in a competitive market will be determined by the lowest costs means of moving the product from one location to another. If

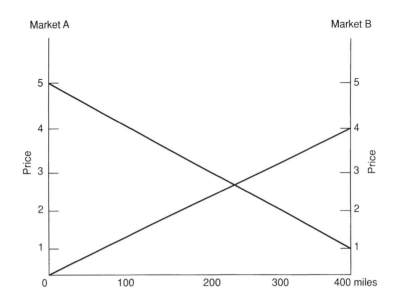

Fig. 4.11. Determination of the boundary between two markets A and B given prices in each consuming centre and transportation costs.

trucks are the lowest cost alternative, their rates will determine the differentials, but if rail rates are lower then differentials will be based on rail transportation. It is not unusual to find that the lowest cost method of transport varies with the distance that the product is to be moved. This is illustrated in Fig. 4.12. This figure shows that trucks are the least cost mode of transportation between a and b, then rail costs are lowest for b to c, and water (barges) will have the lowest costs for distances greater than c. Another thing to note about Fig. 4.12 is that total transportation costs do not increase by a factor of two when distance doubles. There is a fixed cost of transportation associated with loading, unloading and administrative costs and variable costs which vary with the distance moved. Figure 4.12 shows that the fixed costs associated with each transportation method vary, as do the variable costs for distance moved.

Interregional price relationships

Most agricultural commodity markets are much more complex than the examples above where surplus-producing areas ship to one market or producers have a choice of shipping to one of two markets. Instead, the typical situation is for a number of supply regions to be shipping to numerous consuming markets. These consuming markets can be thought of as population centres and surrounding areas while the supply regions will vary by commodity, depending on such things as climate and comparative advantage in

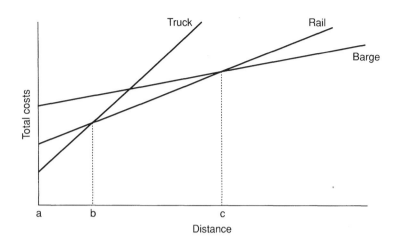

Fig. 4.12. A hypothetical illustration of differences in total costs of transportation by mode.

production of the agricultural products. USA orange production, for example, is concentrated in Florida, Texas, Arizona and California while wheat production is centred in the Northwest and Plains states, and the majority of corn production is in the Midwestern states. Geographical price relationships arising from these more complex markets can be analysed by use of spatial equilibrium or trade models. Given supply and demand conditions in each market and transfer costs, these models enable the analyst to determine the quantities of product shipped from each region to every other region and the net prices that will exist in each region under a 'least cost' trading pattern.

The operation of a trade model can be illustrated by a simple two-region example. This is explained in some detail because it forms the analytical basis for the international trade model used in the next chapter. Each of the two regions produces and consumes a commodity. The supply and demand functions for the commodity in each region are known and can be drawn as in Fig. 4.13:

$$\text{Demand: } P_a = 12 - QD_a; \quad P_b = 20 - QD_b$$
$$\text{Supply: } P_a = QS_a; \qquad\quad P_b = 4 + QS_b$$

where QD refers to the quantity demanded, QS to the quantity supplied, and the subscripts are for the two markets, a and b. With no trade permitted, the price in each of the regions is independent and is affected only by supply or demand changes in that region. Without trade, one can solve for the equilibrium price and quantity in each region by setting supply equal to demand. The equilibrium for region A is $P_a = 6$ and $Q_a = 6$, while for region B the equilibrium is $P_b = 12$ and $Q_b = 8$.

Now let us consider what will occur if trade is permitted. First, we will examine the case in which transfer costs are zero, using excess supply and

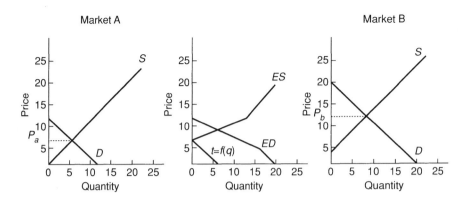

Fig. 4.13. A simple two-region trade model.

demand curves. An excess supply for a region is the amount by which the region's quantity supplied exceeds the quantity demanded at each price, or $ES_a = QS_a - QD_a$. At the isolated equilibrium price, this quantity is zero, while at higher prices, it is the horizontal difference between the supply and demand curves at that price. While an excess supply can be derived for each region, only the excess supply for the low price (surplus) region will enter our trade solution. In a similar manner, the excess demand is the amount by which the region's quantity demanded exceeds the quantity supplied at each price, or $ED_b = QD_b - QS_b$. At the isolated equilibrium price the quantity is zero, while at prices below the equilibrium, the quantity will be the horizontal difference between the supply curve and the demand curve at each price. Only the excess demand for the high price (deficit) region will enter our trade solution.

To express the excess supply and demand in relation to price, we need to have the supply and demand functions in the quantity-dependent form:

$$\text{Demand: } QD_a = 12 - P_a; \quad QD_b = 20 - P_b$$
$$\text{Supply: } QS_a = P_a; \quad QS_b = -4 + P_b$$

Substituting the quantity-dependent equations in the excess supply and demand equations gives:

$$\text{Excess supply: } ES_a = P_a - (12 - P_a) = -12 + 2P_a$$
$$\text{Excess demand: } ED_b = 20 - P_b - (-4 + P_b) = 24 - 2P_b$$

The excess supply and demand curves can be plotted in the trade sector diagram. If the transfer cost is zero, a solution may be obtained by setting $ES_a = ED_b$, where ES_a is the quantity exported from region A to region B and ED_b is the quantity imported into region B. In this illustration $ES_a = ED_b$ when price equals 9, and the excess supply is equal to 6. Region A would produce 9 units, consume 3 units and export 6 units, while region B would produce 5 units, consume 11 units and import 6 units from region A.

The addition of a transportation cost for shipments between the two regions will reduce the quantity of product shipped, with shipments decreasing as transportation costs increase. Referring back to the equilibrium prices in the two regions, P_a = 6 and P_b = 12, it is easy to see that there will be no incentive to ship a product between the two regions if the cost of transportation is equal to or greater than the unit price difference of 6. Let us look at transfer costs ranging between 0 and 6 per unit of product. To begin, suppose there is a positive transfer cost of t = 2. To solve for prices in each region and the amount of product traded, simply find the value of Q in the trade sector for which the vertical distance between the excess supply curve and the excess demand curve is equal to t = 2. This can be done by inspection or algebraically by adding transfer costs to the excess supply function (in the price-dependent form) and solving for a new equilibrium. The new excess supply is:

$$P = 6 + 1/2\,Q + 2, \text{ or } P = 8 + 1/2\,Q$$

The new equilibrium quantity is Q = 4, and at this point P_b = 10 and P_a = 8. At these prices region A produces 8, consumes 4 and exports 4, while region B produces 6, consumes 10 and imports 4 from region A.

The line in the middle panel of Fig. 4.13 labelled $t = f(q)$, shows the quantity imported and exported for all positive values of t. It is obtained by relating the vertical distance between ES_a and ED_b to Q (the equation for the difference between excess demand and supply is $t = 6 - Q$). It shows, for example, that if t > 6, trade ceases since the transfer cost exceeds the price difference without trade. If t = 4, the quantity traded would be 2 and so on. The maximum trade will occur where the transfer cost is zero (t = 0) and is Q = 6.

The trade model just presented illustrates the principles outlined at the beginning of this section on geographic price relationships, i.e. price differences between two regions that trade with each other will be equal to the cost of transportation, and the maximum price difference between two regions that do not trade will be equal to the cost of transportation. This assumes, of course, that there is a homogeneous product, a competitive market structure with full information on prices and supplies, and no barriers to trade. Since these assumptions are often violated in real world commodity markets, one may find price differences that are greater than the cost of transportation with no trade occurring, or price differences that differ from the cost of transportation when there is trade.

Using the trade model

Even though the two-region trade model is a gross simplification of the real world, it does provide some insights into how changes in supply, demand and transfer costs can affect prices, quantities produced and consumed and quantities traded for each region. You can briefly consider the operation of the model by increasing or decreasing demand in one of the regions, or considering the impact of a production subsidy, or by changing the transportation rate. One can also use the model to evaluate the impacts of tariffs or changes in the exchange rate between countries.

The simple two-region graphic model is of limited use for analysis of the common problem involving many producing and consuming regions. For example, every major city in Europe can be considered to be a consuming region and every supply point for a farm commodity as a shipping region. There is an aggregate supply function for each shipping region and each city has an aggregate demand function for a given product. The economic problem is to determine the equilibrium solution for the amounts of a product that will be produced and shipped from each supply point to each demand point and the prices at each point. However, this kind of detail is not available to a typical researcher, in even the most ambitious empirical interregional trade study. What the analyst usually does is specify geographic regions, choose one (or more) demand and supply points in each region, and then develop the supply and demand relationships for those points as well as the transfer costs from each central supply point to each demand point.

PRICES AND PRODUCT QUALITY

While commodities are often regarded as homogeneous, almost all demonstrate significant variability in product dimensions that are important to buyers. Sellers, to improve returns and communicate with buyers, sort their commodity into lots with similar characteristics, which can be easily described to buyers. Standards spell out the commodity characteristics, attributes, magnitudes, dimensions and tolerances that create the subdivisions designated as 'grades' that form the basis for sorting a heterogeneous commodity into more homogeneous groupings. Commodity grades are often viewed as equivalent to quality, with separate grades reflecting different quality levels. Alternatively, grades may reflect different bundles of characteristics, that determine end-use possibilities, as distinct from quality. Thus, grades and grade standards convey information about a commodity that facilitates communication between buyers and sellers.

The historical development of grades and standards was based on the growth of markets and the requirements of market participants. To expand the boundaries of their markets from local to regional to national to international areas, sellers had to describe their products accurately to buyers who could not inspect them physically. As markets developed, individual descriptions evolved into a set of common descriptive terms that were formalized into voluntary grades and standards backed by the employment of graders. A lack of uniformity with the voluntary approach often led to government-sponsored regulations and third party grading as a means of protection for both buyers and sellers. Hill (1990), in a discussion of the development of grades and standards in the USA grain industry, illustrates some of the issues involved in moving from the descriptive terms of individual traders, to a voluntary industry programme, to uniform government grades and standards. In the USA, grades and standards tend to be utilized most for marketing channel transactions between producers, packers, processors and retailers. With the exception of a limited number of commodities (including beef, eggs, and butter), very few grades carry through to the retail level and are recognized by consumers. Thus, while grades and standards can be very important in facilitating trade in the marketing channel, consumers tend to rely much more on brands as an indicator of quality.

An economic framework for viewing grades

Grades and standards provide a structure for improving the flow of information in commodity markets. With the backing of government, grades and standards reduce transactions risks, extend the physical boundaries of markets, and increase both economic and productive efficiency. Grades can eliminate the need for physical inspection and provide a basis for market news and price comparisons. Buyers' needs are communicated to sellers through prices; sellers can adjust production and marketing practices to take advantage of price differentials.

Grades and standards may have one or more economic impacts. Grades may increase demand, resulting in increased sales at a given price and/or increased prices. Cost reductions through improvements in physical operating efficiency and improvements in pricing efficiency result in reduced marketing margins, with benefits in a competitive economy accruing to both producers and consumers in the form of higher returns at the producer level and lower prices at the retail level.

Buyer benefits

At the consumer or intermediate buyer level, the sorting of commodities into grades allows buyers to select the particular quality characteristics which they prefer and are willing to pay for, thus increasing utility. This is best illustrated by using indifference curve analysis (introduced in the previous chapter) as presented by Freebairn (1967). Figure 4.14 is an indifference map for two buyers, I and II, who have the same purchasing power, and for two grades of a commodity (a and b), where the two grades are regarded as closely related but different products. The two grades are supplied in the proportions shown by the line *ST*, and before grading, buyers are restricted to the expansion path *ST*. Given a budget line (income constraint) *CC'* representing equal budgets for the two buyers and the relative prices of the two grades, before grading each buyer will purchase a combination of the two grades I and II shown by point *J* on the expansion path *ST*. Neither buyer is maximizing utility and the ratio of their marginal utilities is clearly not equal at point *J*. When the commodity is graded and buyers can select the preferred combination of grades a and b, each buyer will maximize utility by moving to the point where their budget line is tangent to an indifference curve, and in so doing the ratio of their marginal utilities are equalized. Buyers I and II will purchase the combinations of

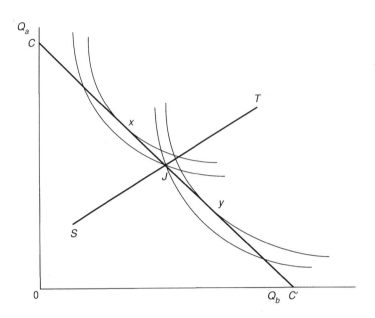

Fig. 4.14. Possible gains in utility from introduction of a grading system (Freebairn, 1967).

grades a and b shown at the points *x* and *y* respectively. Since grading enables each buyer to move to a higher indifference curve, it is clear that satisfaction has increased and each is 'better off'.

Efficiency aspects of grades and standards

Market information is necessary for achieving maximum marketing efficiency[3]. Thus, grades and standards are an important determinant of technical and pricing efficiency. Technical efficiency in commodity marketing is concerned with minimizing costs for a given level of output (or maximizing output for a given level of costs) in the performance of the marketing functions. This is equivalent to stating that the firm is operating on the frontier production function in performance of each of the physical activities such as transportation, storage, packing, and processing, as well as exchange activities. Pricing efficiency deals with the optimum combination of resources and the profit-maximizing level of output under competitive conditions. It is concerned with the manner in which prices are established, how prices guide production and consumption decisions, and the economic allocation of both inputs and outputs. An analytical framework for evaluating the contribution of grades and standards to pricing and technical efficiency in commodity marketing follows.

Measurement of marketing efficiency

Farrell (1957) linked the relationships between technical, pricing and economic efficiency in a framework for measurement of efficiency. His framework, which is useful in differentiating sources of efficiency, is illustrated in Fig. 4.15. The curve *S'S* is a unit isoquant, representing the minimum quantities of two inputs A and B required to produce one unit of output, as explained in the previous chapter. Any combination of the two inputs on the isoquant, such as represented by a firm operating at either point Q or Q', lie on the production function and are technically efficient. There is only one point on the isoquant, however, that is price efficient, i.e. where the inputs are used in their least-cost combination. This point, which is found with the isocost line *CC'*, turns out to be point Q', where the marginal rate of technical substitution is equal to the price ratio of the two inputs. A firm operating at point Q' is both price efficient and technically efficient; likewise a firm operating at point P is neither price efficient nor technically efficient. Farrell (1957) shows that the efficiency of a firm at point P (or any other point on the map) can be measured as a ratio by extending a ray from the origin to the point, as shown by the line OP in Fig. 4.15. The firm operating at point P could maintain its production by using less of both inputs and moving to point Q by using the fraction of OQ/OP of each input. Farrell (1957) defines this ratio (OQ/OP) as technical efficiency. The cost of production at point Q' will be OR/OQ of the costs at Q; this ratio is defined as price efficiency. Thus, if a firm operating at point P were both technically and price efficient, its costs would be OR/OP of what they actually are. This measure of overall efficiency (OR/OP) is the product of price (OR/OQ) and technical (OQ/OP) efficiency.

The impact of grades on commodity demand

The indifference curve analysis in Fig. 4.14 can be extended to show that buyers are prepared to offer an aggregate price for a given total quantity of the graded commodity that exceeds the price they are prepared to pay for the same quantity of the ungraded commodity (Freebairn, 1967). Aggregate demand for a commodity may also be increased by shifting the geographic boundaries of the market and by extending the market through increasing the number of end uses of the commodity. This increase in market demand may be the result of one or more of the factors mentioned above.

Reduction in costs of marketing through improvements in physical operating efficiency and improvements in pricing efficiency result in reduced marketing margins, with benefits in a competitive economy accruing to both producers and consumers. A reduction in margins will result in an increase in derived demand at the producer level and an increase in derived supply at the consumer level. This will result in a reduced price at the retail level, an increased price at the producer level, and an increase in the quantity of commodity sold and consumed. A detailed discussion of marketing margins will be found in Chapter 8.

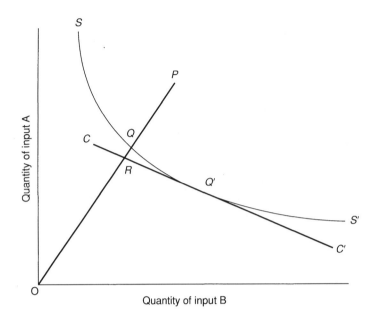

Fig. 4.15. Measurement of technical and price efficiency using the Farrell (1957) approach.

Hedonic prices

Hedonic prices are based on the hypothesis that goods are valued for their utility-bearing attributes (Rosen, 1974). Thus, hedonic prices are the implicit prices of the various attributes embodied in a commodity and the price of the commodity is a function of the amounts of the attributes it contains and the values placed on these attributes. Ideally, the utility-bearing attributes that are important for hedonic prices will be the same attributes used for quality standards and grades.

The application of hedonic price models to agricultural commodities has emphasized the determination of the relative importance of various quality attributes that are included in grades and standards. The empirical approach to estimating hedonic prices has typically been to estimate a regression model. The techniques involved are discussed in Chapter 6, but here a simple model is used, of the form:

$$P = B_0 + B_1 Z_1 + B_2 Z_2 + B_3 Z_3 + e$$

where P is the price of the commodity, the Zs are the quality attributes, the Bs are the coefficients to be estimated and e is the unexplained error. The quality attributes included in the regression model can be those included in determining the grade of the product or others important to buyers. For example, a hedonic price model for malting barley included variables for variety and grade as well as the non-grade variables of per cent protein and per cent plumpness (Wilson, 1984). Once the regression model has been estimated, the estimated coefficients (B_is) can be used to derive marginal implicit prices for the attributes or elasticities of expected price with respect to the attributes. Results of econometric estimates of the marginal implicit values of major characteristics of wheat in world markets indicate that increased protein content was associated with price premiums (0.5% premium for each 1% increase in protein) and that white wheat also received price premiums of almost CAN$16 per tonne (Veeman, 1987).

Estimates of the values of particular attributes can have several uses. Producers may be able to alter their production practices, use of inputs, or varieties to influence attributes that increase product price. The value of particular attributes can provide guidance to plant breeders in variety development and can be a factor in research resource allocations. For the wheat example, Veeman (1987) concluded that total revenue from Canadian wheat exports could be increased by more emphasis on the development of higher yielding varieties that are adapted to the higher moisture regions of the Prairies (even though protein content tends to vary inversely with yield) and that more emphasis on development of white wheats suited to Prairie growing conditions is also merited. Middlemen in the marketing channel may be able to adopt handling methods, containers, storage and processing practices to improve retail level product prices through emphasis on attributes important to final users.

THE IMPACTS OF GOVERNMENT COMMODITY PROGRAMMES

Few governments or politicians appear to be satisfied for any extended period of time with market prices for agricultural commodities. Because of the characteristics of agricultural prices mentioned earlier – extreme variability and persistent low levels – most governments in developed countries feel obliged to develop policies and programmes to stabilize and/or increase the level of agricultural prices. In this section we will examine some policies used by governments to stabilize and increase commodity prices, with attention to some of their economic impacts. Note that governments' intervention in commodity markets is usually designed to accomplish one or more of the following objectives: (i) to raise the average level of prices and incomes; (ii) to reduce price and income variability; (iii) to increase self-sufficiency in food and fibre; and (iv) to change (improve) the allocation of resources.

Policies to be discussed include price supports through government purchases, limiting production or sales, tariffs and import restrictions, consumption or export subsidies, and deficiency payments.

Governments occasionally face the problem of trying to hold down or control rapidly increasing commodity prices, a problem that has usually occurred during wartime, during periods of high inflation, or when there has been a crop failure. The policies that are most often used include price controls and export embargoes. Under appropriate conditions, the government can relax import controls or release stocks of product held in storage.

Price supports

The USA has a long history of government price supports for storable agricultural commodities, including programmes for wheat, corn, rice, oilseeds, wool, cotton, manufactured dairy products (butter, cheese and dried milk) and tobacco. Similarly, the Common Agricultural Policy (CAP) of the European Union supports the market prices of most major agricultural commodities. In the early years of the USA programmes, the support prices were closely linked to parity prices. More recently, the legislation has severed the link to parity prices for most commodities and has given the Secretary of Agriculture considerable discretionary authority for establishing loan rates.

Government purchase programmes

The USA has used a loan programme, administered by the Commodity Credit Corporation, to support the prices of storable commodities such as grains and cotton above equilibrium levels. Under this programme, producers can obtain a nonrecourse crop loan at the support price shortly after the crop is harvested. If the market price increases above the loan rate, the farmer can sell his crop on the open market and pay off the loan. If the market price remains below the loan rate, the farmer simply defaults on the loan and the government takes the crop as security for the loan. In the European Union, the purchase programme has operated with national intervention agencies being obliged to purchase commodities offered to them at a predetermined intervention price.

The economic impact of a price support programme (with government purchases) is shown in Fig. 4.16. Suppose that the government has decided to support the price of a commodity at a price P_s which is above the equilibrium price P_e. To maintain the above equilibrium price P_s, the government will have to purchase a quantity equal to Q_s minus Q_d at a total cost of $P_s(Q_s - Q_d)$.

It is a straightforward matter to demonstrate that the government costs of a price support programme will depend on the price elasticity of demand, the level of the support price relative to the equilibrium price, and the price elasticity of supply if production is uncontrolled. The more elastic the demand and supply curves for the commodity, the more it will cost government to maintain a given level of prices. Because of the potential costs of these purchase programmes, the need for control measures to restrict production is evident.

The final cost of price support programmes to government will depend on how long commodities are held before being resold (or given away) and the price of the commodities that are sold. Government does not have to incur a net cost from a support programme if demand expands by enough

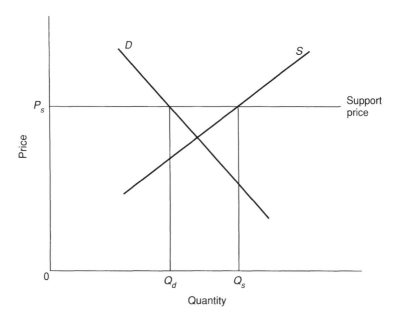

Fig. 4.16. The economic impacts of supporting prices above the equilibrium price.

to cover storage costs or if there are significant price increases as a result of short crops. In practice, however, these programmes can be quite costly.

Limiting production

While government can maintain prices above equilibrium in the short-run with purchase programmes, this approach can become very expensive as producers expand production. In many cases, it will be less costly for government to limit production, or even to pay farmers not to produce, rather than to purchase and store commodities. Two approaches have been used in the USA. The supply of a commodity marketed can be limited by assigning each producer a sales or marketing quota, with penalties for exceeding the quota. A second approach, which has been used extensively, is to limit the use of inputs. In the USA, acreage control or land retirement programmes are most often used, but other inputs, such as milk cows in the 'whole dairy buyout programme' have been removed from production. Similarly, in the European Union, milk production is limited by individual producer quotas, and grain production is limited by a rotational land retirement scheme known as 'set-aside'. While most concede that acreage controls are inefficient, they have been politically acceptable and are easier to administer than sales quotas or limits on other inputs. To be effective, however, significant areas of land must be removed from production since farmers will retire the least productive land first and use increased levels of inputs on their remaining land.

Farmers benefit from reduced production when the demand for their product is price inelastic. This is the case for most farm commodities in the short-run. However, over a period of years the relatively high prices may result in the development of alternative sources of supply or substitute products. There is some evidence that USA cotton producers lost markets to synthetic fibres due to an effective supply control programme which maintained relatively high prices over several years.

The effect of a supply control programme is shown in Fig. 4.17. The total supply is fixed at Q_s, resulting in a new vertical segment of supply that establishes an equilibrium price at P_s. Note that government costs will be lower than with price supports since consumers pay the market price. Total revenue to producers is also less than with price supports, since a smaller quantity is produced and sold. Even with lower total revenue, producers may be better off with the supply controls since their costs are also lower. As one might expect, agricultural business firms that sell inputs to farmers (seed, fertilizer, farm machinery, etc.) and those that are involved in marketing the crop will be adversely affected by supply controls and are generally opposed to such programmes.

The benefits to producers from a successful supply control programme are usually capitalized into the value of the rights to sell or produce the controlled commodity. This is true of any programme in which the rights to obtain higher returns are restricted and negotiable. For example, in some

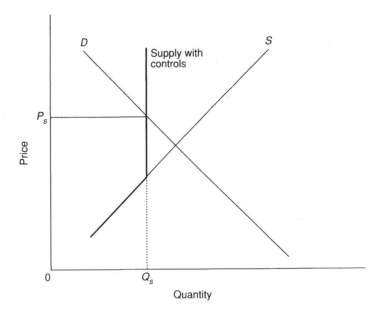

Fig. 4.17. The economic effects of a supply control programme.

EU member states, milk quotas can be sold (or leased), and in the UK in 1996 the quota price was three times the sale price of the milk.

Tariffs and import restrictions

Countries have traditionally used tariffs and quotas to maintain domestic prices at a level above world market prices. The effect of these protectionist policies is to raise internal prices above world prices, thus encouraging domestic production and reducing the market for imported product. The degree to which domestic production and consumption are affected by import restrictions depends on the slopes of the demand and supply curves in the importing nation. The potential impact of an import duty on domestic production, consumption and imports is shown in Fig. 4.18. The more complicated international implications of agricultural trade policy mechanisms are explored in the next chapter. Here, however, we assume that the supply of imports is perfectly elastic at the world price P_w. Without a tariff or other protection, domestic production will be equal to Q_1 and consumption will be equal to Q_4. If a tariff equal to the difference between P_w and P_t is imposed on imports, such that the domestic price rises to P_t, producers will eventually increase output to Q_2 and consumers will reduce purchases to Q_3. These are the production and consumption effects of a tariff.

We can see from Fig. 4.18 that consumers will pay the full amount of the tariff (if import prices remain the same), and imports will decrease. Tariffs have an advantage of providing a source of government revenue (as long as

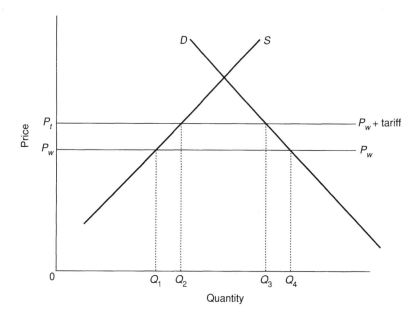

Fig. 4.18. The effects of a tariff on domestic production, consumption and prices.

there are any imports). Note that the same effect on prices could be obtained by use of a quota to restrict imports, but the government would get no revenue (unless government establishes a charge for the quotas (import licences)). The benefits of the higher prices would accrue to the exporting countries which were offered quotas. This was the case with the USA sugar programme until the early 1970s, when the programme featuring quotas was allowed to expire. During the period that quotas were effective, major sugar exporters retained lobbyists, lawyers and other specialists in Washington, DC, whose charge was to maintain or expand their individual quotas. There was, for example, an intensive lobbying effort to obtain additional quota when Cuba's sugar quota was reallocated.

Consumption or export subsidies

Surpluses of farm products can be avoided, even with prices above free market levels, if the government subsidizes domestic consumption or exports. Producers typically prefer these programmes, which increase demand, to those which restrict production. Some of these programmes, such as food stamps or school lunches, may also have substantial non-farm legislative support. The impacts of these programmes on agricultural producers are likely to be uneven and are typically rather small in total. Producers of commodities with high income elasticity (meat, cheese, snack foods, and some fruits and vegetables) will gain if increased spending goes to upgrading diets. Others, such as those producing rice, dry beans or wheat, will gain very little. Marketing firms will gain as consumers buy more marketing services.

The USA and the EU have used export subsidies (or very generous credit terms) to reduce surpluses of farm products. Export subsidies are regarded as unfair by other exporting nations, because of their obvious costs and impacts.

Deficiency payments

With a deficiency payment the government is responsible for making up the difference between the price guaranteed to farmers and the market price. This was the way agricultural commodity prices were supported in the UK before membership of the EU. When compared with a price support programme, consumers benefit from increased supplies at lower market prices, and storage, handling and disposal problems are eliminated. Government expenditures, however, become a visible part of the budget and maintaining political support may become difficult.

An advantage of deficiency payments to support prices is that they avoid the loss of markets to substitute products or to imports and exports of other suppliers. The economic effects of a deficiency payment programme can be illustrated by Fig. 4.19. Assume that the price guaranteed to producers (P_s) is above the equilibrium price. The total supply will become a vertical line at the point where the guaranteed price intersects the supply schedule (Q_s). The total amount supplied will be purchased at a price P_c where the vertical supply is equal to demand. Government payments will be equal to the difference between the guaranteed price to producers (P_s) and the market clearing price (P_c) multiplied by the total production.

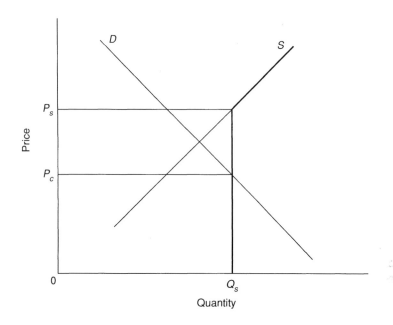

Fig. 4.19. The economic effects of supporting prices with deficiency payments.

Government price controls

Price controls establish a maximum price for a given commodity or groups of commodities to prevent rapid price increases that are not politically acceptable. The approach taken is usually to fix prices at a given level at a specific point in time, i.e. prices can go no higher than they were on 1 October, 19—. As noted, price controls are usually attempted during periods of rapid inflation or during wartime when shortages of 'critical' items develop. Thus, prices are fixed at a given level but demand is increasing. This can be illustrated by Fig. 4.20. Suppose that the market was initially in equilibrium with demand curve DD and supply curve SS resulting in price P_c and quantity Q_e. A ceiling price is established, with the price to buyers fixed equal to $P_{c'}$ which was initially an equilibrium price. As demand increases to $D'D'$, buyers will attempt to increase their purchases of the commodity to Q_c but sellers will continue to offer quantity Q_e of the commodity for sale, resulting in a shortage equal to $Q_c - Q_e$. The experience with most programmes has been for the shortage to increase over time.

What happens when prices are not allowed to perform the rationing process? Something must take the place of price – it may be government ration coupons which act as a new form of currency available in very restricted amounts. Without ration coupons, the rationing may be on the basis of first come – first served, with queues taking the place of price, or a black market may be formed, or it may be who you know determining whether or

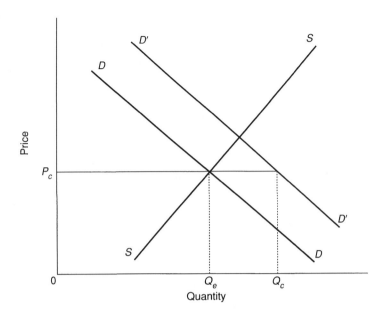

Fig. 4.20. The economic effects of price ceilings with increasing demand.

not you obtain the good at the controlled price. Firms take all kinds of uneconomic (economic??) actions to get around the controls, and some may even close in the short-run to avoid the controls. For example, if a percentage mark-up is allowed each time a product changes ownership, a product may be bought and sold many times without ever leaving a given warehouse. Items may be exported and then imported to get around the rules. Many other actions will be taken depending on the way in which the rules are established for each commodity.

CONCLUDING REMARKS

A knowledge of the many factors affecting agricultural commodity prices is very important in the study of food marketing problems, performance, and opportunities. This chapter has reviewed the aggregation of individual consumers' demand and firms' supply functions into market relationships. While the traditional demand determinants (tastes and preferences, income, and prices of related goods) together with the supply shifters (prices of inputs, prices of related goods, and technology) affect equilibrium market prices, other factors can also have important impacts. Price variations through time, as related to storage, seasonal shifts in supply and demand, and production cycles were analysed. Spatial price relationships, as related to geographic market areas, were outlined and price relationships with and without trade were presented in terms of a simple two-sector trade model. Product characteristics, as measured by grades and standards, are also related to prices and to the efficient operation of markets. Price differences stemming from product characteristics were explained using a hedonic pricing model. Finally, some of the results and problems encountered with government intervention in agricultural markets were examined.

NOTES

1. Ezekiel (1938) published one of the first papers on the cobweb model. Waugh (1964) provides a good summary of the model together with references.
2. Marshall (1920) stated: 'Thus the more nearly perfect a market is, the stronger is the tendency for the same price to be paid for the same thing at the same time in all parts of the market: but of course if the market is large, allowance must be made for the expense of delivering the goods to different purchasers; each of whom must be supposed to pay in addition to the market price a special charge on account of delivery.'
3. Freebairn (1973) suggests two requirements must be satisfied for the introduction of uniform grading to increase efficiency through improved information. First, a significant portion of buyers must differentiate between units of a commodity on the basis of some characteristics (i.e. all units of the commodity are not regarded as perfect substitutes). Second, it is necessary to establish market participants' general acceptance of the grade standards as a useful measure of the characteristics that they consider important.

FURTHER READING

Tomek and Robinson (1990) cover each of the major topics in this chapter. Additional discussion of price movements over time can be found in Goodwin (1994). Gardner (1987) and Helmberger (1991) each illustrate and discuss the impacts of government intervention in commodity markets.

REFERENCES

Bressler, R.G. Jr. and King, R.A. (1970) *Markets, Prices and Interregional Trade.* John Wiley and Sons, New York.

Ezekiel, M. (1938) The cobweb theorem. *Quarterly Journal of Economics* 53, 255–280.

Farrell, M.J. (1957) The measurement of productive efficiency. *Journal of the Royal Statistical Society* 120, 253–281.

Freebairn, J.W. (1967) Grading as a market innovation. *Review of Marketing and Agricultural Economics* 35, 147–162.

Freebairn, J.W. (1973) The value of information provided by a uniform grading system. *Australian Journal of Agricultural Economics* 17, 127–139.

Gardner, B.L. (1987) *The Economics of Agricultural Policies.* The Macmillan Publishing Co., New York.

Goodwin, J.W. (1994) *Agricultural Price Analysis and Forecasting.* John Wiley and Sons, Inc., New York.

Helmberger, P.G. (1991) *Economic Analysis of Farm Programs.* McGraw-Hill, Inc., New York.

Hill, L.D. (1990) *Grain Grades and Standards: Historical Issues Shaping the Future.* University of Illinois Press, Urbana, IL.

Houck, J.P. (1984) Market: A definition for teaching. *Western Journal of Agricultural Economics* 9, 353–356.

Marshall, A. (1920) *Principles of Economics,* Book II. Macmillan, London.

Rosen, S. (1974) Hedonic prices and implicit markets: Product differentiation in pure competition. *Journal of Political Economy* 82, 34–55.

Tomek, W.G. and Robinson, K.L. (1990) *Agricultural Product Prices,* 3rd edn. Cornell University Press, Ithaca, NY.

Veeman, M.M. (1987) Hedonic price functions for wheat in the world market: Implications for Canadian wheat export strategy. *Canadian Journal of Agricultural Economics* 35, 535–552.

Waugh, F.V. (1964) Cobweb models. *Journal of Farm Economics* 46, 732–750.

Wilson, W.W. (1984) Hedonic prices in the malting barley market. *Western Journal of Agricultural Economics* 9, 29–40.

5 International Trade in Agricultural Commodities

Heinz Ahrens

Professur für Agrarpolitik und Agrarumweltpolitik, Landwirtschaftliche Fakultät, Martin-Luther-Universität Halle-Wittenberg, Emil-Abderhalden Str. 20, 06108 Halle (Saale), Germany

INTRODUCTION

The aim of this chapter is to describe and to analyse developments in international trade in agricultural commodities[1]. First, major determinants of this trade, policy and non-policy, are treated, within the framework of a simple model. Second, against the background of this analysis, some basic trends in international trade in agricultural commodities, which have taken place since the 1970s, are discussed. In doing so, the causes of, and need for, an international liberalization of this trade are elucidated. Finally, some perspectives on international trade in agricultural commodities will be pointed out.

DETERMINANTS OF INTERNATIONAL TRADE IN AGRICULTURAL COMMODITIES

International trade in agricultural commodities is the result of international specialization. The latter is influenced by policy and non-policy determinants. A complete analysis would require a multi-country, multi-product general equilibrium model. In the following, a partial equilibrium approach will be used to illustrate the impact of policy and non-policy determinants on agricultural trade.

The model, which essentially develops the interregional trade model illustrated by Fig. 4.13, comprises two 'countries', country A and the 'rest of the world', R, trading in one particular agricultural commodity. Domestic producer and consumer prices are assumed to be the same. Country A is a 'large country' – that is, exports or imports are large enough to have an influence on world prices. The same is, of course, true for the rest of the world. As transportation costs are neglected, world price and domestic price are identical under free trade.

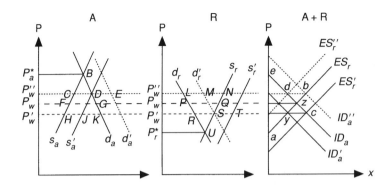

Fig. 5.1. Free trade.

Free trade

Assume that in Fig. 5.1 both countries A and R, for which the supply and demand curves for the agricultural commodity, s_a, d_a, s_r and d_r, are given, pursue a policy of self-sufficiency. Domestic prices will, therefore, be at P_a^* and P_r^*. When trade is allowed, P_a will fall and P_r will rise to the common price P_w (P_w is derived in the right-hand part of the figure, from the intersection of A's import demand curve ID_a and R's export supply curve ES_r at z). At this level A's imports (equivalent to the distance FG) are equal to R's exports (PQ). For A, consumers' surplus (as defined in Chapter 3) rises by the area $P_a^*BGP_w$ while producers' surplus falls by $P_a^*BFP_w$, the net welfare effect thus amounting to $+BGF$. For R, consumers' surplus is reduced by $P_wPUP_r^*$ while producers' surplus rises by $P_wQUP_r^*$ so that the net welfare gain is $+PQU$. In the right-hand part of the figure, total gain in world welfare is represented by the area 'eza'.

In this system, world price and trade flows are influenced by shifts of the supply and/or demand curve as discussed in Chapters 3 and 4. In short, an increase in supply may be due to: (i) an expansion of the area under cultivation; (ii) an improvement in climatic conditions; (iii) a release from stocks; or (iv) productivity growth. Any increase in demand may be caused by: (i) population growth; (ii) rising income (assuming a 'normal' income elasticity of demand) resulting from the growth of the Gross Domestic Product (GDP); (iii) a rise in the price of a competing product or a fall in that of a complementary one; or (iv) a change in consumer preferences.

What are the consequences of changes in these market forces?

1. If there is an increase in A's domestic supply, from s_a to s_a', resulting in a shift of A's import demand curve to ID_a', the common price falls to P_w' where A's reduced quantity of import (JK) and R's reduced quantity of export (RS) are the same. Country R experiences a welfare loss of $PQSR$, and its foreign exchange earnings fall because of the smaller export quantity and lower world price.

2. In the case of an increase in A's domestic demand, from d_a to $d_a{}'$, the common price rises from P_w to $P_w{}''$ and the quantity traded to $CE = LN$. Country R experiences a welfare gain of $LNQP$. The value of the trade flow rises because of both the larger export quantity and the higher world price.

3. If there is an increase in R's domestic supply, from s_r to $s_r{}'$, leading to a shift of R's export supply curve to $ES_r{}'$ the common price falls to $p_w{}'$ where A's higher quantity of import (HK) is equal to R's increased quantity of export (RT). A experiences a welfare gain amounting to $FGKH$. R's foreign exchange earnings and A's foreign exchange cost may rise or fall.

4. In the case of an increase in R's domestic demand, from d_r to $d_r{}'$, the common price rises from P_w to $P_w{}''$ and the quantity traded falls to $CD = MN$. A experiences a welfare loss of $CDGF$. The value of the trade flow may rise or fall.

In this system of free trade, three things are worth noting. First, on the basis of the common price $(P_a = P_r = P_w)$, both countries produce at the same (marginal) cost. Second, productivity gains in one country lead to a geographical redistribution of part of world production, in favour of that country: production migrates to low-cost locations. Third, if a change in supply or demand in one country creates a world scarcity or glut, this is absorbed by both countries via a rise or fall in domestic price.

Protection

Agricultural policies influencing trade may broadly be grouped in two categories: positive or negative protection. In the simplest case, where protection consists of fixing the agricultural commodity price above or below world price, the effects can be shown as follows.

Positive protection

Protectionist importer Figure 5.2 starts from the system of free trade presented in Fig. 5.1. In the upper part, we now assume that country A raises its price to $P_a{}'$ so that imports fall to BC. As the import demand curve shifts from ID_a to $ID_a{}'$ where it is perfectly inelastic with respect to world price, R's price and world price fall to $P_w{}'$. For A, this means a twofold reduction in the foreign exchange cost of imports (quantity effect and price effect). The net change in surplus amounts to $-BCGF$; on the other hand, A gains the additional budget income from the import levy, $BCED$. R witnesses a net welfare loss of $PQML$, and a twofold loss of foreign exchange earnings. The world as a whole suffers a welfare loss of 'dzy'.

As it is assumed that the two countries' supply curves have the same slope, protection by A has no effect on total world production but merely brings about a change in the latter's geographical distribution, away from

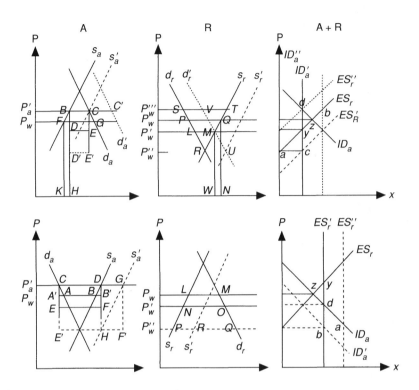

Fig. 5.2. Positive protection.

R, towards A. But the redistribution of *WN* = *KH* from R to A leads to an increase in aggregate (variable) cost: while in R variable cost falls by *QNWM*, in A it rises by *BHKF*. As a result of protection, A produces at higher marginal cost than R.

What if market forces are subject to change?

1. If there is an increase in A's domestic supply, to s_a', the reduction in A's import demand (in this case to zero with a shift of the import demand curve to ID_a'') drives down world price (to P_w''). For A, revenue from the import levy falls to zero (with a less drastic fall in import quantity, it might, however, fall less, stagnate or even rise; the result is indeterminate). For R, the result is a (in this case: total) loss in foreign exchange earnings and a loss of overall welfare (−*LMR*). As a result of this process, the gap between the (marginal) cost at which the two countries produce the commodity widens.

It should be noted that these effects of A's production increase on the volume and value of trade, world price, and R's welfare are much stronger than under the system of free trade (Fig. 5.1). This is due to the fact that A does not allow its own price to react to the glut caused by itself, so that the burden of 'absorbing' the glut by a higher domestic quantity demanded and a lower domestic quantity produced is passed entirely on to the rest of the world.

2. In the case of an increase in A's domestic demand, to d_a', the system works the other way round. As A's price does not react to the scarcity created by itself, the rise in world price (to P_w''' where $BC' = ST$) and the positive effects on R's welfare (+*STML*) and foreign exchange earnings are stronger than they would otherwise be. The end result is the same as the one that would occur under free trade.

3. If there is an increase in R's domestic supply, to s_r', again the fall in world price (to P_w'') is particularly pronounced because of the fixity of A's price (P_a'). Refusing to increase its import quantity, A benefits greatly from the fall in world price, in terms both of foreign exchange cost and revenue from the import levy (+*DEE'D'*). In this case, paradoxically country R does not benefit from its productivity increase: at the new lower world price the quantity exported by it is no higher than before (*RU* = *LM*), and foreign exchange earnings have fallen. To put it differently, the system prevents the normal free trade logic according to which competition induces production to move to locations where productivity increases. Protection neutralizes trading partners' advantages from productivity gains via artificially deep falls in world and domestic prices that force producers to produce a lower-than-normal quantity (at artificially low marginal cost).

Another effect of A's price inflexibility is this: if A's production of the commodity is subject to annual fluctuations, e.g. owing to changes in climatic conditions, A itself does not contribute to the absorption of these 'shocks' but passes them on to the rest of the world. This means that world price fluctuates more widely than under free trade: protection has a destabilizing effect on world price. Similarly, fluctuations of production in R are not absorbed by A; the latter's stability of domestic price increases the instability of international and R's domestic price. In this respect the policy of price support is more harmful than that of an *ad valorem* tariff which also drives a wedge between domestic and international price but allows the former to fluctuate together with the latter[2].

Protectionist exporter In the lower part of Fig. 5.2, country A is assumed to be a protectionist exporter. Fixation of a domestic price at P_a' results in an increase in export supply (from *AB* to *CD*). To be able to sell it on the world market at the new price P_w', A has to pay export subsidies amounting to *CDFE*. Since there is a net gain in surplus of *CDBA*, A experiences a loss of welfare of *DFECAB*. The latter consists of two elements: the welfare loss that would occur if A were a small country having no impact on world price (*CAA'*+ *DB'B*), and the 'terms of trade effect' resulting from the fact that large country A does in fact depress world price (*B'FEA'*). A's foreign exchange earnings are boosted by the quantity effect but reduced by the price effect of the protectionist policy. For importing R, the welfare effect is positive (+*MONL*) while the foreign exchange effect depends on the relative weight of the fall in world price and the increase in the quantity imported. World welfare decreases by '*ydz*'.

What are the consequences of changes in these market forces?

1. If there is an increase in A's domestic supply, to s_a', the increase in A's export supply (to *CG*) has another depressing effect on world price (fall to P_w''). The cost of export subsidies rises (to *GF'E'C*). Again, the effects of this increase in supply on the volume of exports/imports, world price, etc. are stronger than under the system of free trade (Fig. 5.1). Again this is due to the fact that A's price does not react to the glut created by A, so that the burden of adaptation of demand and production to the new situation is shifted entirely to R. However, since R is an importer, the beneficial effects on the consumer outweigh the detrimental ones on the producer so that A experiences a net welfare gain (+*NOQP*).

2. If there is an increase in R's domestic supply, to s_r', A refuses to reduce its export quantity *CD*, with the result of a fall in world price (to P_w''). For A, there is an unwelcome rise in the cost of export subsidies and a reduction of total welfare (by *FHE'E*) as well as a loss of foreign exchange earnings. For R, the result is a reduction, not of the quantity imported (*RQ = NO*) but of the price of imports so that at least there is a certain saving in foreign exchange. Here again, the system prevents the normal free trade logic according to which competition induces production to move to locations with productivity increase. Protection neutralizes trading partners' advantages from productivity gains via artificially pronounced falls in their producer price which make profit-maximizing producers produce at artificially low marginal cost.

Clearly, if there are annual fluctuations of production in A, the former's price support policy adds to the instability of world price and R's domestic price.

Negative protection

Importer   In the upper part of Fig. 5.3 it is assumed that importing country A fixes its domestic price at P_a'. The import demand curve shifts to ID_a', and import quantity increases to *ED*. As a result, R's price and world price rise to P_w'. For A, the foreign exchange cost of imports rises both under the quantity and price aspect. Another drawback for A is that imports need to be made cheaper to the consumer via import subsidies (*CDEB*). There is a net welfare loss of *CDGFEB* of which *CHJB* is the terms of trade effect. For country R, export quantity, export earnings and welfare increase, the latter by *TQPS*. The world as a whole suffers a welfare loss amounting to '*baz*'. The geographical redistribution of world production from A to R causes an increase in variable cost of (+*TNWQ−FHKE*):

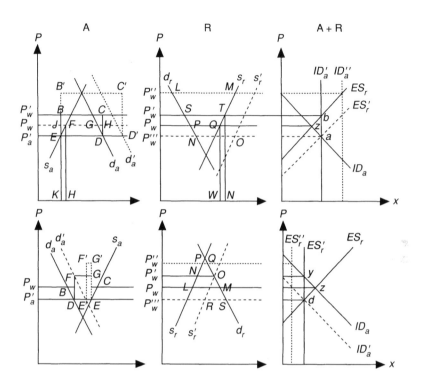

Fig. 5.3. Negative protection.

1. If there is an increase in A's demand to d_a', the rise in A's import demand (to ID_a'') raises R's price and world price (to P_w''). As a result, A's import cost and import subsidies increase (the latter to $C'D'EB'$).

2. In the case of an increase in R's domestic supply, to s_r', world price and R's price are driven down to $P_w''' = P_a'$. For A, this is most welcome: the cost of import subsidy and the foreign exchange cost of imports both go down. This effect offsets the initial effect of A's policy of negative protection. However, this is a coincidence; any additional change in market forces would again reveal the overly great and destabilizing effects of this kind of protection on the world market.

Exporter In the lower part of Fig. 5.3 it is assumed that exporting country A fixes domestic price of the commodity at P_a' and introduces a (variable) tax on exports. Export quantity falls from BC to DE. As a result, world price and R's price rise to P_w'. For A, foreign exchange earnings may go up or down. The revenue from export tax amounts to $GEDF$. The rest of the world suffers a welfare loss ($OMLN$), and so does the world as a whole ('yzd'). The geographical redistribution of world production from A to R causes an increase in the variable cost of (the same quantity of) world production.

What are the consequences of these changes in the market forces?

1. If there is an increase in A's domestic demand to d_a', the fall in its export supply to $E'E$ raises world price and R's price (to P_w''). R suffers a welfare loss of $QONP$.

2. In the case of an increase in R's domestic supply, to s_r', if A keeps its price unchanged, the result will be a pronounced fall in world price to $P_w''' = P_a'$ so that the free trade situation is reached. However, for A, this means that income from the export tax falls (to zero); this represents a strong incentive for the government to reduce domestic producer price still further (with the result that there will be another reduction of export quantity).

EVOLUTION OF INTERNATIONAL TRADE IN AGRICULTURAL COMMODITIES

Growth in the 1970s

Determinants

During the 1970s, most countries in the world pursued agricultural policies that implied positive or negative protection.

The majority of OECD countries granted positive protection to major agricultural commodities. Typical policy instruments were those discussed above: support prices combined with variable import levies and variable export subsidies; or tariffs, import quotas, deficiency payments and other direct income support. The major motive behind these policies was the desire to slow down the negative income effects resulting from (i) the workings of market forces (increase in productivity, stagnation of demand, fall in prices) and (ii) the limited mobility of those employed in agriculture. In affluent societies the agricultural sector is relatively small so that the system can 'afford' the extra burden to the taxpayer; and the share of agricultural products in the household's budget has become so small that the effects of price support on real incomes may not be felt too much, either. In some OECD countries, society holds a 'romantic' view of agriculture. Furthermore, the agricultural 'lobby' is sometimes successful in convincing the public that the conservation of the farm structure serves major social objectives such as the upkeep of certain rural regions, or the conservation of environmental quality[3].

In many developing countries, agricultural policies boiled down to negative protection. Agricultural producers were taxed by ceiling prices and export taxes, often via parastatal marketing boards and state trading organizations; and currency overvaluation. Behind these policies, several motives can be detected. First, low prices for staple food like rice, wheat and maize serve the interests of the masses of the poor (who have a tendency not to starve quietly). Second, low food prices, practised in the big cities, in particular in the capital, are in the interest of the 'ruling elite', benefiting its families and friends, partisans, students, bureaucrats, soldiers, etc. (Streeten, 1983). Third, under the development strategy of industrialization, low food prices serve as a tool to change the internal terms of trade in favour of industry (transfer of real income from agriculture to industry) and to keep down industrial wages. Fourth, exports are a convenient source of taxation. Since most policies of protection, positive or negative, have the effect of driving a wedge between the international and domestic prices, the relative difference between the two prices, or the Nominal Coefficient of Protection (NCP) can be used as a crude measuring rod: NCP = (P_d/P_w). At the end of the 1970s, typical NCPs were in the following ranges (IBRD, 1982):

NCP	Examples
1.5–2.0	EC (wheat, beef); Japan (rice)
1.0–1.5	Japan (beef); EC (rice)
0.66–1.0	Egypt, India (wheat); Argentina, Kenya (beef); India, Senegal (rice); Mexico, Kenya (cotton)
0.33–0.66	Argentina, Bangladesh (wheat); Brazil, Cameroon (rice); Egypt, Pakistan (cotton); Brazil, Tanzania (coffee)

All in all, during the 1970s agricultural policies pursued by most industrialized countries had the effect of depressing world prices while those practised by many developing countries worked in the opposite direction. These effects were reinforced by the evolution of market forces: in OECD countries, demand for major agricultural commodities grew slowly or stagnated because of slow population growth and low income elasticities of demand, while at the same time there were high rates of productivity growth (which in turn was partly the result of policies of protection: high prices, investment subsidies, public investment in agricultural production and marketing infrastructure, etc.).

At the same time, in many developing countries the demand for agricultural commodities increased considerably because of high rates of population growth, substantial increases in income – in particular in the middle-income countries – and high income elasticities of demand; while production growth lagged behind, partly as a result of adverse farm financial conditions of the previous decade and the continuing policy-induced disincentives to producers. Some Less Developed Countries (LDCs) downplayed former self-sufficiency objectives in the face of political pressure for dietary improvements. For several LDCs the satisfactory growth of foreign exchange earnings from exports of primary products and other commodities provided sufficient foreign exchange for the financing of rising food imports; others easily resorted to international lending, facilitated by the expansion of the world money supply (expansion of the Eurodollar market, creation of new reserve assets, recycling of petrodollars by the international banking system (USDA, 1987)).

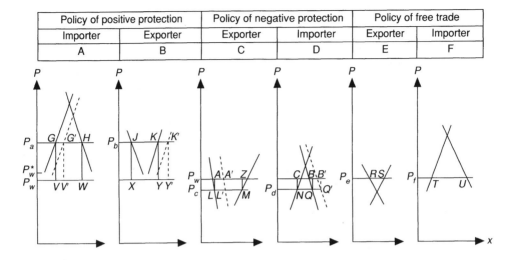

Fig. 5.4. Impact of agricultural policies and market forces on agricultural trade in the 1970s (schematic representation).

Theoretical analysis

In Fig. 5.4 an attempt is made to show, in a stylized way, the consequences of these policies and market trends for international agricultural trade. The world is assumed to consist of six trading partners. Under completely free trade, world market price would be at P_w^*. However, OECD-type countries A and B fix a support price at P_a and P_b and LDC-type countries C and D fix a ceiling price at P_c and P_d. The net effect is to drive down world price and domestic prices of E and F to $P_w = P_e = P_f$.

We now assume that technical progress in A and B shifts supply curves to the right while in C and D growing demand shifts the relevant curves to the right; the net effect on world price being zero. What are the consequences? For country A, the quantity and foreign exchange cost of imports decline; at the same time, government revenue from the import levy goes down. For country B, export quantity and foreign exchange earnings rise but expenditure on export subsidies also goes up. To country C, the decline in the quantity exported means a loss of foreign exchange earnings and a loss of revenue from the export tax. Country D is faced with a rise in the quantity and foreign exchange cost of imports while at the same time having to spend additional amounts on import subsidies[4]. The situation of country E and F remains unchanged.

This evolution will be particularly welcome to country B: it can sell a growing quantity of exports to country D, more than is lost on the market of A, at a world price that is not subjected to downward pressure (which is positive with respect both to foreign exchange earnings and the cost of export subsidization). For country C, the situation is less satisfactory, and it may tend to put the blame on countries A and B for their world price depressing agricultural policies. In contrast, country D appears to be a beneficiary of these policies, with respect to the foreign exchange cost of imports and to that of import subsidies. Free trader E is also harmed by A's and B's agricultural policies.

If one combines countries A and B into the group 'OECD' (in the sense of typical OECD countries) and countries C and D into the group 'LDCs' (typical LDCs) this process has the following effects on trade balances and trade structures:

1. The OECD experiences an improvement while the LDCs witness a deterioration in the (net) balance of trade.
2. For the OECD, the share in total world exports rises while that in total world imports declines; and vice versa for the LDCs.
3. Under the assumption of proportionality, the three exporting countries redirect their exports, away from the OECD importing country (A), towards the LDC importing country (D). At the same time the three importing countries change their source of supply, away from the LDC exporting country, towards the OECD exporting country (B).

These are the basic effects for a commodity where OECD imports and LDC exports decline so much in absolute terms that total (OECD and LDC) imports and exports remain constant (stagnation of world trade). It is clear that these effects would be the same for a commodity where, for example, OECD imports and LDC exports do grow in absolute terms, but less than LDC imports and OECD exports. However, in the latter case world trade would expand.

The evolution of trade

During the 1970s, the volume of agricultural trade increased considerably. The value expanded from US$50 billion in 1970 to more than US$225 billion by 1980. The number of countries that depended regularly on imports or exports for more than 5% of their food supplies or markets grew from fewer than 25 in 1970 to more than 40 by 1980 (USDA, 1987).

The observed trends of agricultural trade during the 1970s largely followed the pattern indicated above: for several commodities exports by OECD (LDC) countries followed a trend of positive (negative) growth while for imports it was vice versa. For other commodities, LDC imports and OECD exports rose faster than OECD imports and LDC exports.

Aggregating all agricultural commodities into one group, trade was characterized by the following trends (OECD, 1984): (i) OECD exports rose faster than those of LDCs, and LDC imports grew more than OECD imports; and (ii) OECD exports and LDC exports to LDCs rose faster than those to OECD countries, and OECD imports and LDC imports from OECDs rose at a faster pace than those from LDCs.

For the individual commodities, the consequence was (*cf.* Table 5.1):

1. The share of OECD countries in world exports of most agricultural commodities rose considerably, largely at the cost of LDCs[5].

Table 5.1. OECD and LDC shares in world exports and imports of major agricultural commodities, 1967–80 (source: OECD, 1984).

	Share in exports (%)				Share in imports (%)			
	OECD		LDC		OECD		LDC	
	1967	1980	1967	1980	1967	1980	1967	1980
Meat	62.1	78.9	22.3	10.4	86.3	69.1	5.1	20.4
Dairy products	91.0	95.4	1.5	1.5	75.2	57.6	19.8	33.5
Foodgrains and products	68.7	79.4	14.4	14.4	28.5	19.9	50.8	50.8
Feedgrains	63.8	88.7	25.1	6.7	87.8	47.5	4.4	18.2
Oilseeds and products	85.4	67.9	10.7	31.5	81.5	62.3	9.0	22.1
Fruit	34.7	34.7	49.1	49.1	81.5	77.3	9.0	14.8
Vegetables	68.1	76.7	12.1	12.1	63.6	63.6	23.7	33.5
Sugar	14.2	27.3	67.3	67.3	62.5	39.3	18.6	30.8
Tropical beverages	2.1	5.7	97.1	91.7	83.9	78.6	8.4	11.1
Agricultural raw material	64.2	67.5	24.4	20.8	78.2	71.9	10.7	14.2
All above commodities	54.3	65.0	34.2	27.3	72.6	59.8	15.9	22.6

2. OECD shares in world imports went down for most commodities while those of LDCs increased.

As regards world prices, for major agricultural commodities these followed no clear upward or downward trend. At the same time, the evolution was characterized by considerable short-term fluctuations in world markets. For wheat, in 1973–74 decreases in world supply and increases in world demand led to a 'world food crisis' which was reflected in an extraordinary rise in international prices of wheat, coarse grains and other commodities (OECD, APMT, 1994). Major determinants were: (i) unfavourable weather conditions in important wheat-exporting countries (including a drought in the USA and Canada in summer 1974) as well as in wheat-importing countries (socialist countries like the former USSR and the People's Republic of China, and developing countries like Egypt and Bangladesh); and (ii) the world oil crisis that gave an additional push to international wheat prices in late 1973.

Problems in the 1980s

Determinants

During the first half of the 1980s, supply and demand conditions on world agricultural markets underwent considerable changes. While in most OECD countries agricultural policies and the long-term trends of demand and supply remained more or less the same for many agricultural commodities, major changes took place in many developing countries: production growth increased; demand growth decelerated; and negative protection was reduced. As a result, LDCs' exports were boosted while import growth slowed down.

These developments were brought about by a variety of factors. As regards the supply side, earlier infrastructural and other investments, made during the 1970s, began to bear fruit, and current investment in agriculture increased. At the same time, technological advances materialized. The 'green revolution' spread. Regarding the demand side, LDCs' per capita incomes began to stagnate during the world recession of the early 1980s. Furthermore, declining world primary product prices and rising expenditures on oil imports and on the external debt service[6] made it necessary for many a developing country to cut down on imports, including those of agricultural commodities. Another important factor was the tightened supply and rising cost of credit on the international markets.

LDCs' agricultural and trade policies contributed to the expansion of exports and stagnation of imports of agricultural commodities. Amongst them were: (i) the dismantling of disincentives to agricultural production like ceiling prices or export taxes; (ii) the introduction of restrictions on imports (state trading, import licencing, tariffs, import quotas); and (iii) the introduction of incentives to exports, such as subsidies, or tax premia for export-oriented production.

In addition, in a growing number of LDCs, structural adjustment programmes agreed upon with the International Monetary Fund (IMF) and/or the World Bank, included trade-improving policies like currency devaluation and anti-inflation programmes. All in all, in a number of developing countries, the negative protection of agriculture was reduced, or abolished, partly even replaced by positive protection.

In some developing countries, a special feature behind the dismantling of negative protection to agriculture was a reorientation of the overall development strategy towards the agricultural sector since the strategy of industrialization had not fulfilled its objectives and the neglect of agriculture was having serious consequences on rural employment and incomes, thereby contributing to one of the gravest phenomena observed in LDCs, namely massive rural to urban migration.

The effects of the aforementioned factors on LDC's production are reflected in some global figures: the index of per capita food and agricultural production which had risen from 100 in 1975–77 to 104 in 1979–81 for LDCs and to 108 for developed market economies, went up to no less than 113 by 1984–86 in LDCs while rising to 111 in developed market economies. However, some regions, in particular sub-Saharan Africa, did not share this success story (FAO, MBS, 1987).

Theoretical analysis

These trends and their consequences are illustrated in Fig. 5.5 in which the initial situation (straight lines) is identical with the final one in Fig. 5.4. We now assume technical progress to take place in every country. Furthermore, we assume that countries A through D leave domestic price unchanged. As a consequence, world price falls to P_w'. What are the consequences for the individual countries?

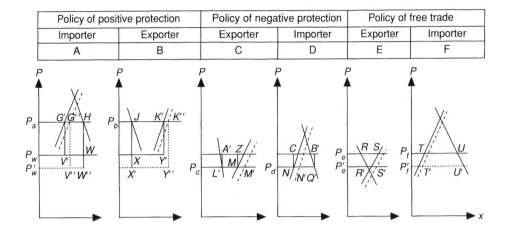

Fig. 5.5. Impact of agricultural policies and market forces on agricultural trade in the first half of the 1980s (schematic representation).

Importing countries For country A the quantity imported goes down (to $G''H$), and even more so does the foreign exchange cost of imports (simultaneous fall in world price). Even more favourable are the consequences for country D. As in the case of country A, import quantity falls (to $N'Q'$), and there is a twofold reduction of the foreign exchange cost of imports. At the same time, the fall in world price has a positive budget effect (the subsidy on food imports going down to zero). For country F, falling world price has the advantage of lowering the foreign exchange cost of import per unit. On the other hand, the fall in domestic price, with the resulting effect on the quantity of import, works in the opposite direction.

Exporting countries For country B the consequences will not be at all welcome. While the quantity exported rises (to JK''), foreign exchange earnings may even fall. Furthermore, the cost of export subsidies experiences a twofold increase (to $JK''Y''X'$). Similarly for country C, the only difference is that for it, the negative budget effect consists of a reduction of income from export taxes (to zero). To avoid this, domestic price could be lowered, but this might be self-defeating to the extent that it would reduce the quantity exported. For country E, the fall in domestic price tends to counteract the effect of the productivity increase on the quantity exported; because of this and of the fall in world price, foreign exchange earnings may even decline.

In this scenario, the (positive) nominal coefficients of protection increase for A and B. The (negative) NCPs of C and D decrease. Of course the fact that they go down to zero is a coincidence – with a more marked decline in world price and/or under the assumption of the introduction of policies raising domestic price, the negative protection would have turned into a positive one. Countries A through D, by keeping their prices constant, do not contribute to the absorption of the worldwide glut. That is why world price and E's and F's domestic prices have to fall further than would otherwise be necessary; they fall until E and F have entirely absorbed the effect of the artificially strong expansion of production quantity in A through D. In spite of their productivity increases, E's export quantity falls and F's import quantity expands strongly. Seen from the point of view of their producers: agricultural incomes are boosted by technical progress (downward shift in the marginal cost curve) but this is more than offset by the artificially steep fall in domestic price.

This evolution implies an additional distortion of the geographical distribution of world agricultural production. In high-cost A and B, production increases more than it should while in low-cost E and F it undergoes an artificially pronounced decline. Country C and D expand their production more than they would under free trade, but this serves to partly make up for the distortion inherited from the 1970s.

The evolution of trade

The evolution of trade in agricultural commodities reflected the general tendency of glut. In the first half of the 1980s, unprecedented stock accumulations took place and world agricultural trade stagnated at falling prices (USDA, 1987; Basler, 1988).

As Table 5.2 shows, the high growth rates for the value of total agricultural exports by developed market economies experienced during the 1970s fell considerably during the first half of the 1980s. This was partly due to falling world prices (see below). At the same time, the agricultural export performance of developing countries improved, in particular for Asian and Latin American countries. The trend was accompanied by a considerable reduction in LDCs' import growth. Statistics by OECD (1989) show that the annual nominal growth rate of OECD agricultural exports to other OECD countries fell from 13.4% (average 1970–1980) to −0.7% (1980–1986) while for exports to LDCs it fell from 21.5% to −4.3%. As can be seen from Fig. 5.6, as a consequence of the worldwide glut real world prices for major agricultural commodities fell considerably during the first half of the 1980s.

Table 5.2. Growth rates of agricultural trade, 1970–1985 (% per annum) (source: FAO, 1987).

	Exports*		Imports*	
	1970–80	1980–85	1970–80	1980–85
Developed market economies	5.7	0.7	2.0	2.5
Developing countries	1.8	3.3	8.0	1.7
Africa (sub-Saharan)	−2.2	0.4	6.6	1.6
Near East/North Africa	−2.6	−0.1	12.7	6.1
Asia	4.0	4.4	5.4	−1.8
Latin America	2.7	3.8	8.3	−3.2

* Expressed in value terms.

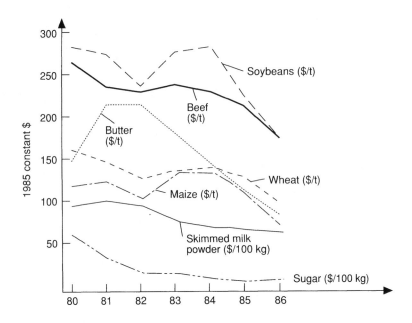

Fig. 5.6. Real world prices for agricultural products, 1980–86 (source: OECD, APMT, 1992). Prices in dollars, per tonne or per 100 kg, are deflated by World Bank index of manufacturing unit value, 1985 = 100.

Agricultural trade conflicts and liberalization under the GATT

Introduction

It is clear from Fig. 5.6 that the fall in world prices during the first half of the 1980s was detrimental to all exporting countries, whether characterized by positive, zero or negative protection. Who is to be blamed? On the one hand, the change in world price is due to the worldwide productivity increase; and each country has the right to become more productive. On the other hand, the intensity of the change is the result of A's, B's, C's and D's policies. The difference being, however, that while A and B raised their positive protection, C and D reduced their negative protection. In the final analysis, A and B are the 'culprits' responsible for the low *level* of world price. While the former partly closes its frontiers to exports by B, C and E, the latter indulges in unfair competition on the markets of A, D and F. Assume that 'country' B is the aggregate of two countries, one (called 'EU') with a relatively high, and the other (called 'USA') with a relatively low degree of positive protection. In this case the latter will pass the blame on to the former, in particular if the EU's system of protection was there first and was the cause of the USA defensive-support policies.

Table 5.3. Agricultural trade of the USA and the EC 1979–1981 and 1985–1987, three-year moving averages (billion US$) (source: Fröhlich, 1989).

Period	USA total		EC total*		USA–EC bilateral	
	Export	Import	Export	Import	USA export	USA import
1979–81	37.8	20.5	24.0	40.7	8.9	3.0
1980–82	38.6	20.5	25.2	39.2	9.0	3.3
1981–83	37.2	20.6	24.3	36.8	8.9	3.6
1982–84	35.3	21.7	23.3	36.2	8.4	4.1
1983–85	32.9	23.5	23.6	35.7	7.8	4.7
1984–86	30.2	25.6	24.4	36.9	7.0	5.2
1985–87	28.0	26.6	26.2	39.1	6.7	5.4

* Excluding intra-EC trade.

The conflict between the USA and the EC

One of the countries which was hardest hit by the fall in world agricultural prices was the USA. During the 1970s, the USA share of the world agricultural market had increased considerably: while world trade had increased fourfold, USA exports had expanded sixfold (USDA, 1987). As can be seen from Table 5.3, during the first half of the 1980s USA agricultural exports fell considerably. As at the same time USA agricultural imports increased, it was felt that these developments were largely caused not only by the appreciation of the dollar, which weakened the USA competitive position, but also by the high degree of protection practised by major world trade competitors.

The country on which criticism was focused most was the European Union, then called the European Community (EC). Its Common Agricultural Policy (CAP), the system of producer minimum prices, variable levies and variable export subsidies (so-called 'export restitutions') (OECD, 1987b) had been criticized by politicians and scientists in less protectionist OECD countries and LDCs for many years (Sampson and Snape, 1980; Buckwell *et al.*, 1982; Ahrens *et al.*, 1983; Ahrens, 1985; Mathews, 1985; Stoeckel, 1985; Dicke *et al.*, 1988; Rosenblatt *et al.*, 1988; Wissenschaftlicher Beirat, 1988). In the words of Gupta *et al.* (1989, p. 38):

> There are three major effects on other countries that have been attributed to the CAP. First, pricing policies and protection in the EC have led to excessive production, reducing imports and expanding exports. With growing output in other parts of the world and with no corresponding shifts in demand, the rising farm output in the EC has depressed world prices. This effect has been exacerbated by subsidized exports of EC surpluses. ... Second, ... the EC's variable imports and export levies have tended to insulate the EC markets from external price fluctuations, thereby amplifying the variability of world commodity prices. ... This means that the rest of the world must adjust more to any quantity shifts. Third, the greater instability of prices renders the incomes of primary producers and exporters unstable. When combined with risk-averse behaviour by farmers in production, this instability causes farmers to contract output, thereby lowering their incomes.

(a)

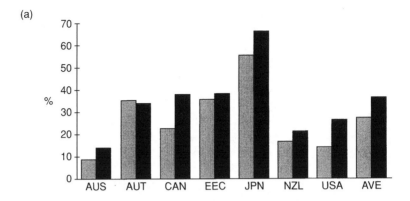

(b)

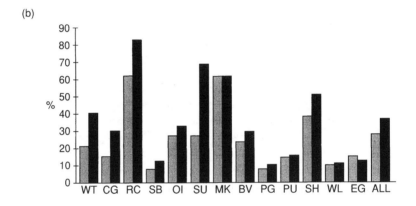

Fig. 5.7. Net percentage PSEs, 1979–81 and 1984–86. (a) by country, (b) by commodity (source: OECD, APMT, 1988). For each country or commodity, the first column refers to 1979–81, the second to 1984–86. AUS = Australia; AUT = Austria; CAN = Canada; JPN = Japan; NZL = New Zealand; AVE = OECD average; WT = wheat; CG = coarse grains; RC = rice; SB = soybeans; OI = other oilseeds; SU = sugar; MK = milk; BV = beef and veal; PG = pig meat; PT = poultry; SH = sheep meat; WL = wool; EG = eggs.

In spite of the decline of major world prices, EC agricultural exports increased during the first half of the 1980s. Since at the same time imports went down, the EC's balance of agricultural trade improved. Under the influence of the EC's protectionist system, US agricultural exports to the EC declined while imports from the EC rose (*cf.* Table 5.3).

Agricultural protection by the EC was higher than that by the USA. This was clearly shown by the OECD's calculations of a new measure of agricultural protection, developed by Josling in two FAO documents (FAO, 1973, 1975), the Producer Subsidy Equivalent (PSE). Based on the definition of total Producer Subsidy Equivalent:

$$PSE = Q\,(P_d - P_r) + D - L + O$$

with Q = quantity of production; P_d = domestic producer price; P_r = reference price (world price at border); D = direct payments; L = levies on producers; O = other payments from the government budget, net percentage PSE is calculated as PSE (%) = (PSE / ($QP_d + D - L + O$) 100 (OECD, 1987a).

As Fig. 5.7 shows, at the beginning of the 1980s the EC's average (all commodities) net percentage PSE was more than twice that of the USA (37% against 16%).

The USA system of support to agricultural producers of grain, oilseeds and cotton, as outlined in the previous chapter, contains several elements of price and income support, in particular nonrecourse commodity loans, inventory and financial activities by the Commodity Credit Corporation (CCC), and direct cash transfers for deficiency payments (grain and cotton). In contrast to the EC, the USA practised a system of acreage reduction according to which grain and cotton farmers, to be eligible to the price and income support, have to comply with supply control programmes. These are aimed to compensate the production incentives generated by the price and income support programmes (OECD, 1987c). Meanwhile, the EC has introduced a similar system[7].

During the first half of the 1980s, under the impact of falling world prices, PSEs rose for almost all OECD countries but the increase was particularly marked for the USA. This was largely due to the fact that, as a reaction to the EC's (and other countries') high degree of protection that hindered USA exports, the USA themselves introduced policies to enhance the competitiveness of USA farm exports[8]. While in the EC average net PSE rose by only three points, the increase was 12 points for the USA (Fig. 5.7).

However, this meant an additional burden to the USA budget. While the EC system of protection focused on support prices thereby passing the major part of the burden to the consumer, the USA system of agricultural support, dominated by deficiency payments and export subsidies, passes the major part of the burden to the taxpayer. As can be seen from Table 5.4, the total cost to the latter increased threefold during the first half of the 1980s. Total cost, both to the taxpayer and consumer, which had amounted to one half of that of the EC at the beginning of the 1980s, rose to the same level within a few years.

Table 5.4. Cost of agricultural policy: USA and EC, 1979–1981 and 1984–1986 (billion ECU) (source: Fröhlich, 1989).

	USA		EC	
Item	1979–81	1984–86	1979–81	1984–86
Cost to the taxpayer	19.4	59.4	21.1	30.4
Cost to the consumer	9.6	20.3	36.7	49.8
Budget revenue	0.3	1.2	0.8	0.6
Total cost	28.7	78.5	57.0	79.6

Against this background, the USA were motivated to prevent a continuation of the export subsidization race where world price is continually being driven down, export subsidies go up for every exporting country, and no one is better off in the end. The USA expected 'that its farm exports would benefit from a more liberal agricultural trading environment, in which both import barriers and export subsidies would be reduced' (USDA, 1987).

The Uruguay Round of negotiations

Since the middle of the 1980s, the problem of world agriculture can be described as follows: as the factors of production employed in world agriculture become more productive whereas effective demand for agricultural commodities grows relatively slowly, part of the factors must move out of the agricultural sector. The question is: where; which factors (labour, land, capital, inputs); and how (what mechanism should determine the where and the which)?

In the international community there is a certain consensus that in the long-run the market mechanisms should decide. Against this background, the USA successfully urged the international community to include agriculture into the Uruguay Round of GATT negotiations on trade liberalization, launched in September 1986 in Punta del Este, Uruguay.

It is not our intention here to analyse in detail the interesting history of the various controversies experienced under these negotiations. However, in brief:

1. While the USA insisted on a relatively rapid and complete reduction of agricultural protection, the EC (and some other countries) were more hesitant (in the beginning, it even made suggestions that came close to introducing a system of 'managed trade' where markets are shared on the basis of international agreements on import and export quantities, irrespective of comparative advantage). Behind this was the fear that under a system of free trade, European agriculture and the cultural heritage connected with it would be jeopardized.

2. The Cairns group of 14 'free trader' exporting countries (Canada, Australia, New Zealand[9], Argentina, Brazil, Uruguay, Chile, Colombia, Indonesia, Malaysia, Philippines, Thailand, Hungary, Fiji Islands), basically supported the USA position but suggested a compromise regarding the speed of liberalization.

3. The emerging positive protection by LDCs, in particular Newly Industrializing Countries (NICs), was discussed. According to calculations by USDA (1987), in the period 1982–1984, PSE was in the range between +50 and +75% for: wheat in Taiwan, South Korea and Brazil; corn in South Korea and Mexico; barley, beef, rice and soybeans in South Korea.

4. Major net food-importing LDCs like Egypt and Bangladesh expressed fears about the future cost of food imports and availability of food aid as a consequence of liberalization-induced rises in world agricultural prices.

Finally, based on the so-called Blair House Accord of November 1992, a compromise was reached in December 1993. Its main provisions are:

1. Domestic support: (i) reduction by 20% (base period: 1986–1988) of all domestic support measures in favour of agricultural producers (expressed in terms of Aggregate Measurement of Support: AMS) in equal instalments from 1993 to 1999; with the exception of the so-called 'green-box' measures; (ii) allowance of credit for reduction actions undertaken since the year 1986.

2. Market access: (i) conversion of the 1986–88 non-tariff border measures into tariff equivalents ('tariffication'); (ii) reduction by 36% on a simple average basis of tariffs or tariff equivalents in equal instalments between 1993 and 1999 (minimum rate of reduction for each tariff line: 15%); and (iii) maintenance of current access opportunities (access = limitation of tariffs to 32% of normal tariffs) for imported goods and establishment of minimum access opportunities of 3% of domestic consumption and raising them to 5% by 1999.

3. Export competition: budgetary and quantity reduction of export subsidies (base period: 1986–1990) respectively by 36 and 21%. During the transition period (1995–2000), however, the period 1991–1992 may be used as the base period.

As far as the developing countries are concerned, special provisions were granted to them, in particular to the least developed countries. They pertain to the above three reduction commitments, and to the establishment of appropriate mechanisms to ensure that the implementation of the results of the Uruguay Round on trade in agriculture does not adversely affect the availability of food aid.

PERSPECTIVES

Since the middle of the 1980s, the evolution of international trade in agricultural commodities has 'normalized' in the sense that pressures on world prices have eased. For a number of products, world supply and world demand seem to have grown at roughly the same pace in the medium-run. As can be seen from Fig. 5.8, the result has been a certain medium-term constancy, in some cases even a rise in, real world prices (the intermediate high for wheat, maize and soybeans being due to USA drought). At the same time net percentage PSEs have decreased to some extent for most OECD countries and commodities (OECD, 1994).

On the other hand, before the GATT compromise, most OECD countries have not gone far on the way of dismantling their agricultural support programmes. One exception is dairy production where the majority of OECD countries have introduced or intensified supply control measures, with the result of a rise in real world prices, in particular for skimmed milk powder (*cf.* Fig. 5.8). Otherwise the long period of negotiations under the Uruguay Round of GATT was more or less one of hesitation. So far, commodity markets have remained heavily influenced by agricultural policies and support arrangements rather than by market signals.

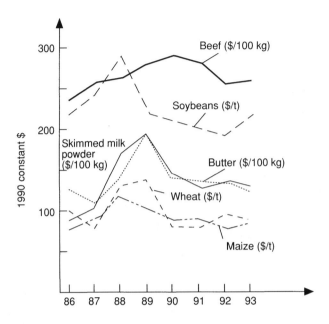

Fig. 5.8. Real world prices for agricultural products, 1986–93 (source: OECD, APMT, 1994). Prices in dollars, per tonne or per 100 kg, are deflated by World Bank index of manufacturing unit value, 1990 = 100.

As far as future developments are concerned, the outlook is subject to considerable uncertainty. On the one hand, the arrangements agreed upon under the Uruguay Round and the continuation of restrictive measures (like CAP reform) that have been taken before, will lead to certain reductions in OECD structural surpluses. This effect may be supplemented by a recovery of world economic growth and a resulting expansion of demand for agricultural commodities.

On the other hand, the restructuring of agriculture in formerly socialist countries will generate productivity increases in the long-run. Similar effects may result from the Eastern enlargement of the European Union. In the long-run, therefore, new pressures may arise for a further dismantling of agricultural protection in a new round of international negotiations.

CONCLUDING REMARKS

From the above analysis it may be deduced that slowly, and with a considerable time-lag, the international community is learning from the adverse effects of trade-distorting agricultural policies. As international trade in agricultural commodities is moving towards what may be called a new medium-term equilibrium, new pressures are building up, thereby necessitating additional adjustments and hopefully resulting in a further expansion of this trade. Whether, when, and to what extent this would be compensated by increases in transportation costs, resulting from the much-discussed environmental strategy of internalizing external costs, is too early to answer.

NOTES

1. In addition to the patterns of trade in agricultural commodities, there is also a significant flow of manufactured food products between nations. The dollar value of this flow is typically greater than that for commodities, but this value includes the cost of processing and manufacturing. Agricultural trade statistics usually do not include the international movement of manufactured food products. Aspects of this trade are covered in Chapter 18.

2. In reality production would be subject to simultaneous shocks in A and R. Model analyses using simultaneous random shocks show that indeed world price fluctuates less widely under a system of free trade (Bale and Lutz, 1979).

3. For a discussion of motives behind positive and negative agricultural protection, cf. Parikh et al. (1988).

4. To avoid the rise in the foreign exchange cost of agricultural imports, country D may seek food aid, which has the additional advantage of making import subsidization unnecessary and even generates additional revenue to the government's budget as the food from food aid is sold to the consumer. This may make it possible to pay the national farmer a producer price above the consumer price and to cover the difference by subsidies.

5. The exceptions pertain to products or product groups where the degree of OECD protection is relatively low (oilseeds), preferential import quotas are granted to LDCs only by OECD countries (sugar) and/or where there is a large tropical element in the product group (fruit, vegetables, tropical beverages).

6. Since the value of the dollar went up sharply, the local currency costs of servicing the debt also increased considerably.

7. In addition, the effect of a deficiency payment, or 'open-ended production subsidy' (USA) on world price is less pronounced than that of a variable export subsidy (EC) (Houck, 1986).

8. As early as in 1982, the USA government established a three-year credit programme to enhance agricultural exports. In 1985, parallel to the Farm Act of that year, the Bonus Incentive Commodity Export Programme (BICEP) that provided a bonus to exporters of USA grain was introduced. This programme was continued as the Export Enhancement Programme (EEP). The major part of the grain sold under this programme was exported to North Africa and the Near East with the declared objective to restrain European exports (USDA, AO, September 1987).

9. As can be seen from Fig. 5.7, net percentage PSEs for Australia and New Zealand were very low in the middle of the 1980s; this was less so for Canada.

FURTHER READING

Basic elements of trade theory as related to agriculture are explained in Ritson's classic book (1977, Chapter 7). A more extensive partial equilibrium analysis of agricultural trade policies is presented in McCalla and Josling (1985) and Houck (1986). For a critical survey of the effects, both domestic and international, of the EEC's system of agricultural protection, the reader is referred to Stoeckel (1985) and Rosenblatt *et al.* (1988). The interests and problems of developing countries in international trade in agricultural products are analysed by Valdés (1987) and summarized, with suggestions for the international negotiations, by Islam (1988). For a profound treatment of major consequences of liberalization of world agricultural trade, on the basis of different quantitative models reflecting the 'state of the art', cf. Parikh *et al.* (1988), Tyers and Anderson (1988) and Goldin *et al.* (1993). As an introduction to the environmental effects of a liberalization of world agricultural trade, Anderson (1992) is recommended. For any student wishing to keep up to date with major facts and trends of international trade in agricultural commodities, the annual publications by OECD (Agricultural Policies, Markets and Trade) and FAO (The State of Food and Agriculture, Commodity Review and Outlook, Production Yearbook, Trade Yearbook) will be most useful.

REFERENCES

Ahrens, H. (1985) *EEC's Common Agricultural Policy and Canadian Agricultural Exports.* Departmental Publication No. 85–1. Department of Agricultural Economics, McGill University, Quebec.

Ahrens, H., Urff, W.v. and Weinmüller, E. (1983) Conséquences de la politique agricole commune sur les pays en voie de développement. *Economie Rurale. Revue Francaise d'Economie et de Sociologie Rurales* 156, 3–10.

Anderson, K. (1992) Effects on the environment and welfare of liberalizing world trade: The cases of coal and food. In: Anderson, K. and Blackhurst, R. (eds), *The Greening of World Trade Issues.* Harvester Wheatsheaf, New York, pp. 145–172.

Bale, M.D. and Lutz, E. (1979) The effects of trade intervention on international price instability. *American Journal of Agricultural Economics* 61, 512– 516.

Basler, A. (1988) Der weltagrarhandel in den ausgehenden achtziger Jahren. *Agrarwirtschaft* 37, 173–179.

Buckwell, A.E., Harvey, D.R., Thomson, K.J. and Parton, K.A. (1982) *The Costs of the Common Agricultural Policy.* Croom Helm, London, Canberra.

Dicke, H., Donges, J.B., Gerken, E. and Kirkpatrick, G. (1988) The economic effects of agricultural policy in West Germany. *Weltwirtschaftliches Archiv* 124, 301–321.

Food and Agriculture Organization of the United Nations (FAO) *Commodity Review and Outlook.* Various issues. FAO, Rome.

Food and Agriculture Organization of the United Nations (FAO) *Monthly Bulletin of Statistics (MBS).* Various issues. FAO, Rome.

Food and Agriculture Organization of the United Nations (FAO) *Production Yearbook.* Various issues. FAO, Rome.

Food and Agriculture Organization of the United Nations (FAO) *The State of Food and Agriculture.* Various issues. FAO, Rome.

Food and Agriculture Organization of the United Nations (FAO) *Trade Yearbook.* Various issues. FAO, Rome.

Food and Agriculture Organization of the United Nations (FAO) (1973) *Agricultural Protection: Domestic Policy and International Trade.* C 73/LIM/9. FAO, Rome.

Food and Agriculture Organization of the United Nations (FAO) (1975) *Agricultural Protection and Stabilization Policies: A Framework of Measurement in the Context of Agricultural Adjustment.* FAO, Rome.

Food and Agriculture Organization of the United Nations (FAO) (1987) *Agriculture: Toward 2000*, revised version. FAO, Rome.

Fröhlich, H.-P. (1989) *Das GATT am Scheideweg*. Deutscher Instituts-Verlag, Köln.

Goldin, I., Knudsen, O. and Mensbrugghe, D. (1993) *Trade Liberalization: Global Economic Implications*. Organization of Economic Cooperation and Development (OECD)/The World Bank, Paris, Washington.

Gupta, S., Lipschitz, L. and Mayer, T. (1989) The Common Agricultural Policy of the EC. *Finance & Development* 1989, 37–39.

Houck, J.P. (1986) *Elements of Agricultural Trade Policies*. Macmillan Publishing Company, New York.

International Bank for Reconstruction and Development (IBRD) (1982) *World Development Report 1982*. IBRD, Washington, DC.

Islam, N. (1988) Agriculture in GATT negotiations and developing countries. In: Miner, W.M. and Hathaway, D.E. (eds), *World Agricultural Trade: Building Consensus*. The Institute for Research on Public Policy, Brussels, pp. 169–189.

McCalla, A.F. and Josling, T.E. (1985) *Agricultural Policies and World Markets*. Macmillan, New York, London.

Mathews, A. (1985) *The Common Agricultural Policy and the Less Developed Countries*. Gill and Macmillan, Dublin.

Organization of Economic Cooperation and Development (OECD) (1984) *Agricultural Trade with Developing Countries*. OECD, Paris.

Organization of Economic Cooperation and Development (OECD) (1987a) *National Policies and Agricultural Trade*. OECD, Paris.

Organization of Economic Cooperation and Development (OECD) (1987b) *National Policies and Agricultural Trade. Study on the European Economic Community*. OECD, Paris.

Organization of Economic Cooperation and Development (OECD) (1987c) *National Policies and Agricultural Trade. Study on the United States*. OECD, Paris.

Organization of Economic Cooperation and Development (OECD) (1989/1994) *Agricultural Policies, Markets and Trade (APMT). Monitoring and Outlook*. OECD, Paris.

Parikh, K.S., Fischer, G., Frohberg, K. and Gulbrandsen, O. (1988) *Towards Free Trade in Agriculture*. Martinus Nijhoff Publications, Dordrecht, Boston, Lancaster.

Ritson, C. (1977) Agricultural trade policy. In: *Agricultural Economics. Principles and Policy*, student edn. Granada Publishing, London, pp. 223–348.

Rosenblatt, J., Mayer, T., Bartholdy, D., Demekas, D., Gupta, S. and Lipschitz, L. (1988) *The Common Agricultural Policy of the European Community – Principles and Consequences*. Occasional Papers No. 62. International Monetary Fund, Washington, DC.

Sampson, G. and Snape, R.H. (1980) Effects of the EEC's variable import levies. *Journal of Political Economy* 88, 1026–1040.

Stoeckel, A. (1985) *Intersectoral Effects of the CAP: Growth, Trade and Unemployment*. Occasional Papers No. 95. Bureau of Agricultural Ecomomics, Canberra.

Streeten, P. (1983) Food prices as a reflection of political power. *CERES, FAO Review* 16(2), 16–22.

Tyers, R. and Anderson, K. (1988) Liberalizing OECD agricultural policies in the Uruguay Round: Effects on trade and welfare. *Journal of Agricultural Economics* 39, 197–216.

United States Department of Agriculture (USDA) *Agricultural Outlook (AO)*. Various issues. USDA, Washington, DC.

United States Department of Agriculture (USDA) (1987) *Government Intervention in Agriculture. Measurement, Evaluation, and Implications for Trade Negotiations.* USDA, Washington, DC.

Valdés, A. (1987) Agriculture in the Uruguay Round: Interests of developing countries. *The World Bank Economic Review* 1, 571–593.

Wissenschaftlicher Beirat beim Bundesministerium für Wirtschaftliche Zusammenarbeit (1988) *Der Einfluß der EG-Agrarpolitik auf die Entwicklungsländer.* BMZ-Aktuell, Entwicklungspolitik. Bundesministerium für Wirtschaftliche Zusammenarbeit, Bonn.

6 Commodity Market Modelling

Philip Garcia and Raymond M. Leuthold

Department of Agricultural and Consumer Economics, University of Illinois at Urbana-Champaign, 305 Mumford Hall, MC-710, 1301 West Gregory Drive, Urbana, Illinois 61801, USA

INTRODUCTION

Commodity models provide a systematic approach for analysing the behaviour of markets. The modelling of agricultural commodity markets has a long and rich history, utilizing a variety of analytical procedures and applied to many commodities[1]. Over time, their analytical procedures have changed to reflect more accurately the nature of agricultural markets. Recently, the development and use of models have become more pervasive due to new and improved data sources, and more accessible computer/statistical programs.

The development of appropriate commodity market models requires a thorough knowledge and understanding of the problem or uses envisaged for the model, the factors (e.g. technical, biological, government and economic forces) influencing decision-makers in the market, the sources of available data, and the current relevant analytical procedures to address the investigation. Hence, it involves a blending of quantitative procedures, economic theory, and understanding of the 'real world' context in which the problem must be addressed.

The uses of agricultural commodity market models are quite varied, but focus on the identification of the economic structure, forecasting and policy analysis. Specification of the appropriate structure is important for estimating elasticities and flexibilities, but takes on added significance when the model is being used for forecasting and policy analysis. Inadequate identification of the structure can make forecasting and policy analysis problematic.

The voluminous literature related to agricultural commodity market modelling makes a comprehensive review of the topic impractical. Here, we address selected issues related to econometric modelling of agricultural commodity markets[2]. Agricultural commodity markets lend themselves to econometric analysis because, at least at the primary level, commodities are fairly homogeneous, and markets are highly competitive. Data also are more readily available from a variety of sources which can be used to characterize market structure and behaviour. Further, by systematically identifying the characteristics of agricultural supply and demand, econometric analysis of commodity markets can provide market participants and policy makers with a clearer vision of the economic environment in which they operate.

Specifically, this chapter presents a relatively non-technical discussion of agricultural commodity market models. First, it provides definitions of commodity market models, detailing their structure and discussing the factors often used to explain supply and demand relationships. This is followed by a brief discussion of model validation and verification procedures. Next, an overview of the uses of commodity market models is provided, focusing on the development and application of planning/policy models, market analysis models and forecasting models. We conclude with a brief assessment of the value of econometric market models in the analysis of agricultural commodities.

COMMODITY MARKET MODELS

Econometric commodity models quantify the relationships that explain economic behaviour of a market or system of markets. Econometric agricultural commodity models are mathematical representations of markets, often comprised of supply, demand, inventory and market-clearing relationships. Variables which influence these relationships are specified, based on economic theory, and knowledge of the underlying factors and constraints that influence decision-makers. Econometric procedures consistent with the structure of the model and the assumptions of the underlying error processes are used to quantify the relationships.

To assist in defining the nature of commodity models, consider a simplified three-equation specification of supply, demand and an identity:

$$Q_D = \beta_0 - \beta_1 P + \beta_2 PS + \beta_3 Y + \varepsilon_1 \tag{6.1}$$

$$Q_S = \alpha_0 + \alpha_1 P + \alpha_2 PO + \varepsilon_2 \tag{6.2}$$

$$Q_D = Q_S \tag{6.3}$$

The model posits that demand of the commodity (Q_D) is related linearly to its own price (P), price of substitutes (PS), income (Y), and an error process (ε_1). Supply (Q_S) is assumed to be related linearly to its own price (P), price of alternative outputs (PO) which compete for the use of the resources, and an error term (ε_2). The system is closed by a market-clearing identity.

Equations 6.1–6.3 represent the system of relationships and factors which influence price and quantity of the commodity. They are the structure of the model, with Eqns 6.1 and 6.2 referred to as behavioural equations, and Eqn 6.3 as an identity. An important aspect in understanding the structure of commodity models is identifying which variables are determined by the structure and which variables are determined outside the structure. Here, price and quantity are determined simultaneously by the model and are referred to as endogenous or dependent variables. Other variables such as income, and price of alternatives in consumption and production are viewed as determined outside of the model and are referred to as exogenous or independent variables[3]. The precise effects of the variables on the structure are quantified by the parameters or coefficients (αs and βs) of the system. The error terms influence demand, supply and price, and are stochastic in nature, assumed to be influenced by random occurrences, e.g. weather and omitted variables.

Equations 6.1–6.3 with their estimated parameters can be used to solve for equilibrium price and quantity values given the values of the exogenous variables and the error process. The solutions are generated either graphically or mathematically (Intriligator, 1978). The graphic solutions in price and quantity space can be generated assuming the expected value of the error terms are zero, and incorporating the values of the independent variables in the intercepts (Fig. 4.3, p. 54, is an example of this). Mathematically, the solutions are based on finding the simultaneous solution for price and quantity expressed in terms of the exogenous variables and the error terms. These expressions are referred to as the reduced form, with particular prices and quantities calculated given specific values of the exogenous variables and the error terms.

Two other forms of econometric commodity models are predominant in the literature: recursive and single equation models. The recursive framework specifies that quantity and price are determined sequentially through time, and assumes one-way causality from independent and predetermined variables to the dependent variable. Equations 6.1–6.3 can be modified to form a recursive structure, Eqns 6.3–6.5, with the assumption that quantity supplied (Eqn 6.4) is a function of lagged predetermined prices:

$$Q_{St} = \delta_0 + \delta_1 P_{t-1} + \delta_2 PO_{t-1} + \varepsilon_4 \tag{6.4}$$

$$P_t = \Phi_0 - \Phi_1 Q_{Dt} - \Phi_2 PS_t + \Phi_3 Y_t + \varepsilon_5 \tag{6.5}$$

As quantity available for consumption is predetermined by the supply, price clears the market, in effect, becoming the dependent variable in the demand relationship (Eqn 6.5). This price then feeds back to supply (Eqn 6.4), influencing production in the subsequent period.

Single equation models consider only one aspect of the larger system, usually either specifying supply or demand relationships. In the context of our model, the quantity demanded (Eqn 6.1) can be specified as a single equation model assuming that prices are either exogenous to the market or predetermined. Similarly, supply (Eqn 6.4) can be considered a single equation model when prices are predetermined or exogenous to the structure. This type of model assumes one-way causality from the right-hand side variables to the dependent variable.

Several of these concepts can be developed more formally in the context of a linear structural model:

$$y\Gamma + xB = \varepsilon \tag{6.6}$$

where Γ is a $g\times g$ matrix of coefficients of endogenous variables; B is a $k\times g$ matrix of coefficients of predetermined variables; y and x are row vectors of the g endogenous and k predetermined variables respectively; and ε is a row vector consisting of g stochastic disturbance terms[4]. The structural form consists of g structural equations, each of which has an independent role and meaning in the model (e.g. a demand function), and when combined with the values of the parameters determine the endogenous variables in terms of the predetermined variables and the error terms. In effect, the structure represents the model builder's view on how the market works.

The structural form is solved, expressing the endogenous variables as explicit linear functions of the exogenous variables and the stochastic error terms, generating the reduced form. Postmultiplying Eqn 6.6 by Γ^{-1} and solving for y yields:

$$y = -x B\Gamma^{-1} + \varepsilon\Gamma^{-1} \tag{6.7}$$

Introducing the $k\times g$ matrix of reduced-form coefficients Π and the $1\times g$ reduced-form error disturbance vector v, where $\Pi \equiv - B\Gamma^{-1}$, and $v \equiv \varepsilon\Gamma^{-1}$, the reduced form is written as:

$$y = x\Pi + v \tag{6.8}$$

The reduced form determines the probability distributions of the endogenous variables, given the exogenous variables, the coefficients, and the probability distributions of the stochastic terms. The coefficients in Π represent the change in the endogenous variables given a change in the exogenous variables.

When dynamics explicitly enter the structure in the form of lagged endogenous variables, it is possible to solve for the final form. The final form expresses current endogenous variables as functions of base values and all relevant current and lagged exogenous variables and the error terms. Consider the following equation where the structural form includes only one lag:

$$y_t\Gamma + y_{t-1}B_1 + x_tB_2 = \varepsilon_t \tag{6.9}$$

where y_{t-1} is a vector of lagged endogenous variables. The lagged endogenous and exogenous variables are grouped as predetermined variables of the system. The B in Eqn 6.6 also has been partitioned to conform with the partitioning of the predetermined variables. The reduced form is solved as in Eqn 6.8, which in turn is used to solve iteratively for the final form:

$$y_t = y_0\Pi_1^t + \sum_{t-1}^{j=0} x_{t-j}\Pi_2\Pi_i^j + \sum_{t-1}^{j=0} v_{t-j}\Pi_i^j \tag{6.10}$$

where $\Pi_1 \equiv -B_1\Gamma^{-1}$, $\Pi_2 \equiv -B_2\Gamma^{-1}$ and $v \equiv \varepsilon_t\Gamma^{-1}$. The development of the final form permits the calculation of multipliers which are used in policy analysis (Labys, 1973; Intriligator, 1983). For large models that may include non-linearities, the final form frequently is not calculated. In these instances, the model typically is solved through the use of numerical solution procedures, such as Gauss–Seidel, to obtain equilibrium market determined values. The procedure orders equations temporally and successively iterates values of the endogenous variables until a solution is found (Hallam, 1990).

Equation 6.6 represents the general linear simultaneous model. An important consideration in simultaneous models is the ability to extract information of the structural parameters from the reduced form, the identification problem. In essence, the system needs to be identified such that sufficient information exists to determine structural parameters B and Γ from the reduced-form parameters Π. This can be accomplished in several ways, but the most common is through the imposition of zero restrictions, implying that certain variables appearing in particular equations do not appear in other equations[5]. An example of an appropriately identified system is the recursive model. A recursive framework reflects one-way causality in y which means that in Eqn 6.6 Γ is upper triangular, and also that the covariance between the error terms in any two equations is zero. The relationships among the equations reflect the sequential flow in the endogenous variables through time[6].

Whether a structure is best specified in a simultaneous, recursive or single equation framework depends on a variety of considerations, but frequently is influenced by the modeller's knowledge of the system and purpose of the model. An important consideration is the temporal unit of observation (e.g. weekly, monthly or quarterly data) relative to the biological or adjustment lags in the system. As the lag structure increases in length relative to the temporal unit of observation, important variables can be viewed as predetermined by past occurrences and, hence, the structure can be specified in a recursive framework. Because of the relatively long biological and agronomic lags in agricultural production, the relevance of recursive systems in agricultural markets has long been recognized (Waugh, 1964; Hallam, 1990; Tomek and Myers, 1993).

Another consideration is the structure of the demand. Even when supply is predetermined by past occurrences, the quantity available for use may be allocated to different end users based on relative price relationships (e.g. the allocation of grains to inventories, to processing for human consumption, feed, or export markets). In this case, price in the current period and quantities allocated to the various markets are determined simultaneously. Clearly, the nature of the causality in the model is highly dependent on the underlying characteristics of the market and requires the analyst to develop a comprehensive understanding of its structure for appropriate specification. In general, as the model becomes richer in the details of demand, the degree of simultaneity increases. Hausman-type tests have been used to help to determine the direction of causality (Thurman, 1987), but in larger models, the knowledge of the modeller often proves more useful.

While less common today, single equation models still are used particularly when aspects of either supply or demand are assessed in considerable depth, e.g. the incorporation of risk in supply models (Traill, 1978). Single equations also are used as pricing models generated by reduced forms of larger structural models.

SPECIFICATION OF SUPPLY AND DEMAND RELATIONSHIPS

Understanding and accurate specification of supply and demand relationships are the keys to developing useful market models. For example, agricultural supply is heavily influenced by the agronomic and biological nature of the commodities. Hence, it is important to understand the stages of growth and development of the commodity. It also is important to identify the physical and economic factors that influence output at each stage of the production process. In addition, because of the nature of agricultural production, lags exist between the decision to increase output and its sale. This raises issues about the dynamic specification of expectations and risk, as well as the importance of adjustment processes through time. Further, agricultural supply has been traditionally influenced by changes in technology and the use of government programmes to stimulate or control production. These characteristics, where important for a particular commodity, need to be incorporated into the specification of supply.

Supply

The modelling of supply often begins with the decomposition of output into yield, an output/unit of area or animal, and units of area (e.g. hectares) or animals (livestock numbers)[7]. Yields are modelled as a function of changing technology as specified by time trends, changing input/output price relationships, rates of input application such as fertilizer or pesticides, and weather factors during important plant or animal development stages in the growth cycle (Dixon *et al.*, 1994). Crop yield models also have been specified assuming a non-symmetric stochastic disturbance term, instead of explicitly entering weather variables into the exogenous set of factors (Gallagher, 1986). The effect of changing price ratios on selected crop yields has been somewhat variable (Houck and Gallagher, 1976; Brandt *et al.*, 1992; Dixon *et al.*, 1994), but may be more pronounced in perennials, where as harvest approaches price changes appear to affect the intensity of the harvest and the quality of the crop. Input/output price relationships also have been moderately important in explaining rates of grain feeding and subsequent changes in weight per animal slaughtered (Skold *et al.*, 1988). Typically, the driving force in crop yield models over time has been technology, which shifts out the production function. In a particular year, random variables, such as weather, often greatly influence the level of crop yields.

Acreage and livestock number relationships are assumed to be influenced by relative profitability often measured by output and input price variables. In addition, these relationships are specified to include the effects of adjustment processes, expectations, risk, and government programmes. Adjustments in agricultural production can occur because permanent price changes may not be immediately recognized. Similarly, changes in investment may take considerable time to implement completely and may be hampered by adjustment costs. The specific adjustment processes and the role of expectations frequently depend on the characteristics of the data and the nature of the production process. The shorter the unit of temporal observation relative to the biological or agronomic lag, the increased likelihood that adjustments and expectations will play a more important role in specifying the supply relationship. Adjustment processes are most commonly measured through the use of lagged dependent variables either in a partial adjustment framework or the use of some other weighted lag structure. In addition, because of the asymmetry in investment decisions with increasing and decreasing profitability, acreage equations have been modelled permitting differences in response to increasing and decreasing prices (Traill *et al.*, 1978).

Price expectations are important in specifying agricultural supply because differences in time between input decisions and marketing decisions can be substantial. Hence, the expected harvest price may influence input/output mix decisions at planting. Price expectations have been modelled assuming that expectations are an extrapolation of previous prices. In this context, a variety of procedures: naive expectations, adaptive expectations, and polynomial representations of previous prices have been utilized (Askari and Cummings, 1976; Hallam, 1990). Researchers also have used linear and non-linear rational expectations models which base expectations on the underlying supply and demand structure of the commodity market, and forecasts of the exogenous variables (Goodwin and Sheffrin, 1982; Shonkwiler and Emerson, 1982; Holt, 1992). In addition, futures market prices have been used as expected prices under the assumption that an efficient futures market uses all available information to formulate price (Gardner, 1976). Clearly, because of its ease of implementation and relatively plausible findings, the adaptive expectations model has been the most used approach historically. Interestingly, tests for the appropriateness of the various price expectations hypotheses frequently have not resulted in the acceptance of one of the models at the rejection of others (Shonkwiler, 1982; Antonovitz and Green, 1990)[8].

Agricultural economists have been sensitive to the role of risk, and government programmes in influencing supply response. Risk, measured as some deviation from price expectation, has been incorporated rather successfully into supply analysis (Just, 1974; Traill, 1978). Input and output price risk also has been successfully estimated (Hurt and Garcia, 1982), as have down-side deviations from expectations as a risk measure (Tronstad and McNeill, 1989). More recently, econometric and time series procedures have been combined to identify the effect of the changing nature of the unexpected price variability on agricultural supply (Holt and Aradhyula, 1990). Finally, the influence of government production programmes that limit production and establish support prices have been investigated through various procedures. These include the use of dummy variables (Lidman and Bawden, 1974), the use of composite policy variables such as effective support price and effective diversion payment rates along with output price (Houck and Ryan, 1972), and the generation of revenue-cost variables to measure the revenue from programme participation (Lee and

Helmberger, 1985; Brandt *et al.*, 1992; Chen and Ito, 1992). The revenue–cost approach has been used to estimate different parameters during alternative government regimes (Lee and Helmberger, 1985) and within the context of switching regression procedures (Chen and Ito, 1992). In a related context, the effects of government programmes which truncate the probability distribution of expected producer price also have been successfully examined (Holt, 1992).

Often, the quantification of acreage and livestock number response relationships has been disaggregated spatially or in terms of the production cycle depending on the availability of the data and desired uses of the model. Crop acreage response has been disaggregated by states, crop reporting districts and other spatial areas in efforts to generate relatively homogeneous production zones (Taylor *et al.*, 1993; Park and Garcia, 1994). Within these production zones, acreage response functions for various commodities also have been estimated with constraints which force the shares of the total crop land used by the different commodities to equal unity (Taylor *et al.*, 1993). Perennial crops have been disaggregated, temporally and with respect to the stages of the growth cycle. Hence, plantings, available stock of trees for production and culling of trees from the production are modelled separately (French *et al.*, 1985).

For animal livestock relationships, a detailed understanding of the production process is used to model output. Frequently, livestock numbers are specified at important stages of animal growth process. The effects of key economic variables are then modelled at each stage. For example, the number of hogs slaughtered has been decomposed into the slaughter of barrows and gilts, and sows, allowing the sows slaughtered to be influenced by long-run breeding inventory as well as short-run marketing considerations. Slaughtered barrows and gilts primarily are a function of lagged pig crops (i.e. lagged farrowings times the pigs saved per litter). Sow slaughter is affected by expected profitability of future production of barrows and gilts and lagged farrowings. Sow farrowings, in turn, are influenced by expected profitability and adjustments in the production process (Stillman, 1985). This type of disaggregation permits the modelling of changes in the breeding herd as well as marketing decisions, and allows elasticities of price response to differ across the production process[9]. Additionally, biological constraints imposed at various stages of production also have been more formerly incorporated into estimation procedures through restrictions on the estimated parameters (Johnson and MacAulay, 1982; Skold, 1992) or through the choice of functional form (Chavas and Klemme, 1986). These constraints limit livestock numbers at any stage of production from exceeding the feasible dimensions of previous stocks.

Demand

Somewhat different, but equally interesting, issues arise in specifying demand where several modelling procedures exist. Demand models are categorized as *ad hoc* or partial specifications, and demand-systems specifications. Partial specifications, modified versions of Eqns 6.1 and 6.5, typically are developed within the context of a specific commodity supply and demand model. These partial specifications frequently do not conform to all the tenets of consumer behaviour (e.g. Slutsky and homogeneity conditions), but are structured to include the fundamental determinants of demand. In specifying these relationships, decisions must be made concerning the nature of causality between price and quantity, which other prices or quantities to include, and whether income and prices should be deflated. In addition, the analyst must determine which functional forms to use, and whether dynamics (either because of habits or other adjustments) play an important role in explaining demand. Finally, the relevance of seasonality and changes in structure of demand through time need to be considered.

In livestock markets, demand often has been modelled at retail or wholesale (Stillman, 1985; Devadoss *et al.*, 1993). Here, it is not uncommon to encounter either linear or loglinear price or quantity per capita dependent relationships. Price indexes are used to deflate nominal prices and income or specific indexes are generated to reflect the movement of important substitutes or final product prices. Often, empirical specifications include seasonal dummy variables, and slope and intercept shifters to permit structural change.

Empirical work on the demand for grains and soybeans has focused on disaggregating the total demand into its various components (Roy and Ireland, 1975; Brandt *et al.*, 1992; Devadoss *et al.*, 1993). For example, for feed grains the primary components of demand are food use, feed use, seed use, stocks and exports. For soybeans, the components are industrial crush, other domestic uses, exports, and stocks. The demand for soybean meal and oil also is disaggregated into domestic use, exports and stocks. Consumer theory, producer theory of derived demand, and inventory theory then are used to identify the variables influencing the various uses. Each component is modelled as a function of its own price and sets of shifters that affect disappearance in each category. Restrictions derived from economic theory are not imposed on the structure of the markets. For example, feed use may be determined by its own price, the price of competing feed products, and livestock prices and numbers. Estimation procedures depend on the complexity of the market structures, but it is clear that price and use are highly dependent.

Partial specifications particularly are useful for investigating characteristics of a particular commodity or set of related commodities (Labys, 1973; Hallam, 1990)[10]. For example, the analyst may be interested in investigating structural change or the effects of promotional programmes for specific commodities. Modelling within the context of one commodity, or set of related commodities, may permit the analyst more flexibility in determining the exact effects of the programmes, or structural break. Similarly, the analyst may be interested in generating a parsimonious price forecasting model. In this case, complete demand specification may lead to imprecise price forecasts.

Recently, demand-systems specifications have become more popular for modelling demand. Specific functional forms which aggregate over individual commodities to form a small number of commodity groups are used to estimate the system of demand equations subject to constraints placed by economic theory. Clearly, the Almost Ideal Demand System (AIDS), which uses a semilogarithmic form and expresses consumption shares as a function of logged prices and logged expenditure share divided by a price index, is the most commonly used[11]. This formulation has been employed to generate elasticities, examine differences in groupings of commodities, and examine structural change in the consumption of meats (Eales and Unnevehr, 1988; Moschini and Meilke, 1989; Green and Alston, 1991). In addition, inverse AIDS formulations have been specified to reflect the price determination process often consistent with agricultural commodities, and dynamics have been included to capture the changing nature of demand through time (Eales and Unnevehr, 1993). While useful in generating elasticity estimates of commodity groupings, the demand-systems specifications frequently provide little other information about the demand characteristics associated with the individual commodities. In addition, because of the difficulty in specifying the other factors and the structure influencing demand, the theoretical constraints imposed are often rejected in statistical testing.

The modelling of demand for agricultural commodities is complicated by several issues. First, many agricultural commodities have multiple end users and frequently the scope of markets, particularly for grains, is international. In a competitive market, this implies that the demands are interrelated, prices and quantities are determined simultaneously, and that market structures must allocate the commodities in space, form and time. For internationally traded commodities, two related concerns emerge. In competitive markets, clearly, trade flows are affected by foreign as well as domestic factors and policies. Conceptually, price differentials allocate commodities. In an international context, it becomes important to identify effectively the foreign determinants of demand which alter excess supply and demand curves. In imperfectly competitive markets, successful modelling may require the identification and quantification of the nature of competitive structure, the importance of the market participants, and the reactions of market participants to market activities of their competitors (Behrman, 1978; Paarlberg and Abbott, 1986; Love and Murniningtyas, 1992).

Second, the demand for inventories, particularly for speculative commodities, is challenging to specify. The holders of stocks may be numerous and their motivations varied. Often, data are not sufficiently detailed to model accurately the behaviour of various market participants. When data are available, private inventories have been specified as functions of production and previous stocks, which reflect precautionary and transaction stock demand, and expected price relationships, which represent the speculative dimensions of inventory activities. Expectations have been specified using extrapolative procedures such as naive, adaptive expectations, and, more recently, using rational expectations (Labys, 1973; Miranda and Glauber, 1993). The extrapolative procedures for generating expectations can be readily formulated, but lack some realism. The rational framework, which is more appealing because expectations are based on the underlying market structure, is more difficult to quantify, often requiring the combined use of econometric and stochastic dynamic programming procedures (Miranda and Glauber, 1993). In a somewhat similar vein, expected price/inventory relationships have been modelled within the context of cash and futures markets (Subotnik and Houck, 1982; Giles *et al.*, 1985). In this case, inventory decisions are influenced by changes in the basis (i.e. the difference between cash and futures). Incorporating this incentive for speculative storage activity adds realism to the model at the cost of added complexity, the specification and quantification of futures market activities. The modelling of stock behaviour is further complicated by the presence of government stock programmes which are common for many agricultural commodities. Government stock programmes are varied, can change frequently, but often are designed to establish price supports or to stabilize prices within a band. Frequent changes in government programmes make econometric estimation difficult due to concerns with structural change. The establishment of price supports or price bands imply that the probability distribution of expected price is bounded as the support levels are enforced by the government's commitment to purchase stocks (Holt, 1992).

Finally, the analysis of demand in agricultural markets is complicated because consumers' activities, particularly for livestock, often are measured at retail or wholesale, while producers' activities are quantified at farm level. This implies the existence of a marketing margin (Chapter 8) which may be viewed as either absolute or a constant per cent or some combination of both (George and King, 1971). In addition, cost and size variables often need to be included to identify how marketing margins change over time and in response to changes in input usage (Lyon and Thompson, 1993). Effectively incorporating marketing margins in commodity analysis is more challenging as issues of market power, uncertainty, and an inventory behaviour are directly addressed (Wohlgenant, 1985; Schroeter, 1988; Schroeter and Azzam, 1991; Koontz *et al.*, 1993).

VALIDATION AND VERIFICATION

An important aspect of any commodity modelling effort is that the analyst should have confidence in the model. Prior to using, the model builder must validate, or verify, the model, a procedure of examining the significance of model parameters and the accuracy of model forecasts (Labys and Pollak, 1984). Needless to say, all models have errors, whether they come from structural misspecification, measurement and data errors, inappropriate functional form, or estimation procedures and problems. Model validation, then, is an examination of whether the model specified and estimated corresponds closely to, or represents accurately, the assumed structure.

An outside reviewer initially will evaluate a model according to whether the model structure corresponds to the objectives and designed use of the model. This assessment involves comparison with alternative specifications that exist in the literature, and, of course, will receive different treatment based on the evaluator's knowledge of the market being modelled as well as the literature. It is important to note that reasonable models for the same sector or commodity can have varying structures.

Beyond this evaluation, in its simplest form, validation of the empirical content of the model involves checking the signs and magnitudes of the parameters against prior expectations, and evaluating the goodness of fit of parameters and equations. This latter evaluation involves checking the level of significance of each parameter with its t-statistic, and examining the R^2 of the equation or performing F-tests on whether the equation provides significant explanation. Models often need to be modified after initial estimation because these simple checks may present results that are unsatisfactory and do not fit the underlying theory or assumptions.

Although this chapter does not discuss estimation procedures of commodity models in any detail, the analyst must use appropriate techniques. For a simple single equation and many recursive models, ordinary least squares is the common technique of estimation. However, for simultaneous equation models, more advanced econometric procedures may need to be used, ranging from seemingly unrelated regression to various multivariate regression techniques such as two-stage least squares, three-stage least squares and maximum likelihood procedures (Judge *et al.*, 1985; Maddala, 1992). The techniques or procedures used often are dictated by the nature or characteristics of the data and whether these data and their resulting errors correspond to the assumptions of the model.

The analyst always faces the possibility of misspecifying an equation or model because it may be impossible to define or identify the correct underlying behavioural assumptions that one is attempting to model. The true specification (model) may never be known. This is specification error, and it results from uncertainty about the correct mathematical and statistical relationships. The causes can be numerous, including variable selection, functional form, data error and incorrect specification of how the stochastic term enters the regression equation. The most common specification error is to omit a relevant explanatory variable (usually because of lack of data, or knowledge), or to include irrelevant explanatory variables. The omission problem is most serious because it can lead to biased estimates of parameters. Similar biases occur with employing an incorrect functional form, or a trend existing in one of the explanatory variables. The extent of the bias depends on the relationship between the omitted variable with the included variables and the dependent variables. Sometimes the analyst willingly omits a variable when its inclusion causes more econometric problems than its exclusion, as in cases where strong linear relationships between independent variables, i.e. multicollinearity, exists.

Arriving at the final model is an iterative process involving trade-offs between the perceived true model specified from economic theory and industry knowledge against model alterations based on econometric results. The analyst needs to use the significance and sign of parameters and performance of goodness-of-fit tests as an initial guide. Additional tests include the Durbin Watson test for autocorrelation, Goldfeld–Quandt or Breusch–Pagan multiplier test for heteroskedasticity, Hausman test for measurement error, Chow test for structural change, examination for multi-collinearity, and tests for misspecification (Maddala, 1992; McGuirk *et al.*, 1993). Final models often end up looking quite different from the originally hypothesized model. However, the analyst must be careful and use good judgement in modifying the model from the original hypothesized structure. To be driven entirely by statistical fit and econometric procedures may result in a final model being so misspecified from the true model that the model provides no insights and is useless.

The validation procedures just described are for the most part parametric tests. Model builders also may employ several non-parametric techniques in evaluation. Some are discussed in the next section under forecast model evaluation, such as the various mean-squared error measures and turning point tests. In addition, there are runs tests, rank correlations of predicted and actual changes, tests of randomness, and sensitivity analysis (Labys and Pollak, 1984). Finally, in some cases it is appropriate to evaluate the model in a benefit/cost framework (Garcia *et al.*, 1988; Leuthold *et al.*, 1989).

APPLICATIONS

The usefulness of a model depends on its objectives and how it is constructed. But how it is constructed also depends on how it is going to be used. Commodity market models may be constructed for use in economic policy and planning analysis, for market analysis which involves structural parameter estimation and the derivation of elasticities, or for forecasting purposes. These groupings are not mutually exclusive, but rarely can a model be constructed in a framework that satisfies adequately all of these purposes.

Economic planning and policy models

Planning analysis is performed at all levels of economic activity. Normally, this is considered an activity of government agencies, like central planning boards, but corporations and various market organizations also develop planning models. These entities could be involved with planning in the energy, agriculture or metals and mining sectors. Different among these agencies is the level of aggregation, ranging from firm level, to sector analysis, to country or national planning. Intermixed in the planning may be domestic and export commodity sectors. These models also are built with analyst knowledge beyond simple economics, including also political, technological, legal and institutional considerations and relationships. That is, non-price factors may play an important role in the models.

For policy models, the analyst identifies outcome variables which signify the objectives of economic policy, and control variables which are the instruments of such policies. By systematically varying the levels of control variables, the impact of policies on the outcome variables can be observed.

Control variables are normally exogenous variables (instruments) generated independently of the model. Outcome variables are a subset of the endogenous variables whose values are solved in the model. The control and outcome variables are linked by the model structure. A reduced-form example of this type of model was shown earlier as Eqn 6.8 where outcome variables are in the vector y_t, and the vector x with its associated matrix of coefficients could be subdivided into two parts, non-policy exogenous variables (vector z_t), and variables representing the control exogenous variables, or policy instruments (vector r_t). In the *instrument-target approach* it is assumed that there are desired levels of the target (outcome) variables, and it is also assumed that the number of instruments equals the number of targets (Intriligator, 1978, 1983).

An alternative to the *instrument-target approach* is the *social-welfare function approach* that allows for trade-offs among the endogenous (target) variables. This is done by assuming an objective function which is maximized or minimized subject to the constraints of the model. Examples include maximizing social welfare, minimizing price variability, or variations of optimal control. Frequently, this function is a quadratic loss function (Intriligator, 1978; Dixon and Chen, 1982).

A third and most common approach to policy evaluation is the *simulation approach* which relaxes the specific assumptions of exact targets and welfare functions. The estimated reduced-form model (Eqn 6.8) is used as the base to study the behaviour of a system. The model can be simulated successively over time or space, or both, observing alternative combinations of policy and subsequent endogenous (outcome) variables. Such model simulation can be done to observe what might have happened in the past using historical data, or how new policies will affect outcome variables in the future. In this latter situation, the analyst may need to forecast future values of certain exogenous variables. Alternatively, rather than inputing control variables as given or deterministic, these variables can also be varied stochastically, permitting the analysis to incorporate risk and uncertainty. One could also imagine using these models in a game theoretic context, analysing how changes in certain control (policy) variables impact, say, export programmes.

One must be cautious, however, in interpreting policy model evaluations, owing to the *Lucas Critique* (Lucas, 1976). This critique argues that individuals are rational, and in anticipation of the effects of new policies, will modify their behaviour accordingly. As a result, Lucas argues that simulation results cannot provide useful information on the consequences of alternative policies (Hallam, 1990). The real questions are whether this critique applies when modest changes are made in policy, which is the most common case, or whether parameter estimates can be modified to reflect the nature of the changes.

Market (structural) analysis models

Market analysis models are designed to depict how a market is structured and behaves. These models centre on quantification of relationships among the variables of interest within the system.

Market analysis models are typically disaggregated and detailed. They begin usually with equations representing the supply or production sector, or equations representing the demand or consumption sector, or both. An example is the model represented by Eqns 6.1–6.3. This model may represent a market at the world level, or be disaggregated by country or region in an international context, or be disaggregated further to represent states or subregions within one country or state. The number of regions or countries being represented dictate the size of the model. Considered simultaneously with this level of spatial aggregation is the appropriate level of time aggregation, i.e. whether the model will be estimated with annual, quarterly, monthly or even more disaggregated data. As discussed earlier, time aggregation, lags and simultaneity of equations are highly interrelated. Some variables that are endogenous when using annual data may become exogenous, or predetermined, and lagged when using disaggregated data due to normal biological, production or information lags.

The model is not complete with simply demand (consumption) or supply (production) equations. Linkages must be made among the various equations, which might come through equilibrium conditions, pricing equations, trade or transportation linkages, storage equations and constraints from institutional, regulatory or technological considerations. Some models concentrate on a single commodity or product, while others link various products or sectors in a multicommodity framework, such as the grain and livestock sectors (Arzac and Wilkinson, 1979). Identification of exogenous and endogenous variables will vary considerably as more sectors are added to a model and the interrelationships become more complex.

Earlier discussion in this chapter alludes to the many decisions and assumptions the model builder must make while constructing a model. These include the level of spatial and time aggregation, the number of commodities or sectors, which variables are exogenous or endogenous, time lags, and whether the model should be a single equation, a set of simultaneously related equations, or of a recursive framework. Uncertainty about the future on either the supply side or demand side of the model will require assumptions about the formation of expectations. Finally, an important decision facing the model builder before empirical estimation is to determine the functional form of the model and its individual equations.

A principal function of market models is to generate structural parameters, and hence obtain elasticities or flexibilities. Elasticities and flexibilities tell us how market participants react to changing conditions of the market, usually changes in prices, quantities or income. This is basic information about the market and the behaviour of its actors.

Parameters of simultaneous equation models are often not directly obtainable or observable during the estimation, and a reduced-form version of the model is derived. This is usually done by algebraic manipulation of the structural model, and the procedure varies depending on the dynamics of the model. Reduced-form parameters, or impact multipliers, tell us the immediate effect of changes in exogenous variables on endogenous variables. These have the interpretation of comparative statistics results, the effect of each exogenous variable on each endogenous variable. Parameters from a dynamic model (Eqn 6.10) can be used to trace out the effect of a change or shock, either for one period or permanent, of an exogenous variable on all endogenous variables over time, giving a time path of the endogenous variables leading back to equilibrium. In order to study this dynamic behaviour, one must generate the total multipliers from the estimated parameters, which then account for direct and indirect effects over time (Labys, 1973). Of course, to be able to perform this evaluation and analysis, the model must be stable, conditions ascertained from the reduced-form parameters.

Forecasting models

Models built for forecasting purposes usually start as market analysis models. That is, the model is specified according to proper economic theory and market conditions. Forecasts can then be generated from either the structural model if it is single equation or recursive, or from the unrestricted or restricted reduced-form model if its structure is simultaneous. However, the model builder's objectives, and, hence, evaluation of the proper specification of the model may differ from the above models. In a model designed for market analysis, it is critical that the parameters (and the subsequently derived elasticities and flexibilities) be unbiased. This is because the model is being used to study participant behaviour, structural change or policy issues; misspecification can lead to erroneous implications and policy.

However, in the case of forecasting models, the prime objective may be the accuracy of the model's forecasts, and not whether it satisfies all of the important economic and econometric assumptions. There can be trade-offs between model specification and model performance (forecasting accuracy). Consequently, a model designed for forecasting may be more parsimonious in both the number of variables and equations. A simple structure often is preferred. In addition, the model or its forecasts may be combined with a non-economically specified model based strictly on the behaviour of the time series being predicted, such as a Box–Jenkins ARIMA model.

Important to the structure of forecasting models is the timing of when information is known. The analyst must not incorporate variables into the model, the values of which will not be known at the time that actual forecasts are made. If there is such a variable in the model, then its value must be forecast independently and, hence, treated as exogenous in the model.

The analyst first estimates the model using historical data and appropriate econometric procedures. Forecasts are created by incorporating future or known values of exogenous variables into the model followed by the necessary algebra of multiplying these values by their parameters. This procedure assumes that the statistical relationships and parameters will continue to hold in the future, i.e. there is not a structural change. If these exogenous (explanatory) variables are known or assumed, the forecasts are considered unconditional, or *'ex-post'*. When the explanatory variables must be forecasted, the forecasts are conditional, or *'ex-ante'*. If the underlying model is simultaneous, the forecasts are generated from the reduced-form equations. The forecasts are termed 'unrestricted' when all exogenous variables are included in the forecasting equations. If only those explanatory variables which are significant remain in the model, i.e. insignificant variables are dropped in subsequent estimations, then the forecasts are termed 'restricted'. The latter situation is most common, using current system information.

An example of a short-run *'ex-ante'* forecasting equation is:

$$\hat{y}_{T+1} = y_T \hat{\Pi}_1 + \hat{x}_{T+1} \hat{\Pi}_2 + \hat{u}_{T+1}$$

The left-hand side, vector y_{T+1}, are the forecasted values of the endogenous variables at time $T+1$. In the forecasting function are y_T, current values of the endogenous variables, x_{T+1}, predicted future values of exogenous variables, and other predetermined variables of the model. The final term, u_{T+1}, is the future disturbance term, accounting for errors due to omitted variables, measurement problems, incorrect specification, etc. It may represent known shocks, or be in an ARIMA framework as noted below.

The true test of the model is to forecast beyond the period of fit, or out-of-sample. In-sample forecasts should necessarily perform well because the regression has already been fitted over these data. Out-of-sample forecast errors may come from inaccuracies within the model itself or, in the case of *'ex-ante'* forecasts, the errors could come from inaccuracies in forecasting explanatory variables. In the case of dynamic models where forecasted variables are being fed back into the model as known, or lagged endogenous variables, a time path of forecasts can be projected for several periods. However, errors in early forecasts may compound further out in time. If explanatory variables are being assumed, the analyst may have to attach some probabilities to the values, thereby presenting a range, or scenario, of forecasts.

Most forecasts from econometric models now are done in association with Box–Jenkins ARIMA time series models (Allen, 1994). There are several methods of incorporating this time series information. One procedure is to use time series techniques to analyse the error terms of the econometric model and then make appropriate adjustments to the econometric model based on this structural information. Another procedure, and maybe more common, is to model independently the (historical) data series being forecast using Box–Jenkins procedures, then to forecast out-of-sample with this ARIMA model. These forecasts can be applied separately, or combined with the econometric forecasts. Combined forecasts are called 'composite' forecasts, meaning forecasts from two or more models are combined. Several procedures exist for weighting and combining forecasts, but often simple averages of independent forecasts perform the best. The justification for combining alternative modelling approaches is that each model reflects different information underlying the historical data series. The analyst then compares forecasts in terms of their accuracy, selecting the best model(s). ARIMA forecasts are now almost always included in the analysis because they often outperform the econometric forecasts, or at least the composite forecast outperforms the individual forecasts. ARIMA models perform the best as short-run forecasts[12].

Forecasting models are judged quantitatively and qualitatively with non-parametric tests. The standard quantitative evaluation is to examine after the fact the forecast errors, or the difference between predicted and actual values. For several compelling statistical and evaluative reasons, these errors are squared, summed and then averaged, called the mean-squared errors. Alternative forecast models can then be compared, or evaluated, using either these mean-squared errors, their root mean-squared errors, root mean-squared percentage errors, or other statistics derived from these measures, such as the Theil inequality, or U, coefficient. A standard base in many cases is to compare the forecasting model to some naive forecast. A common procedure is to use this period's price as the forecast of the next period's price, or naively to forecast no change in price.

An additional evaluation of the forecast model is to examine turning points, or the direction of the forecasts. For some models, knowing whether the forecast variable, such as price, will increase or decrease, is more important than evaluating the size of the error. Turning point evaluation can be done in a contingency table format, as shown in Naik and Leuthold (1986), or by using Henriksson–Merton (1981) timing tests, as demonstrated by McIntosh and Dorfman (1992). Finally, forecasts may be evaluated in terms of their economic value, especially when prices are involved. Trading simulations may be conducted based on model forecasts, attaching economic value to the forecasting model (Gerlow *et al.*, 1993).

CONCLUDING REMARKS

Agricultural economists have numerous modelling procedures available for the analysis of commodity markets. Here, we have discussed selected issues related to the development and use of econometric market models, the most commonly utilized procedure for agricultural commodity analysis. The development of an appropriate econometric market structure follows an iterative process of problem identification, specification, estimation and validation. It requires the identification of the numerous economic and non-economic factors which influence market behaviour, the assembly of data that reflect the underlying theoretical models of producer and consumer behaviour, and the estimation of the parameters with statistical procedures that are consistent with the structure of the model and its errors.

Commodity market models have been used extensively in analyses of policy, market structure, and forecasting. Policy analysis permits the assessment of the effectiveness of the use of alternative control variables in achieving economic objectives. The analysis of market structure provides characterizations of commodity markets, and the development of market elasticities and flexibilities. Forecasting is useful to assess likely outcomes of market processes, given the market structure and forecasts of exogenous variables.

While subject to difficulties in their development and use, econometric market models provide a systematic and useful paradigm for understanding the workings of agricultural commodity markets. The specification of a commodity market model facilitates the development of an in-depth understanding of the complex set of factors which influence agricultural supply, demand and inventory relationships. The verification of models provides an opportunity to assess the consistency of a preconceived understanding of the market structure and behaviour with the reality of the data. Modifications of the structure to more closely reflect market activities permit the development of alternative hypotheses of market behaviour and a richer understanding of the market environment. Further, using market models can provide information to decision-makers which facilitates the decision process by more clearly defining the structure of the market in which they operate, the consequences of alternative actions, and the likely movements of market–determined variables.

NOTES

1. The literature on commodity modelling is quite extensive. Labys (1973) and Labys and Pollak (1984) provide useful taxonomies and flow diagrams of the various modelling procedures. Labys and Pollak (1984) and Hallam (1990) present rather detailed discussions on econometric modelling of agricultural commodity markets. Recent modelling developments can be found in the rather large number of agricultural economics journals, many of which are referenced here. Two articles with extensive bibliographies are Tomek and Myers (1993) on commodity market modelling and Allen (1994) on forecasting.
2. Due to the extensive literature, it is not possible to identify all the important contributions here. The selection is influenced by our reading of the literature and our own work.
3. While not specified in the above model, lagged endogenous variables often appear in commodity models and are referred to as predetermined, their magnitude determined by previous market activities.
4. The notation follows Intriligator (1983).
5. A more detailed discussion of conditions necessary to ensure identification can be found in most econometric texts (e.g. Intriligator, 1978).
6. In this situation, each of the structural equations is identified, and ordinary least squares provides consistent and asymptotic estimates of the parameters.
7. Generally, the units of area have focused on acreage planted. See Brandt *et al.* (1992) for a discussion of the importance of using acreage harvested for certain crops.
8. In part, this may reflect data deficiencies, difficulties in specification of the complete market structure, limited power of the testing procedures, or difficulties in generating exogenous forecasts in rational formulations.
9. Skold (1992) considers an alternative procedure to model the addition to breeding inventories as well as the disinvestment in the breeding stock specified by Stillman (1985).

10. Livestock and grain sectors also have been linked in various modelling efforts (Arzac and Wilkinson, 1979; Offutt and Blandford, 1984; Devadoss *et al.*, 1993). In such frameworks, livestock prices and numbers influence feed use and prices. In turn, feed and livestock prices with appropriate lags affect livestock production.
11. See Eales and Unnevehr (1993), Buse (1994) and Hahn (1994) for a discussion of the limitations of estimating linear approximations to the AIDS model.
12. Recent successful forecasting models have used economic theory to suggest the correct variables, but then the data are combined and estimated within a vector autoregressive (VAR) framework (Allen, 1994).

FURTHER READING

Allen, P.G. (1994) Economic forecasting in agriculture. *International Journal of Forecasting* 10, 81–135.

Hallam, D. (1990) *Econometric Modelling of Agricultural Commodity Markets.* Routledge, London.

Labys, W.C. (1973) *Dynamic Commodity Models: Specification, Estimation and Simulation.* Lexington Books, Lexington, MA.

Labys, W.C. and Pollak, P.K. (1984) *Commodity Models for Forecasting and Policy Analysis.* Croom Helm, London.

Maddala, G.S. (1992) *Introduction to Econometrics*, 2nd edn. Macmillan Publishing Co., New York.

Taylor, C.R., Reichelderfer, K.H. and Johnson, S.R. (eds) (1993) *Agricultural Sector Models for the United States.* Iowa State Press, Ames, IA.

REFERENCES

Allen, P.G. (1994) Economic forecasting in agriculture. *International Journal of Forecasting* 10, 81–135.

Antonovitz, F. and Green, R. (1990) Alternative estimates of fed beef supply response to risk. *American Journal of Agricultural Economics* 72, 475–487.

Arzac, E.R. and Wilkinson, M. (1979) A quarterly econometric model of United States livestock and feed grain markets and some of its policy implications. *American Journal of Agricultural Economics* 61, 297–308.

Askari, H. and Cummings, J.T. (1976) *Agricultural Supply Response: A Survey of the Econometric Evidence.* Praeger & Co., New York.

Behrman, J.R. (1978) International commodity market structures and the theory underlying international commodity market models. In: Adams, F.G. and Behrman, J.R. (eds), *Econometric Modeling of World Commodity Policy.* Lexington Books, Lexington, MA, pp. 9–45.

Brandt, J.A., Kruse, J.R. and Todd, J. (1992) Supply, demand, and the effects of alternative policies on the US oats industry. *American Journal of Agricultural Economics* 74, 318–328.

Buse, A. (1994) Evaluating the linearized Almost Ideal Demand System. *American Journal of Agricultural Economics* 76, 781–793.

Chavas, J.P. and Klemme, R.M. (1986) Aggregate milk supply response and investment behavior on US dairy farms. *American Journal of Agricultural Economics* 68, 55–66.

Chen, D.T. and Ito, S. (1992) Modeling supply response with implicit revenue functions: A policy-switching procedure for rice. *American Journal of Agricultural Economics* 74, 186–195.

Devadoss, S., Westhoff, P., Helmar, M., Grundmeier, E., Skold, K., Meyers, W. and Johnson, S.R. (1993) The FAPRI modeling system: a documentation summary. In: Taylor, C.R., Reichelderfer, K.H. and Johnson, S.R. (eds), *Agricultural Sector Models for the United States.* Iowa State Press, Ames, IA, pp. 129–150.

Dixon, B.L. and Chen, W.H. (1982) A stochastic control approach to buffer stock management in the Taiwan rice market. *Journal of Development Economics* 10, 187–207.

Dixon, B.L., Hollinger, S.E., Garcia, P. and Tirupattur, V. (1994) Estimating corn yield response models to predict impacts of climate change. *Journal of Agricultural and Resource Economics* 19, 58–68.

Eales, J.S. and Unnevehr, L.J. (1988) Demand for beef and chicken products: separability and structural change. *American Journal of Agricultural Economics* 70, 521–532.

Eales, J.S. and Unnevehr, L.J. (1993) Simultaneity and structural change in meat demand. *American Journal of Agricultural Economics* 75, 259–268.

French, B.C., King, G.A. and Minami, D.D. (1985) Planting and removal relationships for perennial crops: an application to Cling peaches. *American Journal of Agricultural Economics* 67, 215–223.

Gallagher, P. (1986) US corn yield capacity and probability: estimation and forecasting with non-symmetric disturbances. *North Central Journal of Agricultural Economics* 8, 109–122.

Garcia, P., Leuthold, R.M., Fortenbery, T.R. and Sarassoro, G.F. (1988) Pricing efficiency in the live cattle futures market: Further interpretation and measurement. *American Journal of Agricultural Economics* 70, 162–169.

Gardner, B.L. (1976) Futures prices in supply analysis. *American Journal of Agricultural Economics* 58, 81–84.

George, P.S. and King, G.A. (1971) *Consumer Demand for Food Commodities in the United States with Projections for 1980.* Giannini Foundation Monograph 26. University of California, Division of Agricultural Sciences, Berkeley, CA.

Gerlow, M.E., Irwin, S.H. and Liu, T.R. (1993) Economic evaluation of commodity price forecasting models. *International Journal of Forecasting* 9, 387–397.

Giles, D.E.A., Goss, B.A. and Chin, O.P.L. (1985) Intertemporal allocation in the corn and soybean markets with rational expectations. *American Journal of Agricultural Economics* 67, 749–760.

Goodwin, T.H. and Sheffrin, S.M. (1982) Testing the rational expectations hypothesis in an agricultural market. *The Review of Economics and Statistics* 64, 658–667.

Green, R. and Alston, J.M. (1991) Elasticities in AIDS models: a clarification and extension. *American Journal of Agricultural Economics* 73, 874–875.

Hahn, W.F. (1994) Elasticities in AIDS models: Comment. *American Journal of Agricultural Economics* 76, 972–977.

Hallam, D. (1990) *Econometric Modelling of Agricultural Commodity Markets.* Routledge, London.

Henriksson, R.D. and Merton, R.C. (1981) On market timing and investment performance. II. Statistical procedures for evaluation forecasting skills. *Journal of Business* 54, 513–533.

Holt, M.T. (1992) A multimarket bounded price variation model under rational expectations: Corn and soybeans in the United States. *American Journal of Agricultural Economics* 74, 10–20.

Holt, M.T. and Aradhyula, S.V. (1990) Price risk in supply functions: an application of GARCH time series models. *Southern Economic Journal* 57, 230–242.

Houck, J.P. and Gallagher, P.W. (1976) The price responsiveness of US corn yields. *American Journal of Agricultural Economics* 58, 731–734.

Houck, J.P. and Ryan, M.E. (1972) Supply analysis for corn in the United States: Impact of changing government programs. *American Journal of Agricultural Economics* 54, 184–191.

Hurt, C.A. and Garcia, P. (1982) The impact of price risk on sow farrowings, 1967–78. *American Journal of Agricultural Economics* 64, 565–568.

Intriligator, M.D. (1978) *Econometric Models, Techniques, and Applications*. Prentice-Hall, Inc., Englewood Cliffs, NJ.

Intriligator, M.D. (1983) Economic and econometric models. In: Griliches, Z. and Intriligator, M.D. (eds), *Handbook of Econometrics*, Volume I. North-Holland Publishing Co., Amsterdam, pp. 181–221.

Johnson, S.R. and MacAulay, T.G. (1982) Physical Accounting in Quarterly Livestock Models: An Application for US Beef. Working Paper. University of Missouri-Columbia, Columbia, MO.

Judge, G.G., Griffiths, W.E., Hill, R.C., Lutkepohl, H. and Lee, T.C. (1985) *The Theory and Practice of Econometrics*, 2nd edn. John Wiley & Sons, New York.

Just, R.E. (1974) An investigation of the importance of risk in farmers' decisions. *American Journal of Agricultural Economics* 56, 14–25.

Koontz, S.R., Garcia, P. and Hudson, M.A. (1993) Meatpacker conduct in fed cattle pricing: an investigation of oligopsony power. *American Journal of Agricultural Economics* 75, 537–548.

Labys, W.C. (1973) *Dynamic Commodity Models: Specification, Estimation and Simulation*. Lexington Books, Lexington, MA.

Labys, W.C. and Pollak, P.K. (1984) *Commodity Models for Forecasting and Policy Analysis*. Croom Helm, London.

Lee, D.R. and Helmberger, P.G. (1985) Estimating supply response in the presence of farm programs. *American Journal of Agricultural Economics* 67, 193–203.

Leuthold, R.M., Garcia, P., Adam, B.D. and Park, W.I. (1989) An examination of the necessary and sufficient conditions for market efficiency. *Applied Economics* 21, 193–204.

Lidman, R. and Bawden, D.L. (1974) The impact of government programs on wheat acreage. *Land Economics* 50, 455–474.

Love, H.A. and Murniningtyas, E. (1992) Measuring the degree of market power exerted by Government Trade Agencies. *American Journal of Agricultural Economics* 74, 546–555.

Lucas, R.E.J. (1976) Econometric policy evaluation: a critique. In: Brunner, K. and Meltzer, A. (eds), *The Phillips Curve and Labour Markets*. North-Holland Publishing Co., Amsterdam, pp. 19–46.

Lyon, C.C. and Thompson, G.D. (1993) Temporal and spatial aggregation: alternative marketing margin models. *American Journal of Agricultural Economics* 75, 523–536.

Maddala, G.S. (1992) *Introduction to Econometrics*, 2nd edn. Macmillan Publishing Co., New York.

McGuirk, A.M., Driscoll, P. and Alwang, J. (1993) Misspecification testing: a comprehensive approach. *American Journal of Agricultural Economics* 75, 1044–1055.

McIntosh, C.S. and Dorfman, J.H. (1992) Qualitative forecast evaluation: a comparison of two performance measures. *American Journal of Agricultural Economics* 74, 209–214.

Miranda, M.J. and Glauber, J.W. (1993) Intraseasonal demand for fall potatoes under rational expectations. *American Journal of Agricultural Economics* 75, 104–112.

Moschini, G. and Meilke, K.D. (1989) Modeling the pattern of structural change in US meat demand. *American Journal of Agricultural Economics* 71, 253–261.

Naik, G. and Leuthold, R.M. (1986) A note on qualitative forecast evaluation. *American Journal of Agricultural Economics* 68, 721–726.

Offutt, S.E. and Blandford, D. (1984) The impact of the Soviet Union upon the US feed/livestock sector – an assessment. *Journal of Policy Modeling* 6, 311–340.

Paarlberg, P.L. and Abbott, P.C. (1986) Oligopolistic behavior by Public Agencies in international trade: The world wheat market. *American Journal of Agricultural Economics* 68, 528–542.

Park, W.I. and Garcia, P. (1994) Aggregate versus disaggregate analysis: Corn and soybean acreage response in Illinois. *Review of Agricultural Economics* 16, 17–26.

Roy, S.K. and Ireland, M.E. (1975) An econometric analysis of the sorghum market. *American Journal of Agricultural Economics* 57, 513–516.

Schroeter, J. (1988) Estimating the degree of market power in the beef packing industry. *Review of Economics and Statistics* 70, 158–162.

Schroeter, J. and Azzam, A. (1991) Marketing margins, market power, and price uncertainty. *American Journal of Agricultural Economics* 73, 990–999.

Shonkwiler, J.S. (1982) An empirical comparison of agricultural supply response mechanisms. *Applied Economics* 14, 182–194.

Shonkwiler, J.S. and Emerson, R.D. (1982) Imports and supply of winter tomatoes: an application of rational expectations. *American Journal of Agricultural Economics* 64, 634–641.

Skold, K.D. (1992) *The Integration of Alternative Information Systems: An Application to the Hogs and Pigs Report.* CARD Monograph 92-M4. The Center for Agricultural and Rural Development, Iowa State University, Ames, IA.

Skold, K.D., Grundmeier, E. and Johnson, S.R. (1988) *CARD Livestock Model Documentation: Pork.* Technical Report 88-TR 4. The Center for Agricultural and Rural Development, Iowa State University, Ames, IA.

Stillman, R.P. (1985) *A Quarterly Model for the Livestock Industry.* USDA Technical Bulletin 1711. USDA, Washington, DC.

Subotnik, A. and Houck, J.P. (1982) A quarterly econometric model for corn: a simultaneous approach to cash and futures markets. In: Rausser, G.C. (ed.), *New Directions in Econometric Modeling and Forecasting in US Agriculture.* North-Holland Publishing Co., Amsterdam, pp. 225–255.

Taylor, C.R., Reichelderfer, K.H. and Johnson, S.R. (eds) (1993) *Agricultural Sector Models for the United States.* Iowa State Press, Ames, IA.

Thurman, W.J. (1987) The poultry market: Demand stability and industry structure. *American Journal of Agricultural Economics* 68, 322–333.

Tomek, W.G. and Myers, R.J. (1993) Empirical analysis of agricultural commodity prices: a viewpoint. *Review of Agricultural Economics* 15, 181–202.

Traill, B. (1978) Risk variables in econometric supply response models. *Journal of Agricultural Economics* 29, 53–61.

Traill, B., Colman, D.R. and Young, T. (1978) Estimating irreversible supply functions. *American Journal of Agricultural Economics* 60, 98–106.

Tronstad, R. and McNeill, T.J. (1989) Asymmetric price risk: an econometric analysis of aggregate sow farrowings, 1973–86. *American Journal of Agricultural Economics* 71, 630–637.

Waugh, F.V. (1964) *Demand and Price Analysis: Some Examples from Agriculture.* USDA Technical Bulletin 1316. USDA, Washington, DC.

Wohlgenant, M.K. (1985) Competitive storage, rational expectations, and short-run food price determination. *American Journal of Agricultural Economics* 67, 739–748.

7 Market Structure and Institutions

Paul L. Farris

Department of Agricultural Economics, 1145 Krannert Building, Purdue University, West Lafayette, Indiana 47907–1145, USA

INTRODUCTION

In this chapter we deal primarily with institutions that have become integral parts of agricultural marketing systems. These include private sector business firms of various sizes and types, cooperatives, futures markets and government agencies that provide services, rules or facilities. We consider the salient characteristics, operations and factors affecting perceived general industry success and performance of marketing systems.

The marketing institutions that have evolved in many countries possess similarities that have grown out of basic features of agriculture. Such features are manifested in aspects of production, supply, demand, trade and national policies. As national economies have developed many marketing-related activities have expanded in order to bring together commodities from geographically dispersed production areas, process them in various ways and distribute them in increasingly altered forms to ever more consumers in the cities. Additionally, producing and distributing manufactured inputs and furnishing capital and information services for producers, processors and marketing agencies are seen as increasingly vital to an effectively functioning system.

Coordination and integration of numerous economic activities is taking place more and more through exchange arrangements that reflect supply and demand conditions in markets somewhere in the overall system. Although some national economies have relied on centrally administered allocations and fixed prices to coordinate activities, that trend is declining as many countries are removing government interventions and expanding the role of markets in exchanging goods and services. But because all requirements for free and open markets cannot be achieved and maintained, various degrees of market failure can occur. Such failures can seriously impair economic efficiency, impede the achievement of potential gains and progress and lift up the need for remedial action.

Market failure or faulty performance may be traceable to many causes, including imperfect information, high transaction costs, lags in adjustment to change, inadequate infrastructure, monopolistic elements, or other impediments, such as some government policies and interventions undertaken to achieve diverse goals. Whether market failures can be avoided or ameliorated, and productivity and progress encouraged, are thus important public policy issues, and success can depend upon characteristics of production, manufacturing, distribution and other factors. The types of policy steps considered acceptable may hinge importantly on citizen preferences for particular goals, rules or interventions, as well as ease of achievement.

INDUSTRY ORGANIZATION AND PERFORMANCE

Market classifications

The organization of industry has long been considered of paramount importance in the performance of marketing systems, and scholars have delineated theoretical aspects in some detail. Over a period of many years the four distinct market forms of pure competition, monopolistic competition, oligopoly and monopoly were identified and their differentiating characteristics set forth (Greer, 1984; Scherer and Ross, 1990). Special cases and variations within the market forms have been elaborated. Distinctions are made between pure and perfect competition and, within monopoly, the potential gains from price discrimination can be present or absent. Monopolistic competition contains elements of both monopoly and competition. The interdependent behaviour among firms, which is a differentiating characteristic of oligopoly, is found within a broad array of industry structures.

While the theoretical market forms have been delineated in some detail and their features presented in basic economics textbooks, it has not been easy to apply the conceptual structures empirically. A step towards applicability came in 1940 by Clark, who presented the concept of workable competition. Market forms with elements of monopoly or imperfections might be judged workably competitive if such forms and market behaviour led to performance results considered to be satisfactory.

The next major step is identified with the work of Bain (1968) in the 1950s. He elaborated the structure–conduct–performance approach and presented empirical information within that framework. Bain's ideas provided operational classifications and hypotheses that led to a burgeoning of work by others in the 1960s and later. Numerous studies based on this framework provided a substantial amount of empirical knowledge of food marketing systems. Prior to Bain's statistical analysis approach, historical case studies were the major sources of validation of industrial organization hypotheses.

Intrafirm behaviour

Two other lines of inquiry and thought that were related to ideas of industry organization and performance emerged primarily in the post World War II period. One of these grew out of the work of Barnard (1938), Simon (1958), Cyert and March (1963) and Leibenstein (1966). The emphasis was on intra-firm organization and management. Simon's concept of bounded rationality led to a focus on intrafirm goals and the idea of satisfying replacing opti-mizing as a driving force in management behaviour. In this connection, Leibenstein developed the idea of X-inefficiency, or organizational slack, within organizations. This can occur in a firm if it is shielded at least in part from competitive market forces. What happens is that cost controls become lax, bureaucracies grow, resources employed become excessive and costs rise. Over time, profits may not seem excessive, but prices tend to remain higher, output lower and technological advance less rapid than if competi-tive pressures had been more vigorous. Estimates by one author indicate the effects of X-inefficiency may amount to as much as 10% of costs where con-centration is very high and 5% where concentration is moderate (Greer, 1984). In concentrated markets in food manufacturing and food retailing, X-inefficiency has been estimated at approximately half of the excess of prices over costs (Marion *et al.*, 1979; Parker and Connor, 1979).

In recent times a great deal of interest has been shown in an approach known as new empirical industrial organization. The focus is on intra-industry studies that measure effects of oligopolistic conduct. The inter-dependent actions and reactions of firms may bring about differing results for themselves and for others as they relate to each other by employing various market strategies. An early theoretical example involving two non-collaborating firms with conjectural interdependence was developed by Cournot (Fellner, 1949). Equilibrium was found at a level between pure com-petition and monopoly. Von Neumann and Morgenstern (1944) developed game theory, which provided the basis for much successive work relating to interactive conduct among business firms. Because the range of conduct options that can be employed by some business firms is so broad, and varia-tions in assumptions so many, economic outcomes can be very numerous and highly variable. Whether interaction among firms is non-cooperative or cooperative is critical. Theoretical hypotheses have been elaborated in some detail for various types of interdependent firm behaviour, but empirical analyses and underlying explanations of findings have been limited.

Applicability of market classifications

In general, the classification of market forms within the structure–conduct–performance framework (Table 7.1) is useful in showing where theoretical reasoning and empirical evidence have provided some generalizations and also where knowledge is lacking. The structure of pure competition is one where firms are sufficiently numerous and of such sizes that no one firm nor

Table 7.1. Distinguishing market structure, conduct and performance characteristics associated with theoretical market forms.

Organization and behavioural attributes	Theoretical market forms			
	Pure competition	Monopolistic competition	Oligopoly	Monopoly
Market structure				
Number of firms	Numerous	Numerous	Few	One
Entry conditions	Easy	Easy	Moderate to difficult	Blocked
Product differentiation	Undifferen- tiated	Some differentiation	Variable	Unique product
Market conduct				
Recognition of interdependence among firms	Unrecognized	Unrecognized	Recognized	None
Feasibility of optional strategies	No	No	Yes	Yes
Market performance				
Technical efficiency	High	Moderate	Variable	Variable
Progressiveness	Low	Low	Variable	Variable
Earnings	Normal	Normal	Above normal	Above normal

several in combination can influence market outcomes. Products are homogeneous or standardized and entry is easy. As a consequence, each firm is driven to minimize costs. Earnings are held to competitive levels. Firms tend to operate at output levels where average costs are at a minimum. Technological advance comes primarily from outside the structure, usually, as in agriculture, from public sector research. Classic examples of pure competition are found in agriculture. The theoretical idea of perfect competition requires, in addition, absence of friction in some sense. This implies perfect knowledge about current and future market conditions and the product and perfect mobility and divisibility of production factors and products.

Monopolistic competition, with elements of both monopoly and pure competition, leads to some performance results characteristic of each. Aspects of this are found in some retail structures. Firms are numerous and entry is easy. Products are differentiated, but interdependence among firms is not recognized. Because of product differentiation, the demand curve for a firm's product slopes downward rather than being constantly horizontal as in pure competition. Prices are elevated just enough to pay for the product differentiation. Competition drives individual firm demand down to where average unit costs are barely covered so no excess profits prevail. That point tends to occur at output levels below where average unit costs would be lowest. Thus there tends to be excess resources tied up in this structure.

The interdependent behaviour among oligopolistic firms is found in many different types of industry conditions. Entry may be impeded or difficult, and products may be standardized or differentiated. Oligopoly is widespread throughout the economy and interdependent actions among firms frequently are found in international trade. Types of oligopoly vary widely from structures that are highly competitive to those where numbers of firms are very few and motivations toward spontaneous coordination, coalitions or cartels are very strong. Inferring conduct and performance under oligopoly generally is complicated, not only by several factors related to the wide structural variations within this market form, but also by varying external conditions affecting the industry and managerial characteristics within firms.

Monopoly in its pure form is one firm with a completely unique product and blocked entry. It can set price and output where profits are maximized. If the monopolist can subdivide customers to take advantage of distinct demand segments and charge different groups different prices, total profits can be enhanced further. And if each customer can be forced to pay his or her maximum price, profits can be increased even more. Firms with strong monopolistic characteristics exist, including some that can successfully employ aspects of price discrimination. Where monopoly power is strong, it generally becomes subject to some form of government intervention, as in the regulation of public utilities.

Directions of causation

At this point it is useful to consider more fully the general question regarding directions of causation within the structure–conduct–performance framework. Clearly, at any particular time basic industry conditions and the structures in place have an important bearing on conduct options available and on expected performance. But in a longer run period, basic factors external to the industry can change, transportation developments can cause fixed facilities to be out of position, new technologies can alter economies of scale and impact existing firm and industry operating procedures, and government rules and policies can be modified. Many external developments can open up opportunities for some firms and adversely affect others. As a consequence, the entire structure can become more broad or more concentrated.

Within the framework itself, the dynamics of intrafirm management and conduct patterns can significantly affect structure and performance (Norman and La Manna, 1992). Merger and financial initiatives and decisions made by firms are obviously critical in some circumstances. Selling and buying strategies, pricing, advertising and promotion and product development are also important dimensions. How firms expand and whether they integrate vertically or in some diversified way can, over time, profoundly change industry structures and market performance. Explorations of oligopolistic behaviour employing game theory, both non-cooperative and cooperative, can indicate diverse insights relating to effects of conduct patterns on structure. Thus in analysing organizational issues and roles of various institutions in the marketing system, it is important to examine external influences and the dynamics of internal firm behaviour on changes in structure over time as well as the shorter run effects of structure on performance.

Integration types and implications

A dimension of industry organization that is importantly influenced by conduct, and is increasingly relevant in the agro-food economy, is integration. Types of integration that tie together individual entities are known as horizontal, vertical and conglomerate. In considering these, it is useful to keep in mind both the short- and long-run aspects of structure and conduct.

Horizontal integration refers to the combination of firms that perform similar functions. An example would be the merger of two meat packing firms. The creation of joint ventures or marketing alliances could have similar competitive effects. Horizontal integration, by definition, increases market concentration and, if firms in the industry are few, it can lead to greater control over supply of products produced in the industry and to enhanced market power.

Vertical integration involves the tying together of two or more successive functions in the marketing chain for a commodity. Forms of vertical coordination vary from loose contractual arrangements to outright ownership. A classic example of full integration in USA agriculture is poultry. The process began as feed companies entered into contracts with growers in order to sell more feed. The companies then went into poultry processing. In time some companies integrated into hatching egg production and the development of breeding stock.

Systems of tight vertical coordination have become a widespread development in several commodity subsectors. New technology and management techniques are increasing the interdependence between agricultural production and marketing as opportunities emerge to capture efficiency gains as well as gains that might arise from exercising market power. Efficiency gains can result from closer matching of supplies and demands in vertical systems, developing and marketing new products with exacting quality specifications and enabling production and marketing processes to fit together more closely with lower transaction costs. Generally fewer resources are then required in the overall food chain.

Decision-making within vertical systems is becoming more centralized and increasingly oriented toward the consumer market. Key decisions have tended to gravitate either toward food retailing or food manufacturing, depending on relative degrees of concentration and opportunities to influence consumer purchasing decisions. On the farm input side, the tendency is also toward closer coordination within the marketing channel. Market influence depends on the extent to which products are differentiated or on control of the supply of a strategic input. If manufacturers of farm inputs can impose a minimum level of resale prices or designate exclusive distribution territories, competition may be reduced.

One consequence, aside from possible benefits to large integrating firms, is that alternative mechanisms of exchange and coordination tend to supplant open market transactions. The number and types of selling opportunities available to independent firms or farms tend to decline, become less representative, more inconvenient or costly, or perhaps non-existent. Other disadvantages might be loss of price and quantity information and an increase in variability outside the integrated portion of the subsector. From the standpoint of policy, such disadvantages must be balanced against possible overall gains from more closely coordinated systems, including how such gains might be shared among farmers of different sizes and types, and among marketing firms and consumers. Of course, the question of who receives the gains is importantly influenced by the numbers, sizes and strategies of competitors at various stages in the marketing channel.

Conglomerate integration refers to the branching out of business firms into other lines of activity (Mueller, 1985). Three types are: (i) *product extension*: two or more different but related products are produced (an example might be poultry processing and meat packing or a baking firm might expand the line of bakery products); (ii) *market extension*: a given product is sold in two different market areas, such as milk in two cities distant from each other; (iii) *pure conglomerate*: a firm that is engaged in activities that are unrelated, such as telecommunications and bread baking.

In recent decades, conglomerate growth has been achieved mainly through merger. Causes of diversification into other activities are believed to be numerous, involving a mixture of efficiency, financial, competitive, and growth motivations. Economies of scale in selling appear to be a factor in consumer goods industries. Also, government policy sanctions may deter firms from expanding within industries where they hold large market shares.

Increasing diversification of large firms may have important implications for market performance. Efficiency and productivity gains may be achievable in some situations. However, a large conglomerate firm may have the potential to engage in predatory pricing to eliminate competitors in some markets because it can subsidize its losses in those markets with earnings from other sources. Also, reciprocal dealing among conglomerate firms can impede the entry of new competitors, weaken competition, and increase concentration. If large conglomerate firms confront each other in several industries and markets, they may acquire tendencies toward mutual forbearance or live and let-live competitive behaviour. In addition, firms may be able to influence government rules and policies to enhance their own competitive advantages. The stimulus to compete might then be reduced in order to generate the individual firm benefits potentially achievable.

Multinational food firms

Along with increases in international trade we have seen an expansion of activity by international food firms, particularly in high value-added products. A detailed discussion of the activities of multinational food firms will be found in Chapter 18. A multinational enterprise has been defined as a firm that owns, controls and manages income-producing assets in more than one country (Connor, 1984). Typically multinational firms are very large, and originate from the major industrial countries, especially the United States, Canada and Western Europe. Firm decision-making is decentralized but subject to considerable control through contracts, agreements, technical services and other arrangements. Multinational food firms emphasize mainly food processing and agricultural commodities trading with less emphasis on food service, grocery wholesaling and retailing. However, the extent of foreign direct investment has been rising in all these areas (Connor, 1994).

Multinational firms are typically large enough to benefit from well-placed foreign investments by employing technology effectively and competing with host country firms. They can often introduce new product variations, reduce prices and in some cases increase host country employment. The environment is commonly favourable to the exercise of alternative conduct strategies. Where existing markets within host countries are functioning imperfectly, multinational companies can enhance competition, increase technical progress and expand global productivity. Issues with regard to future host country acceptance of multinational firms are likely to involve their effects on income distributions within and among countries and general concerns about equity, tax policies, traditions and economic development.

Evaluation of industry organizational forms

Several practical measures have evolved as criteria for appraising the performance of industry structures. Most are viewed from the standpoint of general economic welfare and not just in terms of the well-being of a particular group or segment in an industry. Important aspects of performance include profits or price–cost margins, production efficiency, sales-promotion costs, product characteristics, and technological progress. These aspects have been elaborated in various ways, and other dimensions of performance have been suggested.

Factors which influence performance have been the subject of a great deal of empirical investigation even though research in the area of industrial organization has been difficult because of the lack of precise and relevant data. While theoretical reasoning has advanced and provided logical hypotheses, the detailed information required for analytical verification has often been unavailable, particularly on conduct.

Also, evidence shows that there is a wide variation in earnings among individual firms within particular industries. In pure competition, monopolistic competition and oligopoly, some firms do well and others lose out. Even some highly monopolistic firms do better than others, and some have failed. Numerous factors external to the industry or management decisions and strategies within firms may overwhelm structural aspects. These include product differentiation, vertical coordination, market diversification, economies of scale, barriers to the entry of new competitors and intrafirm behaviour.

One point of view is that high profits may come from employing new technology and managing for greater efficiency rather than from market power (Demsetz, 1974). An example where technological advance and economies of scale appear to be associated with rising concentration is in slaughter of steers and heifers in the United States (United States General Accounting Office, 1990). The estimated cost per head, in a 1985 study, declined from US$40.71 for a plant slaughtering 25 per hour to US$22.20 in a plant slaughtering 325 per hour. Four-firm concentration rose from 35.7% in 1980 to 69.8% in 1988.

However, there is considerable evidence that, where concentration is high, one or more performance characteristics have usually been found impaired, often in a significant way. Generally it has been difficult to claim that high concentration was the only cause, since other market conditions were often found to be interrelated with high concentration.

A study of the effects of increasing regional concentration in the USA beef packing industry on fed cattle procurement prices showed cattle prices to be about 3% less in the most concentrated region compared to the least concentrated region. A regional four-firm concentration ratio of 60% or higher was associated with significantly lower prices paid to producers for cattle (Marion and Geithman, 1995).

An illustrative example in food manufacturing shows that with four-firm concentration at 40% and advertising expenditures at 1% of sales, the after-tax profit rate on stockholders' equity was 6.2%. With an increase in advertising to 5% of sales, the profit rate went to 10.7%. With four-firm concentration at 70% and the advertising to sales ratio at 1% the profit rate was 11.5%. But with concentration at 70% and the advertising to sales ratio at 5%, the estimated profit rate became 15.9% (Greer, 1984).

Another study showed grocery store prices 5.4% higher when four-firm concentration was 70% than at 40%. Pretax profit as a per cent of sales was 0.36% with four-firm concentration at 40% and 1.23% at 70% (Marion *et al.*, 1986). The distribution of firms by size within the top four also was found to affect grocery chain profit rates. Prices and profits were higher both as four-firm concentration was greater and the share held by the largest single firm was larger.

Product differentiation effects are especially evident in large food manufacturing firms with strong national brands and few competitors. Such firms are usually in positions to influence product and input prices and to regulate production and flows to market. Large retail food chains that hold significant market shares in individual cities or market areas typically have considerable influence in pricing, in bargaining for food supplies, and in deciding which suppliers get shelf space except, as noted, for certain manufactured items with strong brand preferences.

Economies of scale in selling are particularly large in areas where television advertising is important; such economies are often greater than economies associated with the physical functions of plant operations in processing or in assembly and distribution. These advantages in selling encourage firms to grow to very large sizes, and they tend to increase diversification and conglomeration in food manufacturing firms. Large food manufacturing firms tend to hold most of the leading positions in concentrated industries that sell highly differentiated products (Connor *et al.*, 1985).

However, the market strength of large food manufacturing firms tends to be moderated significantly for some products by the availability of competing private (or 'own') label and generic products at lower prices. An estimated one-third of consumer foods in the USA in 1977 were private label or unlabelled (Connor and Peterson, 1991). Examples for 1980–81 where private label market shares were half or more include canned tomatoes, canned peaches, frozen orange juice and granulated sugar (Connor *et al.*, 1985). At the other extreme, manufacturer brands accounted for 97% of ready-to-eat breakfast cereals. This high market share has been maintained by extensive advertising and product proliferation. Examples where manufacturer brands accounted for 80% or more include coffee, cake mixes, vegetable oil and canned soup. Price differentials between manufacturer and private label brands vary significantly, ranging from minor differences to 30% or more, with an estimated overall average of around 20% (Connor and Peterson, 1991). Chapter 13 explores further the role of own label in product policy.

Research has shown that levels of advertising were found to influence changes over time in seller concentration in the food and tobacco industries. Product classes with no media advertising decreased in concentration. As advertising rose, so did the increase in concentration. This tendency illustrates the effects of a conduct dimension, advertising, on structure. A similar study of all USA manufacturing showed that consumer goods industries appeared more likely to become increasingly concentrated over time, or to resist concentration-deterring forces, than producer goods industries (Farris, 1973a). However, in some food commodity manufacturing industries concentration has increased significantly in recent years. Examples include beef packing, flour milling and oilseed processing.

The possible effects of market structure on technological progress have also received a great deal of attention. Some authors have suggested that large size, and especially high profits, lead to more rapid technological progress because large, high-profit firms can allocate substantial funds to research and development (Galbraith, 1985). Diversified firms might be especially motivated to emphasize research and development, since the potential areas in which they could usefully apply unpredictable results would presumably be much broader than would those for more narrowly specialized firms. Moreover, large firms have both the resources to finance potentially promising new innovations, new ventures, and new products, and the managerial capability and knowledge to launch them.

An opposing view suggests that large size and, in particular, highly concentrated markets, tend to inhibit technological advances. In such markets the attention of firms tends to be focused, instead, on protecting and building up entry barriers to keep out competitors and on advertising and promoting products in order to enhance their uniqueness. Inventive activity often takes place in the smaller or intermediate-size firms (Mansfield, 1981). If an invention is identified as having a large market potential, a larger firm may purchase it or acquire the firm which has developed it. The larger firm then develops and markets the product. The resources required to develop, market test, and launch a new product may be very large, beyond the reach of the firm that produced the initial invention.

In general, evidence indicates that technological progress in most industries involved in the farm and food system does not require huge firms nor highly concentrated markets (Mueller *et al.*, 1982). Moreover, most inventions that have been applied in agribusiness industries have originated in other industries, such as food machinery companies, or in publicly supported research laboratories. Thus the argument that high profits and concentrated markets are essential for technological progress and rising productivity is not strongly supported by empirical observation.

Increasing attention has been given to the effects administered pricing of large firms in concentrated markets might have on achieving the general economic goals of stability and resource employment (Gordon, 1981). Two important aspects of pricing may be involved. One is the tendency of administered price movements to be relatively inflexible, particularly downward. Such pricing behaviour intensifies the bias toward inflation in the general economy and obstructs the attainment of monetary and fiscal policy goals. Adding to these difficulties is the indirect effect of price indexing. Wages and incomes tied to movements in general price indexes tend to create an upward spiral that partly feeds on itself.

The other important aspect is what happens in the labour market. If a firm can pass on higher labour costs through higher prices, it can more easily accommodate demands for higher wages if it does not face strong competition in the market for its products. Over time, labour costs tend to rise more in firms and industries that have the greater power in product markets. The result not only distorts the use of economic resources but also eventually increases the vulnerability of both labour and business in high concentration industries, making it more likely that they will lose their markets to firms with substitute products, new technology, and lower labour costs.

Future work using new empirical industrial organization approaches can be useful, not only in examining hypotheses, but especially in providing detailed firm and industry data for various analyses and insights into structure, conduct and performance. Where underlying causes of conduct patterns can be identified, and some regularity or stability of conduct patterns can be assumed, or inferred, such as the validity of conjectural variations, consequences in terms of effects on structure or performance might be estimated. Obviously, much detailed empirical information about characteristics within firms is required in order to make such estimates. In a review of 12 industry studies, including coffee roasting, tobacco, and food processing in Canada, substantial market power over price was found. A significant cause of high price–cost margins was anticompetitive conduct (Bresnahan, 1989).

A study of pricing behaviour by USA wheat and corn exporting firms found that export price mark-up was positively related to USA exporter seller concentration in foreign markets, but the influence was not large (Patterson and Abbott, 1994). A related study showed that past participation in the USA Export Enhancement Programme was the major factor related to a firm's continued participation in the programme. No influence was observed between a firm's past export market share in export markets and its current participation in the Export Enhancement Programme (Patterson *et al.*, 1996). In an examination of competition policy in military aircraft procurement, it was concluded that even with industry consolidation, rivalries might be preserved to broaden acquisition alternatives (Kovacic and Smallwood, 1994).

Over time it has been estimated that the proportion of national income from industries in the USA considered to be workably competitive rose from 56.4% in 1958 to 76.7% in 1980. The tight oligopoly share declined from 35.6% to 18.0%; that from dominant firm industries went from 5.0% to 2.8% and the monopoly proportion decreased from 3.1% to 2.5%. The tendency toward more competition was believed to have been associated mainly with imports and pro-competitive government policies, such as antitrust actions (Shepherd, 1983).

In food manufacturing, the national four-firm share of industry value of shipments was relatively high, 60% or more, for only 13 of 49 industries in 1992. A slight increase in levels of concentration had occurred between 1987 and 1992 among 47 industries for which concentration ratios were available for both years (Table 7.2). However, national concentration measures may understate the actual concentration in relevant markets. Also, leading firms in some industries do a fairly high share of business for selected processed food items, both nationally and in some local markets.

For grocery store sales in particular metropolitan areas, 41% of 318 SMSAs (Standard Metropolitan Statistical Areas) had four-firm concentrations of 60% or higher in 1982 (Kaufman *et al.*, 1993). A gradual upward trend in average

Table 7.2. Share of value of shipments accounted for by the four largest companies in USA Food and Kindred Products Manufacturing Industries, 1987 and 1992 (source: United States Department of Commerce, Bureau of the Census, 1996).

Four-firm concentration ratio	Number of industries*		Per cent of industries*	
	1987	1992	1987	1992
Less than 40	17	17	36.2	36.2
40–59	17	15	36.2	31.9
60–79	10	11	21.3	23.4
80 and higher	3	4	6.4	8.5
Total	47	47	100.0	100.0

* Industries included are those for which published data are available for 1987 and 1992.

four-firm shares during the 1960s and 1970s was followed by a rapid increase from 58.17% in 1982 to 69.09% in 1987, based on an analysis of 133 SMSAs that were defined the same way in both years (Franklin and Cotterill, 1993). A loosening of federal antitrust enforcement is believed to have been a main factor related to the acceleration in concentration.

Measures of concentration

As we have seen, a commonly used measure of concentration is the share of the total market held by an absolute number of the largest firms in the market, for example, the largest four, eight or 20. The advantages of this concentration ratio are that it is easily constructed and easily understood. However, it covers only a portion of the total size distribution of firms, and the relative sizes of firms are not considered. Also, as has been observed, a large dominant firm share has been found to be influential, so the share of the largest firm is often used in connection with the concentration ratio.

Another measure is known as the Herfindahl–Hirschman Index, which is the sum of squares of market shares of firms, expressed as proportions of total market sales (or assets or employment). This index is highly regarded, but it is very demanding in terms of data. Still another measure, which is also very demanding in terms of data, and seldom used, is the Gini coefficient. It is a measure of inequality in the distribution of firms in an industry from a hypothetical distribution of equal size firms.

Each of the concentration measures has advantages and disadvantages. However, the various measures typically yield comparable results. Such was found in a study of soybean processing, where the share of industry capacity in the four largest firms increased from 35.3% in 1953 to 57.1% in 1984 and the Herfindahl–Hirschman index rose from 0.052 to 0.109 (Farris, 1973b; industry sources).

Obviously, a consideration in the use of any measure is market definition. Appropriate questions relate to whether markets are local, national or international. In USA food retailing, for example, the market has been considered a Standard Metropolitan Statistical Area (SMSA). For most food manufacturing, markets usually refer to the nation as a whole except for local market industries producing some bulky and perishable products. In purchasing livestock, market procurement areas have been defined as particular regions or smaller. Still another problem is product definition, for example, such as a particular type of drug or a range of related products marketed by pharmaceutical manufacturers. Market analysis thus requires that careful attention be given to market definitions.

Public policies toward industrial organization

Just as problems and uneasiness associated with the concentration of economic power have long existed in many countries, so have advocated remedies. Various approaches include providing increased information to customers on market prices, increased use of grades and standards, listing contents and ingredients on product labels and limiting the amount of advertising cost that can be claimed as business expenses for tax purposes. Tax policies affecting investment and depreciation of capital as applied to different kinds of firms have been used to influence changes in industry structures. Incentives can be provided to encourage the entry of new competitors.

Other government policies that might be employed or altered include rules relating to market conduct, such as restricting business practices to prevent vertical price maintenance, predatory pricing, exclusive market areas and unfair advertising. Modification of merger policies affecting horizontal, vertical or conglomerate combinations can have a major influence. In extreme situations, where markets are highly concentrated and barriers to the entry of new firms are very great, either direct regulation of certain patterns of firm behaviour, encouraging or prescribing conditions for greater competitive rivalry or divestiture of assets may be advocated.

Policy decisions should obviously be based on careful analysis of industrial organization and behaviour and evaluation of consequences of alternative choices. In some markets a broadening trend in competition may be occurring and positive benefits are being reflected in behaviour and performance. In other markets the dominant tendency over time might be toward higher concentration and higher entry barriers. In many farm and food system markets, competitive forces are strong and deficiencies can be remedied or mitigated by procompetitive policies that do not include the most drastic measures. Knowledge to formulate appropriate policy options affecting the organization and performance of industry is thus extremely valuable in enhancing the long-term effectiveness, productivity and viability of agricultural market systems.

AGRICULTURAL COOPERATIVES, MARKETING ORDERS AND MARKETING BOARDS

Agricultural cooperatives

An organizational form that has long been a part of agriculture in many countries is the cooperative. Farmers in the USA engaged in a variety of cooperative ventures early in the history of the nation (Torgerson and Ingalsbe, 1984). They did so in order to do collectively what they were unable to do independently within the generally prevailing purely competitive structure of agriculture.

Among the major goals that cooperatives have sought to achieve are the following: (i) improve bargaining power in purchasing farm supplies and selling farm products; (ii) reduce costs of marketing farm products; (iii) obtain products or services which are either costly or not otherwise available; (iv) obtain better market access for cooperative members; (v) improve product or service quality in both farm inputs purchased or commodities marketed; (vi) increase farmers' incomes; and (vii) provide information and education to increase agricultural production efficiency and to enhance quality of life in rural areas.

The distinctive characteristic of a cooperative is its ownership and control by those who use its services. Generally, cooperatives provide services at cost, owners receive benefits proportional to their use, there is democratic control (one member – one vote), and return on equity capital is limited. Typically cooperatives transact business at competitive market prices and periodically return net earnings to members in the form of patronage refunds based on usage.

Cooperatives have structured themselves in different ways as they have grown to take on new responsibilities. Several have become regional, serving a trade territory larger than a local marketing area and sometimes combining with other cooperative associations. Some have become interregional organizations that specialize in particular areas, such as fertilizer manufacturing or transportation. Joint ventures have been formed, some with other cooperatives and some with investor-owned firms to perform specialized functions, such as warehousing or in marketing particular commodities. Some initiatives have been taken in international marketing through cooperative arrangements with other firms.

The philosophical objectives of cooperatives have varied a great deal among associations and over time. During the 1920s strong centralized cooperatives were advocated with emphasis on supply control for orderly marketing. Another point of view was that cooperatives could serve as competitive yardsticks to point up opportunities for increases in efficiency and to enhance competitive performance of other firms. To some, cooperatives were seen as constituting a middle way between an enterprise economy and one in which the public sector dominated. A variant was a view of cooperatives as a system that would coordinate and complement various functional activities to achieve results beneficial to producers. Another idea is that there would be no limits on cooperatives and they would become predominant in all sectors of the economy. Cooperatives have thus been viewed in diverse ways and their roles have varied as economies have progressed, as ideas have evolved and as experience has accumulated.

Cooperatives continue to be a significant organization form within the USA food system and in many other countries. Their major activities are at the farm supply and first handler marketing levels. Cooperatives have integrated backward into refining and production of petroleum products, mining and basic manufacture of fertilizer and feed. They have been particularly successful in achieving substantial savings on some farm supplies and channelling earnings back to producer members. In 1977 the largest 100 marketing cooperatives accounted for about 6% of the food manufacturing value of shipments (Connor *et al.*, 1985). However, cooperatives play a major role in the marketing of milk and some specialty crops where they have integrated forward into food processing, packaging, distribution, advertising and promotion. Although cooperatives are not among the largest food and tobacco manufacturing firms, their market shares were found to be inversely related to price–cost margins in some food and tobacco product markets (Rogers and Petraglia, 1994). The cooperative influence was largest in the more concentrated markets. These findings are consistent with the theoretical competitive yardstick effect.

From time to time questions arise whether a cooperative might use market power to unduly enhance its prices. A study of this issue indicated that price enhancement was positively related to market share and advertising for both proprietary firms and processing cooperatives (Wills, 1985). However, cooperatives with similar market shares and advertising tended to have lower prices than proprietary firms.

Cooperatives have not become important in retail food marketing except in some of the Scandinavian countries, where they account for up to one-fourth or more of the business (Cotterill, 1984), and in the UK, where they were very important earlier in the century. Some relatively small-scale retail cooperatives succeed in particular locations, such as college communities, by providing specialized products and services. Credit unions, however, have become widespread in the area of finance. They are typically organized to serve particular groups, such as individuals associated with universities, corporations or other entities.

Market orders and boards

Market orders and boards are government sanctioned institutions authorized to facilitate collective activities of producers in the marketing of agricultural products. Federal marketing orders in the USA were authorized for milk and selected fruits and vegetables by basic legislation passed in 1937 and subsequently amended. State marketing orders with less extensive operations than federal orders also exist for certain commodities produced within respective states. Marketing boards exist in several other countries.

USA federal marketing orders provide regulations that are binding on all handlers when approved by affected growers. Marketing orders for milk cover market areas, such as a particular city or region, and the producers who ship milk to that market area. Classified pricing plans provide for payments to milk producers according to proportions for fluid and for other uses.

Federal marketing orders for fruits and vegetables are focused on producing areas (Armbruster and Jesse, 1984). Producers of a particular commodity in a certain producing area can approve an order that legally establishes what can be done under the order. Typically the order can place restrictions on qualities or quantities of the produce marketed or provide facilitating services.

In addition to placing restrictions on qualities and quantities of products that can be marketed in particular forms, orderly marketing might include timing of sales, limiting quantities in particular markets and in total, and prescribing packaging and grading services and other such provisions that would be conducive to the improvement of producer prices. Some of the techniques employed include mandatory inspection of products and imports, specifying quantities and qualities to be sold when and where, and providing for standardized packing. Several commodity orders provide for the raising of funds to support advertising, promotion and research.

The representation of producer interests under marketing orders is typically done by producer cooperatives. Marketing orders provide binding regulations that permit cooperatives to achieve benefits for producers that without an order would not be possible. An order, in essence, provides effective group collaboration to strengthen an industry in its marketing activities, furthers communication among members, and increases efficiency and technological progress.

Marketing boards and similar government interventions in other countries typically participate more fully in all functions of marketing than is true of marketing orders in the United States. The Canadian Wheat Board, for example, was formed in 1935 to earn farmers the best possible returns for their wheat and barley. Its vision is to be the world's leader in marketing grain (Canadian Wheat Board, 1992–93). The Board and its accredited exporters are the only sellers of Canadian wheat and barley in international markets. The Board pools returns so that all farmers delivering the same grade of wheat or barley will receive comparable returns at the end of the crop year. Although selling wheat and barley is the Board's main function, it is also involved in a range of supporting activities, such as product and market development, grain delivery and movement coordination and sales follow-up. It communicates to plant breeders market signals on changing world demands for new varieties.

Selling abroad by centralized agencies or governments exists in several other countries, which is fundamentally different from the USA system of marketing through private sector firms. Most USA grain sold abroad is handled by some half dozen large grain trading companies that have subsidiaries, physical facilities, offices and information gathering capabilities in many producing and customer countries. Their knowledge and effective employment of information in buying and selling have a critical influence on their business success and the economic performance of international grain marketing systems.

Other types of institutions that are assuming increasing significance in agricultural marketing are industry organizations. Commodity promotion boards are generating funds through check-off programmes, government assistance and other sources to undertake a variety of activities. These include commodity advertising and promotion in domestic and export markets, research and various activities and services on behalf of industries they represent. Representatives of industry boards collaborate with business firms and research institutions to ascertain changing consumer demand trends, technological applications and possible changes in government regulations and policies that would be favourable to industry goals.

INFRASTRUCTURE FOR AGRICULTURAL MARKETING

Agricultural marketing systems around the world are shaped and guided by numerous individual country characteristics, such as natural resources, size, climate, people and cultural heritage. Also important are national policies, customs, institutions and education.

The infrastructure investment that accompanies and facilitates economic development of a country also plays a major role in shaping the marketing system and its effective performance (Farris, 1983, 1985). Important physical infrastructure components for agricultural marketing are roads, bridges and transport equipment, seaports, storage facilities, electric utilities, processing facilities, water systems and communication systems.

If a country is oriented toward a market economy, as opposed to a centrally directed one, public development of services can encourage and complement the activities of farmers, businessmen and various market participants. Examples include providing basic economic data, marketing information programmes, grading services, auction markets, rules of competitive behaviour, financial availability, education and research. Incentives can be incorporated that will lead to creativity, innovation, competition and dynamic evolution of the marketing system toward higher levels of performance in future years. Public policies can be focused on invigorating and expanding the role of the private sector to do effectively those things it might do best. Consideration can also be given to transferring to the private sector some government activities that might be effectively handled by the private sector.

Because of the nature and size of many infrastructure investment needs, and the fact that potential benefits would be widely distributed, private entrepreneurs are not generally motivated nor financially capable of undertaking some of them by themselves. Thus solving several infrastructure investment needs requires the joint participation and collaboration of public and private sectors.

A major analytical challenge in market system improvement is to identify high priority infrastructure needs, to analyse complementarities between infrastructure investments and development of marketing systems, and to set forth alternatives and plans for appropriate infrastructure development. With limited capital, priorities must be set. A large capital investment like a new wholesale food marketing centre would be expected to last several years. Its location and construction should be preceded by careful analysis. If more than one such investment is planned, the most obvious and urgent facility can be identified and built first and serve as a pilot project. Because a new capital investment project interjects change beyond itself and provides information that is difficult or impossible to anticipate precisely, experience can provide knowledge and ideas for other investment plans to follow.

Changes in industry organization, such as vertical integration among retailers, wholesalers and processors, can have an important effect on the needs for public facilities and public investments in infrastructure. As supermarket chains grow large, they tend to bypass city wholesale markets for portions of their supplies, establish their own contacts in producing areas, and build their own city market distribution centres. Additionally, investments by other industries or by the government in other sectors of the economy can indirectly affect the agricultural marketing system.

Another consideration is technological change. One cannot know exactly how electronic computers and interactive communication systems might alter commodity flow patterns, storage, processing, pricing, or other aspects of food marketing. New processing technology and changes in consumer demand patterns may be induced by durable goods ownership, such as automobiles and microwaves. Institutions serving the economy, including education, banking and finance or communications systems, develop in new ways.

Changes in government farm and food policy may have implications for marketing infrastructure needs. If it were to seem wise for the national government to take steps toward either greater self-sufficiency in food on the one hand, or more free international trade on the other, food diets and economic functions related to food marketing could be changed substantially. International trade policy could alter the size of the total agricultural economy, its enterprise specialization and its export involvement. The encouragement or discouragement of foreign-based multinational firms can have a major bearing on the character of the domestic economy and the food system. Similarly, other government policies could be influential by affecting zoning and land development patterns, locations of facilities, competitive conditions and food marketing organization configurations.

A change in national policy can therefore have major implications for marketing, both as to size and functions required. A dynamic agricultural marketing system can, likewise, have a substantial influence on other sectors of the economy. As a consequence, it is important for marketing policies to be considered an integral component of overall national economic policy. Because optimality of marketing system development over time is affected by many factors of unpredictable influence, public policy should provide for flexibility and for ways to achieve optimum patterns of marketing infrastructure development in the face of uncertain change.

FUTURES MARKETS

Futures markets are special types of markets that enable people to deal in an operationally useful way with prices at a specified time in the future. They are becoming increasingly important in the economies of the USA and several other nations. At these markets one can buy or sell a contract for a precisely defined commodity or financial instrument to be accepted or delivered at a specified future date. The price is established by the interactions of buyers and sellers, or their representatives, in a two-way auction on an organized futures exchange. Because of the value of incorporating the time dimension in business planning decisions, futures markets play a major role in the function and performance of modern marketing systems.

Although organized futures markets have existed for more than a century, their growth has been enormous since the 1970s when the ongoing trading in commodity futures contracts became greatly overshadowed by the phenomenal expansion in contracts for financial instruments. The development of futures options in the 1980s and their wide acceptance and use contributed further to the rapid rise in total business done on futures exchanges. Much use is made by foreign traders of the two major exchanges in the USA, the Chicago Board of Trade and the Chicago Mercantile Exchange (Leuthold and Campbell, 1984). Futures exchanges also exist and are being established in other countries to serve particular needs of those countries.

Characteristics of futures contracts and futures markets

The emergence of futures contracts was a logical development from contracts to buy or sell goods for future delivery dating from medieval times or before. Shortly after the Chicago Board of Trade was founded in 1848, to-arrive, or forward, contracts came into use to enable traders to contract for forward cash sales. These were contracts in which a buyer and seller would agree on the purchase and sale of a commodity at some future time under such conditions as the two determine. Cash forward contracts at the exchange were useful in bringing buyers and sellers together, but they did not deal with some of the needs and risks involved.

Out of this experience and need came the development of futures contracts with standardization of quantity and quality specifications, month and place of delivery and deposits of margin money (Catania *et al.*, 1987). For example, a corn futures contract traded at the Chicago Board of Trade is for 5000 bushels of Number 2 yellow corn. It requires delivery during certain days of a specified month at an approved warehouse. Buying and selling of contracts can take place only in the organized futures exchange. The price is discovered through open outcry in a trading pit, and trading can be done only by members of the exchange either for themselves or for others for a fee. In practice most contracts are settled through offsetting purchases or sales before the delivery month.

In addition to setting forth precise specifications of contracts and providing a facility for trading, the exchanges establish and enforce rules to ensure that trading occurs in an open and competitive environment and that contracts are honoured. The regulatory activities of the exchanges in the USA are shared with an agency of the federal government, the Commodity Futures Trading Commission, which independently monitors and coordinates its activities with the exchanges.

An integral function in the operation of futures exchanges is clearing the trades. At the Chicago Board of Trade, this is done by a separate organization, the Clearing Corporation, owned by exchange members who meet strict financial requirements. The Clearing Corporation takes the opposite side of every trade and each day calculates the gain or loss for each member. Those who gain are credited. Those who lose are debited, and if the loss is above the minimum amount of money on deposit, additional margin money must be deposited. The member then credits or debits the accounts of its individual customers and, if necessary, calls on debited customers for more money.

The margin required is the amount of money buyers and sellers must deposit to ensure performance of the terms of the contract. It consists of two parts, initial and maintenance requirements. The amount varies depending on price and market conditions. If the value of a customer's open position falls below the maintenance level, additional funds must be deposited to comply with the initial requirement. If there is a gain in the value of the open position above the margin requirement, the amount of the gain is credited to the customer and can be withdrawn by the customer if desired.

Futures market participants

Market participants are of two principal types, hedgers and speculators. Hedgers are interested in managing or avoiding risks of adverse price changes on commodities or financial positions they hold or on positions they plan to take in business operations. A hedge is intended as a counter-balancing investment. Speculators have the opposite goal. They take open or exposed positions with the objective of earning a profit on their expectations of price changes favourable to them.

Various types of futures market speculators include scalpers, day traders and position traders. In addition, there are public speculators who are day traders or position traders, but who rely on exchange members to execute their trades. A scalper is a trader who buys or sells and holds positions usually for very short periods with the expectation of capturing small gains from short-run price fluctuations. A day trader focuses on price movements within a day, seeking to gain from expected favourable price changes before the daily close of trading, and an open position will rarely be held over night. A position trader is interested in expected favourable price movements over a period of days, weeks or months.

The two sides of a futures contract are called long and short. A trader who commits to receive delivery is on the long side of the transaction while the one who commits to make delivery is on the short side. Hedging may involve taking either a long or short futures position depending on business operations of the hedger. Speculators may take long or short positions depending on their price expectations.

Market efficiency

The interplay of the various and numerous market participants produces a market environment that, when functioning effectively, approaches the market form of perfect competition. Although a perfectly competitive market has such requirements as complete information available simultaneously at zero cost to all traders, perfect knowledge about price effects of new information and no transaction costs, the operational idea of an efficient market has been found useful. A market is considered to be efficient if price changes over time are statistically a random walk, that is, each price change is a random move. However, there can be allowances or adjustments in an efficient market for seasonal or cyclical price variations or trends in price movements over time.

Market efficiency is particularly dependent on the availability and distribution of market knowledge. A market is considered semistrong efficient if relevant information is publicly available for analysis and use even though not all traders have it or are in a position to use it effectively. But if some pertinent information is privately held by some traders and is not publicly available, insider trading can lead to price manipulation, distort market performance, cause market participants to lose confidence in the integrity of a market and bring about market failure.

A serious concern since the advent of futures trading involves possibilities for the development of 'corners' or 'squeezes'. This can happen if the proportion of the long or short open interest in a particular contract that is held by one or a few traders reaches a critical point. The possession of inside information by a trading firm, in addition to its use in manipulating prices, can contribute toward its ability to build a dominant market position.

A relatively recent example of highly irregular market activity was the accumulating of very large long positions in the July 1989 soybean contract by Feruzzi Finanziaria S.p.A. while the firm also owned more than four-fifths of all cash market soybeans available for delivery (Bailey, 1990). A poor crop in 1988 had caused a relative scarcity of soybeans to be on hand before the new crop would be available, which contributed to the vulnerability of the market. After repeated requests by the Chicago Board of Trade and the Commodity Futures Trading Commission that Feruzzi long futures position be reduced, the Chicago Board of Trade eventually took action to require orderly liquidation of the firm's long futures position.

The relationship of futures to cash prices

The difference between a futures and cash (or spot) price for a commodity or financial instrument at a particular location and time is known as the basis. Because the cash and futures markets tend to be tied together, the fluctuations in the basis are typically narrower than fluctuations in actual cash or futures prices. If basis fluctuations are small, hedgers usually can shift most of the risk of price change to speculators. But if basis fluctuations are large, and especially if they are highly unpredictable, buyers or sellers who hedge may continue to face significant price risk.

Many factors affect the size of the basis, such as supplies and demands by location, quality differences, transportation costs, storage costs and expectations for change in various factors. Basis patterns for commodities with defined harvest and storage periods, like grains, generally reflect a payment for storage between a current date and the maturity date of the futures contract. This is commonly referred to as the carrying charge. The basis may be positive or negative depending on intertemporal expectations of supply and demand. Normally, the cash price at a particular location is below the futures price. The difference, or basis, tends to narrow as time passes. The cash price approaches the futures price because remaining carrying costs for the physical commodity, principally storage costs, become less as the carrying time period is reduced. The cash price at contract maturity is typically somewhat lower than the futures price because of a location difference and possibly other factors such as transaction costs.

If there is a current relative scarcity of a commodity but more supplies are expected to become available in the future, the cash price can be above the futures price. The price differences in this situation are referred to as inverse carrying charges. With low stocks on hand, users may be willing to pay a premium for currently available supplies in order to maintain business operations. The premium is called a convenience yield, which can exist until available supplies become plentiful. Then positive carrying charges are reflected in the basis.

Price trends for a commodity over time, as well as fluctuations in the basis, can have a significant influence on the benefits of futures trading. For example, assume a corn-producing farmer in July expects the corn price to decline by October when he will harvest and sell his corn. He sells December corn futures at US$2.50 per bushel (he takes a short futures position) (Table 7.3). The farmer's declining price expectations turn out to be correct and by 30 October the December corn futures price had declined from US$2.50 to US$2.30 per bushel. The farmer sells his harvested corn for US$2.05 on 30 October, which means that the basis at his local market is 25 cents per bushel. He then buys December corn on the futures market for US$2.30 (he takes a long futures position that offsets his earlier short position) and gains 20 cents per bushel (US$2.50–US$2.30). (This illustration omits transaction costs.) So as a result of his futures trading, he can add 20 cents per bushel to the US$2.05 he received in his local cash market.

In the above example the total of US$2.25 was the amount the farmer received with a 25 cent basis. If the basis would have been less than 25 cents,

Table 7.3. Illustration of farmer use of a short futures hedge for corn.*

Date	Local spot market Price of corn (US$)	December futures market Price of corn (US$)	Basis (US$)
15 July	2.25	2.50 (Sell)	0.25
1 December	2.05 (Sell)	2.30 (Buy)	0.25
Change	−0.20	−0.20	0.00
		(Gain of 0.20 for farmer)	

* By selling corn at the futures market on 15 July for December delivery for US$2.50 per bushel, the farmer protects against an expected price decline for his growing corn between July and harvest time, realizing US$2.25 per bushel. Without the futures market transaction, he would have realized US$2.05 per bushel.

the farmer would have received a higher cash price for his harvested corn and had a higher return. If the basis had been greater, he would have been less well off. But because of the decreasing price trend, his forward pricing was advantageous. By way of contrast, if the price trend had been rising, the farmer would likely have been worse off by pricing corn in the futures market, depending on what changes in the basis might have occurred. In practice, many farmers do not use the futures market directly but indirectly through forward sales to a first handler such as a grain elevator who, in turn, offsets the price risks in the futures market.

The above example illustrates the point that in buying or selling futures, price commitments are made, and the consequences can be favourable or unfavourable depending on subsequent price and basis movements. Usually, because the basis fluctuates through a narrower range than the price of the commodity itself, a hedger avoids large losses but also large gains.

Options

Owing to the desirability of avoiding losses but keeping open the possibility of gains, options on futures contracts have been used widely since the early 1980s. They were used for a time after the USA Civil War and periodically until 1936 when they were banned until their more recent authorization.

An option involves no price commitment, as in a futures contract (Chicago Board of Trade, 1985). It is a way, by paying a premium, to protect against an adverse price move while keeping open the ability to benefit from a favourable price move. A *call* option gives the buyer the right to buy the underlying futures commodity contract at a specific price within a specified period of time; a *put* option gives the buyer the right to sell. The buyer of a call option obtains protection against rising prices without giving up the right to benefit from lower prices; a put option protects against falling prices without giving up the right to benefit from rising prices. Purchasing an option provides a right, but not an obligation. The seller of an option earns the option premium for providing the purchaser the right to that choice. The option seller is then obligated to take an opposite futures position if and when an option is exercised.

Although the basic idea of an option is straightforward, the introduction of some options trading terminology will be useful in illustrating the trading process. As we have seen, *call* and *put* options provide, respectively, rights to buy or sell. The *underlying contract* is the particular futures contract which the option buyer has the right to buy or sell. The *strike price*, or the *exercise price*, is the price at which the buyer has the right to buy or sell. Options have a range of strike prices, each with different premiums. Strike prices are in 25 cent increments for soybeans and 10 cent increments for corn. The *premium* is the market price of the option, which is the most the buyer can lose. *Exercise* refers to the action to acquire a position in the underlying future at the strike price. *Expiration date* is the last day on which an option can be exercised; after that it is worthless.

Premiums are established in the option market for calls and puts. For calls, premiums are higher with lower strike prices because buyers are willing to pay more to buy the underlying commodity at a lower price. For puts, premiums are higher with higher strike prices because buyers are willing to pay more to sell the underlying commodity at a higher strike price.

The considerable attention given to the pricing of options has resulted in additional terminology. *Intrinsic value* refers to the amount of money, if any, that currently could be obtained by exercising an option at a given strike price. A call option has intrinsic value if its strike price is below the futures price. A put option has intrinsic value if its strike price is above the futures price. An option with intrinsic value is said to be *in the money*. An option that is *at the money* or *out of the money* has no intrinsic value. An option's value at expiration, that is its intrinsic value, will be the amount by which it is *in the money*.

When buyers and sellers negotiate premiums in the market, they are concerned with two separate components of the premium, intrinsic value and *time value*. Time value is influenced by the length of time remaining until expiration of the option and the volatility of the underlying futures price. As remaining time decreases, time value declines because the likelihood of an uncertain event in the remaining time becomes less. High volatility of the underlying futures price causes premiums to be higher than if volatility were lower because an option is more likely to be exercised if prices are volatile. Another term of interest to market participants is the *delta*, which is the ratio of the change in the price of the premium to the change in price of the underlying futures. It is typically less than one.

In summary, the advantages of options are setting lower or upper bounds on prices and keeping open the opportunity for favourable price moves. Persons with various price expectations have a wide range of choices of strike prices. The buyer of the put and call options does not have to be concerned about margin calls, and losses are limited to the amount of the premium. Some disadvantages include the premium cost and the fact that basis risk remains.

Economic functions of futures markets

A major economic function of futures markets is to facilitate the efficient management of price risk. A futures exchange brings together those individuals whose business is to take open speculative positions on assets for the purpose of profiting on future price changes of those assets and those business firms, farmers and others who wish to transfer their price risks to others. Speculators seek to develop skills and knowledge in their business and they compete with others who have the same goals. As a consequence, the risk-bearing cost in the economy is minimized, and many processing and marketing firms can carry out their regular operations more efficiently.

Another valuable contribution of futures markets is in pricing. Prices are discovered in open competitive markets where supply and demand information from around the world is focused and interpreted by numerous market participants who have various objectives. Barring monopolistic price manipulative tendencies, which are rare and monitored quite closely, the prices have validity for market users and the prices are widely used as benchmarks for establishing exchange arrangements elsewhere in the overall system. The information provided by futures markets is useful in price discovery processes in cash markets and for use in business planning by numerous firms who must make decisions about future operations.

The financing of business operations is also a function where futures markets contribute in significant ways. The hedging of price risks on inventories, capital assets and production inputs can help stabilize income flows and reduce borrowing costs. Speculators contribute capital that indirectly helps make this possible.

CONCLUDING REMARKS

In this chapter we have become acquainted with the principal organization forms and institutions in the agro-food economy and how they operate. The frameworks, examples and descriptions presented can help guide the analysis of marketing problems of various types and suggest solutions, including public policy changes.

The broad range of public policies that affect agricultural market system development and performance include many relating to agriculture, business firms, institutions, and the roles of private and public sectors in the economy. The increasing trend toward privatization around the world opens opportunities for business organizations and institutions of various kinds, including those in agricultural marketing, to develop and grow. Because agricultural marketing systems are continuing to evolve, the encouragement and consideration of new frameworks, new institutions and new policies that might prove to be workable can contribute importantly to agricultural market system improvement and to help achieve high priority national goals. Formulating appropriate public policies and incentives to encourage innovation, productivity and favourable economic performance will continue to be an important challenge for every country.

ACKNOWLEDGEMENTS

The author thanks John M. Connor and J. William Uhrig for helpful comments.

FURTHER READING

The following selections from the list of references are particularly recommended for further reading: Bain (1968); Catania *et al.* (1987); Chicago Board of Trade (1985);

Connor *et al.* (1985); Farris (1983); Greer (1984); Marion and NC117 Committee (1986); Norman and La Manna (1992); Scherer and Ross (1990).

REFERENCES

Armbruster, W.J. and Jesse, E.V. (1984) Fruit and vegetable marketing orders. *The Farm and Food System in Transition, Emerging Policy Issues*, FS19. Cooperative Extension Service, Michigan State University, East Lansing, MI.

Bailey, F. (1990) *Emergency Action July 1989 Soybeans. The Story Behind the Action.* Chicago Board of Trade, Chicago, IL.

Bain, J.S. (1968) *Industrial Organization*, 2nd edn. John Wiley and Sons, New York.

Barnard, C.I. (1938) *The Functions of the Executive*. Harvard University Press, Cambridge, MA.

Bresnahan, T.F. (1989) Empirical studies of industries with market power. In: Schmalensee, R. and Willig, R.D. (eds), *Handbook of Industrial Organization*, Vol. II. Elsevier Science Publishers B.V., Amsterdam, pp. 1011–1057.

Canadian Wheat Board, Annual Report (1992–93) *The Winds of Change, Challenge and Opportunity*. Canadian Wheat Board, Winnipeg, Manitoba.

Catania, P.J., Keefer, N. and Andrews, B. (1987) *Stock Index Futures*. Chicago Board of Trade, Chicago, IL.

Chicago Board of Trade (1985) *Options on Agricultural Futures*. Chicago Board of Trade, Chicago, IL.

Clark, J.M. (1940) Toward a concept of workable competition. *American Economic Review* 30, 241–256.

Connor, J.M. (1984) Multinational firms in the world food marketing system. *The Farm and Food System in Transition, Emerging Policy Issues*, FS29. Cooperative Extension Service, Michigan State University, East Lansing, MI.

Connor, J.M. (1994) *Time Series Data on US Foreign Direct Investment*. SP94–1. Department of Agricultural Economics, Purdue University, West Lafayette, IN.

Connor, J.M. and Peterson, E.B. (1991) *Market-Structure Determinants of National Brand-Private Label Price Differences of Manufactured Food Products*. SP91–08. Department of Agricultural Economics, Purdue University, West Lafayette, IN.

Connor, J.M., Rogers, R.T., Marion, B.W. and Mueller, W.F. (1985) *The Food Manufacturing Industries: Structure, Strategies, Performance, and Policies.* Lexington Books, D.C. Heath and Company, Lexington, MA and Toronto.

Cotterill, R.W. (1984) Retail cooperatives in the food system. *The Farm and Food System in Transition, Emerging Policy Issues*, FS28. Cooperative Extension Service, Michigan State University, East Lansing, MI.

Cyert, R.M. and March, J.G. (1963) *A Behavioral Theory of the Firm*. Prentice-Hall, Inc., Englewood Cliffs, NJ.

Demsetz, H. (1974) Two systems of belief about monopoly. In: Goldschmid, H.J., Mann, H.M. and Weston, J.F. (eds), *Industrial Concentration: The New Learning.* Little Brown, Boston, MA, pp. 164–184.

Farris, P.L. (1973a) Market growth and concentration change in US manufacturing industries. *The Antitrust Bulletin* 18, 291–305.

Farris, P.L. (1973b) Changes in the number and size distribution of US soybean processing firms. *American Journal of Agricultural Economics* 55, 495–499.

Farris, P.L. (1983) Agricultural marketing research in perspective. In: Farris, P.L. (ed.), *Future Frontiers in Agricultural Marketing Research*. Iowa State University Press, Ames, IA.

Farris, P.L. (1985) Concentration policy for the farm and food system. *The Farm and Food System in Transition, Emerging Policy Issues*, FS45. Cooperative Extension Service, Michigan State University, East Lansing, MI.

Fellner, W.J. (1949) *Competition Among the Few*. Alfred A. Knopf, New York.

Franklin, A.W. and Cotterill, R.W. (1993) *An Analysis of Local Market Concentration Levels and Trends in the US Grocery Retailing Industry*. Food Marketing Policy Center Research Report No. 19. Department of Agricultural and Resource Economics, University of Connecticut, Storrs, CT.

Galbraith, J.K. (1985) *The New Industrial State*, 4th edn. Houghton Mifflin Co., Boston, MA.

Gordon, R.J. (1981) Output fluctuations and gradual price adjustment. *Journal of Economic Literature* 19, 493–530.

Greer, D.E. (1984) *Industrial Organization and Public Policy*, 2nd edn. Macmillan Publishing Co., New York.

Kaufman, P.R., Newton, D.J. and Handy, C.R. (1993) *Grocery Retailing in Metropolitan Areas, 1954–82*. Technical Bulletin 1817. US Department of Agriculture, Washington, DC.

Kovacic, W.E. and Smallwood, D.E. (1994) Competition policy, rivalries, and defense industry consolidation. *Journal of Economic Perspectives* 8(4), 91–110.

Leibenstein, H. (1966) Allocative efficiency vs. X-efficiency. *American Economic Review* 56, 392–415.

Leuthold, R.M. and Campbell, G.R. (1984) Commodity futures markets and food system performance. *Farm and Food System in Transition, Emerging Policy Issues*, FS18. Cooperative Extension Service, Michigan State University, East Lansing, MI.

Mansfield, E. (1981) Composition of R and D expenditures: Relation to size of firm, concentration and innovative output. *Review of Economics and Statistics* 63, 610–615.

Marion, B.W. and Geithman, F.E. (1995) Concentration-price relations in regional fed cattle markets. *Review of Industrial Organization* 10, 1–19.

Marion, B.W. and NC117 Committee (1986) *The Organization and Performance of the US Food System*. Lexington Books, D.C. Heath and Company, Lexington, MA and Toronto.

Marion, B.W., Mueller, W.F., Cotterill, R.S., Geithman, F.E. and Schmelzer, J.R. (1979) *The Food Retailing Industry Market Structure, Profits and Prices*. Praeger Publishers, Praeger Special Studies, New York.

Mueller, W.F. (1985) Large conglomerate corporations in the food system. *The Farm and Food System in Transition, Emerging Policy Issues*, FS44. Cooperative Extension Service, Michigan State University, East Lansing, MI.

Mueller, W.F., Culbertson, J. and Peckham, B. (1982) *Market Structure and Technological Progress in the Food Manufacturing Industries*. NC117 Monograph 11. University of Wisconsin, Madison, WI.

Norman, G. and La Manna, M. (eds) (1992) *The New Industrial Economics, Recent Developments in Industrial Organization and Game Theory*. Billings and Sons, Ltd., Worcester.

Parker, R.C. and Connor, J.M. (1979) Estimates of consumer loss due to monopoly in the US food-manufacturing industries. *American Journal of Agricultural Economics* 61, 626–639.

Patterson, P.M. and Abbott, P.C. (1994) Further evidence on competition in the US grain export trade. *The Journal of Industrial Economics* 42, 429–437.

Patterson, P.M., Abbott, P.C. and Stiegert, K.W. (1996) The Export Enhancement Program's influence on firm-level competition in international markets. *Journal of Agricultural and Resource Economics* 21, 56–67.

Rogers, R.T. and Petraglia, L.M. (1994) Agricultural cooperatives and market performance in food manufacturing. *Journal of Agricultural Cooperation* 9, 1–12.

Scherer, F.M. and Ross, D. (1990) *Industrial Market Structure and Economic Performance*, 3rd edn. Houghton Mifflin Company, Boston, MA.

Shepherd, W.G. (1983) Causes of increased competition in the US economy, 1939–1980. *Review of Economics and Statistics* 64, 613–626.

Simon, H.A. (1958) *Administrative Behavior*. The Macmillan Company, New York.

Torgerson, R.E. and Ingalsbe, G. (1984) The future of farmer cooperatives. *The Farm and Food System in Transition, Emerging Policy Issues*, FS27. Cooperative Extension Service, Michigan State University, East Lansing, MI.

United States Department of Commerce, Bureau of the Census (1996) *Concentration Ratios in Manufacturing*. 1992 Census of Manufacturers MC92-S-2. US Government Printing Office, Washington, DC.

United States General Accounting Office (1990) *Beef Industry Packing Market Concentration and Cattle Prices*. GAO/RCED-91–28. US General Accounting Office, Gaithersburg, MD.

Von Neumann, J. and Morgenstern, O. (1944) *Theory of Games and Economic Behavior*. Princeton University Press, Princeton, NJ.

Wills, R.L. (1985) Evaluating price enhancement by processing cooperatives. *American Journal of Agricultural Economics* 67, 183–192.

8 Marketing Margins in Food Products

Julián Briz and Isabel de Felipe

Unidad de Comercialización y Divulgación Agraria, Departamento de Economía y Ciencias Sociales Agrarias, ETS Ingenieros Agrónomos, Universidad Politécnica de Madrid, Ciudad Universitaria, 28040 Madrid, Spain

INTRODUCTION

This chapter deals with one of the more contentious topics in agro-food marketing. For many years, social and political institutions have been concerned about food price behaviour, and, as a consequence, about the behaviour of marketing margins, which have been used as an index of economic efficiency, bargaining power and market transparency, among other things.

The term 'marketing margin' refers to the part of the consumer's food expenditure which is absorbed by the food marketing sector. Thus, marketing costs are distinguished from production costs. The point of division is typically defined as the 'farm gate'. Expenses occurring before the farm-gate are productions costs, while expenses occurring beyond the farm-gate are marketing costs. A descriptive analysis of this type of data provides a sense of the relative importance of the different cost components of marketing functions. It provides insight into the operation of the food system at a point in time, as well as through time.

An increase in the marketing margin may result from more marketing services. In addition, increasing efficiency in farm production will be reflected by a proportionate increase in the marketing margin. In contrast, policies to protect the farm sector from international competition may increase the cost of agricultural production and thus the 'farmer's share', relative to marketing margins. Marketing margin analysis relates to these policies and gives some indication of the relative performance of the farm sector compared with the typically more concentrated food marketing sector.

The higher bargaining power of economic actors in the food marketing sector may disadvantage farmers by downward price pressure. In the future, increasing price spreads may be accompanied by lower farm price support in developed countries, following the Uruguay Round Agreement in the GATT (General Agreement on Tarriffs and Trade). Farmers and food

© CAB INTERNATIONAL 1997. *Agro-food Marketing*
(eds D.I. Padberg, C. Ritson and L.M. Albisu)

firms will be attempting to increase, or at least maintain, their share of the final value of food products in the increasingly competitive environment. Under a market strategic policy of lower prices, traders try to increase volume of sales rather than sale price, in the context of the whole enterprise and not product by product. Thus we may expect some products priced below cost, with negative marketing margins offset by other products with high margins.

Nowadays, studies on price spreads are a common topic, not only within the academic fora, but also with government agencies, and business and labour union organizations. Special attention needs to be paid to the methodology used because this can have a significant effect on the results obtained. The problem is, of course, that many institutions try to use such studies as arguments in favour of their position and against others. Therefore, objective studies are necessary, avoiding bias and confusion. In this chapter we try to give a general view under a theoretical and practical scenario. After the description of basic ideas, we will deal with methodology issues. This is followed by comparisons of marketing margins between countries and, finally, by some general comments.

HISTORICAL EVOLUTION

Although most countries had been developing instruments to evaluate Food Marketing Margins (FMMs), the first systematic approach was undertaken in the USA in 1921, trying to understand the origins of the differences between prices at consumer and farmer levels. In 1935 the Bureau of Agriculture Economics published the FMM for ten agricultural products. During the decade of the 1950s more similar studies followed (Briz, 1985).

In the European Community, different studies were undertaken by the Statistics Bureau and the Consumer Protection Service, among others, trying to understand price evolution in different member states. Some international institutions, such as the OECD (1978), established specific groups to investigate methodological support for FMM analysis.

A great deal of data is available on marketing margins for some economies but little for others. This information allows us to identify some general patterns. Probably the most robust generalization is that the food marketing margin gradually increases as income increases. Since income has increased in most economies during the latter part of the twentieth century, it may be expected that marketing margins will have increased as well. Table 8.1 shows marketing margins in the USA for the past quarter of a century. During this period, they have increased from about two-thirds of consumer expenditures to more than three-quarters. This growth has not however been a smooth one and, in particular, during the mid-1970s the farmers' share rose, mainly because of high world commodity prices. Thus these changes are the result of several influences, and it is useful to trace some of the developments within the USA food system, which may have contributed to the evolution of FMMs in addition to the influence of rising consumer incomes.

Table 8.1. Marketing margins and farm share of food expenditure in the USA, 1968–1992 (source: USDA, various issues).

Year	Total marketing bill US$ (millions)	Farm value US$ (billions)	Civilian expenditures for food US$ (billions)	Marketing margin (%)	Farm share (%)
1968	65.9	30.9	86.8	68.1	31.9
1969	68.3	34.3	102.6	66.6	33.4
1970	75.1	35.5	110.6	67.9	32.1
1971	78.5	36.2	114.7	68.4	31.6
1972	82.4	39.8	122.2	67.4	32.6
1973	87.1	51.7	138.8	62.8	37.2
1974	98.2	56.4	154.6	63.5	36.5
1975	111.4	55.6	167.0	66.7	33.3
1976	125.0	58.3	183.3	68.2	31.8
1977	132.7	58.2	190.9	69.5	30.5
1978	147.1	68.9	216.0	68.1	31.9
1979	162.8	78.4	241.2	67.5	32.5
1980	179.7	81.1	260.8	68.9	31.1
1981	202.1	82.4	284.5	71.0	29.0
1982	214.1	83.5	297.6	71.9	28.1
1983	229.7	85.3	315.0	72.9	27.1
1984	242.2	89.8	332.0	73.0	27.0
1985	259.0	86.4	345.4	75.0	25.1
1986	270.8	88.8	359.6	75.3	24.7
1987	285.1	90.4	375.5	75.0	25.0
1988	301.9	96.8	398.8	75.7	24.3
1989	315.6	103.8	419.4	75.3	24.7
1990	343.6	106.2	449.8	76.4	23.6
1991	363.5	101.6	465.1	78.2	21.8
1992	371.5	105.3	476.8	77.9	22.1

Farm production went through major adjustments during the 1970s. As we saw in Chapter 5, high world commodity prices, and associated high farm incomes, brought a wave of investment in agriculture production throughout the world. In contrast, in the 1980s, international commodity markets became over-burdened, and both commodity prices and farm incomes were low. However, during both of these adjustments, farms became fewer and larger. Yield increasing and cost reducing technology were introduced.

While there is no way to net out these factors on the production side, it is likely that they contributed to a lowering of the farmer's share and therefore to a relative increase in marketing margins.

Food manufacturing involves quite different functions compared to food distribution (wholesaling, retailing and catering). A great part of value added in food manufacturing is associated with activities in processing plants. These operations are automatable to a much greater degree than with food distribution. However, another major activity in food manufacturing is new product development. It is nevertheless likely that the efficiencies gained from automation offset much of the higher cost of new product development, leaving the activities in food manufacturing contributing little to the increasing marketing margin, implying that it is food distribution which is the major recipient of the growth in FMM.

BASIC CONCEPTS

There are two ways of defining marketing margins which are also referred to as vertical 'price-spreads' or 'mark-ups' (Tomek and Robinson, 1982):

1. The price difference between two marketing stages (either consumer, wholesale, processor or farmer).
2. The cost of the services provided along the marketing channel. These are associated with the so-called marketing of time, space, and transformation of ownership of the product (Caldentey *et al.*, 1993).

We shall now explore some of the concepts and definitions most frequently used.

Absolute marketing margin (AMM)

This is the gap between prices at different marketing levels (farmers, wholesalers, retailers). Thus, $M_1 = P_R - P_F$; $M_2 = P_R - P_W$; $M_3 = P_W - P_F$; where M_1, M_2 and M_3 are Absolute Marketing Margins (AMMs) at different levels, and P_F, P_W and P_R are prices at farmer, wholesale and retail levels.

Prices should refer to a unit of product at the upper (P_R) level, with an equivalent quantity at the lower levels for comparison (P_W and P_F). This

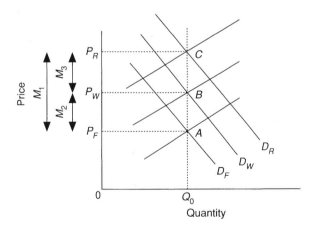

Fig. 8.1. Food margins and prices.

type of margin is commonly used by retailers in many countries. The AMM may be identified by the vertical distance between the equilibrium markets, or supply and demand curves (Fig. 8.1), at different marketing stages (farmer, wholesaler, retailer).

Thus we have a primary demand at retail (D_R) and derived demands at wholesale (D_W) and farmer (D_F) levels. Similarly we may identify primary supply (farmer supply), S_F, and derived supplies at wholesale (S_W) and retail (S_R) levels.

Relative marketing margin (RMM)

Usually, when we refer to the RMM we mean the ratio RMM = AMM$/P_R$, where P_R is the price paid when the product is bought. For instance, the RMM from the farmer to retailer will be RMM$_{FR}$ = M_1/P_R in percentage terms. The use of percentage (or RMM) marketing margin is a frequent practice by wholesalers, adding a given percentage to what they themselves pay for the produce.

Gross marketing margin (GMM)

We obtain the value concept by multiplying the AMM by the quantity marketed, assuming quantity is similar at both price levels. It is represented by the area P_FACP_R in Fig. 8.1. In marketing activities many enterprises use this concept. Formally, the GMM is defined as: GMM = $Q_0(P_R - P_F)$, where GMM = gross marketing margin, P_R = sale price at retail level, P_F = sale price at farmer level and Q_0 = quantity marketed.

Net marketing margin (NMM) and marketing charges (MCH)

Here, the concept takes account of fixed costs, taxes and subsidies. That is: NMM = GMM − FC − T + S, where FC = fixed costs, T = taxes and S = subsidies. Marketing charges refers to the returns to institutions involved in marketing activities. It is the payment of the services of retailers, wholesalers and processors. It is obtained simply by adding subsidies and subtracting taxes from the absolute marketing margin of the business concerned.

Marketing costs

Marketing costs are the return to factors in the marketing process: profits, wages, interest and rents. During the past quarter of a century, farm production costs have progressively become a smaller part of the total cost of food, while marketing costs have increased. Many changes in the food system have contributed to this. First, production is much more specialized in location.

Producing commodities in the location best suited to productive efficiency reduces production costs and may increase marketing (transportation) costs. Tripling the number of food products available to consumers gives more choice and perhaps more satisfaction, but also increases warehousing and distribution costs. The increasing trend to food taken away from home increases food marketing cost. As labour becomes more expensive and labour-saving technology more available, some functions will become more expensive while others will become less expensive. All of these changes have been occurring. A look at the components of the marketing margin may give some indication of the total effect of all the aggregate change. It is, however, not possible to isolate many of the particular observable effects.

The components of the marketing margin in the USA are shown in Fig. 8.2 for 1968 and 1994.

Fixed or flexible margins

Which type of marketing margin should be adopted by entrepreneurs or imposed by government? There are different options and each has advantages and disadvantages. In a very schematic way, three different systems may be considered:

1. *Fixed margins*. Food margins are fixed by an administrative organization at every marketing stage (farmer–wholesaler–retailer). The main advantage is the simplicity to administer and the transparency for all the economic operators. However, it is difficult to control, there is a lack of incentive to invest in new equipment and services, and there is no incentive for an efficient allocation of resources. Finally, it creates discrimination against marginal areas and stimulates parallel markets.

2. *Differentiated margins*. According to geographic areas, seasons of the years, quantity and quality of the products, etc. This gives more flexibility to the system and fosters incentives to special services (transport, storage) marketing channels, farmers in marginal regions, etc. On the other hand, it may be difficult to administer, and be very expensive to control.

3. *Free regulation of margins*. Although there may be some minimum policy price at farmer level, there are no controls on the price spread. Supply and demand establish the price spread. Theoretically there is an optimal allocation of scarce resources. However, as discussed in the previous chapter, special attention may have to be paid to the bargaining power of traders and processors relative to small farmers and consumers.

METHODOLOGICAL ANALYSIS

The study of food marketing margins confronts several problems, particularly in conditions of high inflation or when there is significant variation in the relative price of different products. The criteria for selecting the methodology will be constrained by various factors, for example:

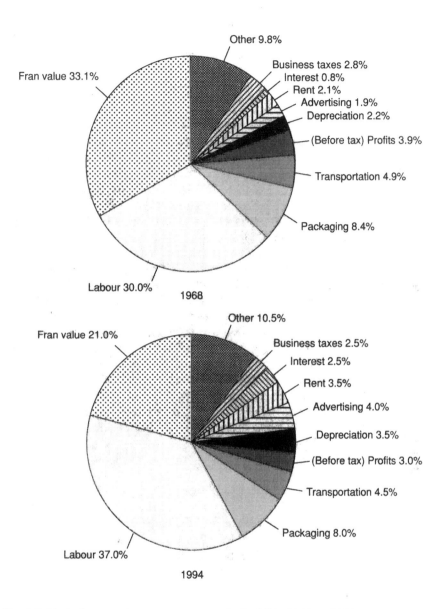

Fig. 8.2. Marketing margin cost components in the USA, 1968 and 1994 (source: Elitzak, 1996).

1. The availability of statistical data and technical or financial support to carry out the study. This is an important point, especially in developing countries.

2. The characteristics of the market products to be analysed, e.g. perishability, seasonality, cycles and trends.

3. The nature of marketing channels – traditional or new distribution systems; number of middlemen, etc.

4. The objectives of the analysis – type of margin to study (absolute, relative, value added, etc.).

Descriptive models

In general, when adopting methodologies to describe food marketing margins, there are two approaches, macro- and microeconomic.

The *macroeconomic approach* aims to identify the role of marketing margins in the general economy. There are two ways of doing this. First, by using National Accounting, we may study the sectors of the country – government, enterprises, households, etc. However, because of the great aggregation used in the system, the usefulness of this method is low for evaluating marketing margins.

Alternatively, Input–Output (I–O) tables have the advantage of focusing the attention on different economic categories. The production system is detailed in the form of, for example, goods and services, final consumption, exports and investment. Thus, I–O tables provide basic information for research on price spreads. However, they face some restrictions because of problems of aggregation due to the diversity of products and enterprises, considerable delays in obtaining data in countries, the assumption of linear production functions and the hypothesis of uniform cost increase along the marketing channel, among other things.

Therefore, once more, we should be cautious in adopting the macro-approach. In contrast, the *microeconomic approach* can give information about the components of the costs, structure and financial situation of enterprises involved in the marketing process. The feasibility of this method depends upon several conditions: (i) the period of analysis, and the possibility of identifying the various elements in the short-, medium- and long-run; (ii) the degree of specialization in the enterprise and the different functions performed; and (iii) the frequency of economic changes along the marketing channel; the sequence, and delay, of changes may influence costs and prices and make more difficult their correct identification.

There are two main methods of microeconomic analysis:

1. The *functional method* tries to analyse the different economic activities associated with each enterprise. Studies of farming, wholesale and retail markets should be combined with behaviour of different factors with significant influence on marketing performance. In order to pursue this method, some countries (e.g. Germany) have selected a sample of enterprises, which provide information about economic activities, products, number of employees, structural costs, subsidies, taxes, benefits and so on. With this method it is possible to analyse the economic or social efficiency of one specific industry. Through historical and cross-section studies, it may be possible to evaluate the effect of public policies, or collusive actions of some entrepreneurs, on the structure and behaviour of the industry. The method would be recommended during periods of serious economic conditions, when society presses for identification of the origins and causes of perceived problems, such as shortages or rapid increases in prices.

2. The *vertical, or product focused, method* is based on the simple idea of following the product through the commercial process, noting prices, quantities, costs, etc. on each transaction. However, the concept of 'product' is not always clear, and frequently it refers to a group of similar products with analogous utilities. For instance, when we talk about meat, we are dealing with a heterogeneous group (beef, pork, lamb, poultry) and different degrees of quality. The method should allow us to identify a unit, or quantity, of product whose price will change along the marketing channel. The problem arises when there is significant transformation, and it is necessary to determine the conversion factors (between raw and transformed products) and when there are by-products. We should pay attention when by-products have a significant value and their prices are independent of the main product. There are many examples of 'main and by-products' with significant relations: soybean, soymeal and soyoil; carcass and hide, etc. Taxes and subsidies may complicate the situation.

Following the methodology developed by the US Department of Agriculture (USDA), Table 8.2 shows how the marketing margins can be calculated when there are by-products.

Table 8.2. Marketing margins: case of principal and by-products analysis.

		Marketing stage	
Concept	Farmer	Wholesale	Consumer (retail)
Quantity*	Q_F	Q_{W1} Q_{W2}	$Q_{R11}, Q_{R12}, Q_{R13}, \cdots$ $Q_{R21}, Q_{R22}, Q_{R23}, \cdots$
Price†	P_F	P_{W1} P_{W2}	$P_{R11}, P_{R12}, P_{R13}, \cdots$ $P_{R21}, P_{R22}, P_{R23}, \cdots$
Value	$V_F = Q_F \times P_F$	$V_{W1} = Q_{W1} \times P_{W1}$ $V_W = \Sigma V_{Wi}$	$V_{R1} = \Sigma Q_{R1i} P_{R1i}$ $V_R = \Sigma V_{Rj}$
Coefficient of transformation‡			
Principal product at wholesale		$a_i = V_{Wi}/V_W$	
Principal product at retail			$b_j = V_{Rj}/V_R$
Absolute Marketing Margin (AMM)			
Farmer to wholesale (principal product)		$\text{AMM}_{FWi} = (V_W - V_F) a_i$	
Wholesale to consumer (principal product)			$\text{AMM}_{WRj} = (V_R - V_W) b_j$

* Q_F = quantity sold at farm level (e.g. liveweight pig sold by the farmer); Q_W = quantity sold at wholesale, here we may introduce different categories at slaughter level (Q_{W1} for carcass and Q_{W2} for by-products to simplify); Q_R = quantity sold at consumer (or retail) level, in this situation we may find a great diversity of products offered to consumer.

† P_F = price at farmer level (received by farmer), in animal products this will be liveweight; P_W = price at wholesale level, in cattle this will be the carcass and by-products prices; P_R = price at consumer (or retail) level.

‡ Standard coefficient for a sector.

Table 8.3. Numerical example of some marketing concepts.

Concept	Product
Selling price (US$ per unit)	2.25
Buying price (US$ per unit)	1.90
Quantity marketed (kg)	560
Absolute Marketing Margin (AMM) (US$ per unit)	0.35
Relative Marketing Margin (RMM)	
In relation to buying price (%)	0.35/1.90 = 0.18
In relation to selling price (%)	0.35/2.25 = 0.15
Gross Marketing Margin (GMM) (US$ per unit)	0.35 × 560 = 196
Taxes (US$ per unit: 0.16)	0.16 × 196 = 31.3
Subsidies (US$ per unit: 0.08)	0.08 × 196 = 15.7
Marketing Charges (MCH) (US$ per unit)	196 − 15.7 + 31.3 = 180.4
Fixed cost (US$ per unit: 0.43)	0.43 × 196 = 84.3
Net Marketing Margin (NMM) (US$ per unit)	180.4 − 84.3 = 96.1
Gross Market Coefficient	196/(2.25 × 560) = 0.15

Comparing prices at different stages

Availability of data often makes it possible to compare prices at different stages. Some corrections may be required due to time-lags in price formation at different levels. Also, there will be problems when we try to compare products with a significant transformation process along the marketing channel.

As a practical example we give in Table 8.3 some hypothetical data for a product, with no transformation process, and maintaining the same quantity marketed.

An extension of comparing prices is to use indices when we aggregate products or services. For instance, we may try to compare the Consumer Price Index (CPI) with the Farmer Price Index (FPI). We need to look for the equivalents on quantities, prices and characteristics of the products at farmer and consumer levels. With all of them we establish 'equivalent indices' and evaluate the 'marketing margin index'.

Evaluation of farmer's share of consumer food expenditure

This is of interest when we wish to evaluate the relative position of farmers in the total economy comparing gross national agricultural value with the total food consumer expenditures. Attention should be paid to foreign trade, in order to calculate its incidence in final food consumption. In general, the participation of farmers in consumer expenditures may be calculated using the formula:

$$FS = \frac{(GAV - BPV)}{CE}$$

where FS = farmer share (in %), GAV = gross agricultural value at farmer level, BPV = by-products value and CE = consumer expenditure.

Algebraic analysis

Empirical research (e.g. George and King, 1971) has frequently shown the marketing margin as a combination of constant and variable factors. As a first approach there are three alternative hypotheses:

1. The margin as function of retail price: $M = a + bP_R$, where M = absolute marketing margin, P_R = retail price and a,b = constant factors.

2. The margin as function of quantity marketed: $M = c + dQ$, where M = absolute marketing margin, Q = quantity marketed on the transaction and c,d = constant factors.

3. The margin as a constant percentage, either of farm or retail price: $M = kP$, where M = absolute marketing margin, P = price and k = constant.

Economic agents use different practices, according to traditions or opportunities. Thus, many wholesalers use a constant percentage (hypothesis 3), and retailers use the absolute margin (hypothesis 1) (Dalrymple, 1961).

The market reaction to prices is measured through supply and demand elasticities. In the same way we can relate them to marketing margins.

The price elasticity of demand:

$$e = \frac{\dfrac{dQ}{Q}}{\dfrac{dP}{P}}$$

analyses the reaction of quantity to price variations. With perishable products, the market depends more directly on quantity as the independent variable, and the concept used is price flexibility $f = 1/e$.

Considering that $P_F = P_R - M_1$, with a simple algebraic operation we arrive at: $f_F = f_R(1 + M_1/P_F)$, where f_F = flexibility at farmer level, f_R = flexibility at retail level, M_1 = absolute marketing margin and P_F = price at farmer level. If $M_1 > 0$, then $f_F > f_R$.

Therefore, as margins usually are positive, consequently price instability at farmer level is greater than at retail market (Houck, 1966). Perhaps another aspect to consider of interest is the margin evolution over the long- and short-run. In the long-run, margins tend to increase, mainly due to greater importance of marketing services, and the farmer share in consumer expenditures diminishes.

In the short-run, some empirical work (George and King, 1971) using quarterly data identified several relationships; considering marketing margin (M) as a linear function of retail price (P_R) and assuming that variations are not the same for all seasons (Fig. 8.3):

1. $M = a + bP_R$. Constant terms (a,b) may be interpreted as no variations between seasons (Fig. 8.3, case I).

2. $M = a_1 + a_2 + a_3 + \ldots + a_i + bP_R = \Sigma a_i + bP_R$. The constants a_i correspond to the different seasons, using dummy variables, with the same slope (Fig. 8.3, case II). Dummy variables (value 1 for one season and zero for others) may be

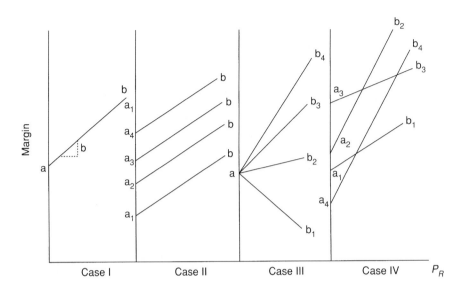

Fig. 8.3. Relationship marketing margin/retail price with seasonal variations.

interpreted as seasonal effect associated with the margin for that specific season, where factors not related to retail price (P_R) change.

3. $M = a + \Sigma\ b_j P_R$. In this case (Fig. 8.3, case III), only the slopes b_j, as dummy variables are affected by seasonality.

4. $M = \Sigma\ a_i + \Sigma\ b_j P_R$. In this general case (Fig. 8.3 case IV), the seasonability influences both groups of elements a_i and b_j.

Static and dynamic models of marketing margins

Marketing margins, by definition, depend upon price evolution at different market levels. In order to understand their performance and foresee their evolution, several authors have developed different models from both the static and dynamic points of view. Gardner (1975) studied the agro-food sector on static hypotheses, considering two groups of inputs: farm and non-farm, with one output at retail level. He analysed the impact on margins of changes in agricultural supply, food consumption or input markets. However, the hypothesis of static equilibrium at different market stages is very restrictive. For that reason, other authors (e.g. Heien, 1980) have used dynamic models for the agro-food sector, considering that market instability is a significant element in price formation, and thus of significance for margins. More recently some authors (e.g. Buzby *et al.*, 1994) have used margin theory and Gardner's model to estimate the impact of banning postharvest pesticide use in the fresh grapefruit industry.

Other empirical work has demonstrated the interaction of price and marketing margin variation through the market stages (producer–wholesaler–retailer), trying to determine the extent of correlation between changes in farm supply, consumer demand, and price spread. Some of the more significant conclusions are: (i) an increase (decrease) in consumer demand usually diminishes (increases) percentage margins; (ii) if farm product supply increases (decreases), this usually produces a decrease (increase) in percentage margins; (iii) government intervention in agricultural markets usually has significant incidence on margin performance; and (iv) market liberalization may diminish marketing margins for some specific products; however, higher value added and more marketing services may offset this.

MARKET FACTORS AND MARKETING MARGINS

Marketing analysis is often based on partial-equilibrium theory (Tomek and Robinson, 1982). Only one factor-margin is considered variable under the *'ceteris paribus'* condition (the other factors being assumed to remain constant). With this hypothesis we may study the influence of changing primary supply, demand or service innovation on food margins.

However, real life is of course more complex. Farmers, middlemen and retailers face political, economic and social actions that interact with marketing strategies. For example, orange producers in Spain pay attention, not only to wholesalers and retailers in the domestic market, but also to European Union regulations on intervention prices, the impact of the GATT Agreement on subsidies or market access, the substitutions effects of other products, and so on.

Interdependence among prices, marketing margins and inflation have been studied trying to identify the role of the different economic factors. Frequently, the performance in one specific industry is a consequence of quantitative (prices, costs, charges) and qualitative (improvement on services, quality, presentation) elements. Upward trends in prices and margins may increase inflation, create economic problems (inflation, unemployment) and have political consequences. Focusing the attention on some of these market factors (Dahl and Hammond, 1977) identified the following:

1. *Marketing input prices.* Labour is an important item, with usually an upward trend. In addition, energy, financial or equipment costs may force margins to change.
2. *Services and value of the food process.* The increasing importance of marketing services affects margins. More elaborate products, better packaging, diversity of product quality, brands, have altered marketing conditions and costs.
3. *Technological changes.* New technologies may increase costs when enterprises have to substitute for the old ones. However, very often in the medium-term, economies of scale and other improvements may provoke lower costs. This is the case, for example, with electronic inventory control,

product identification at retail level with optical instruments, transportation and storage improvement (containers and pallets). Sometimes the cost reduction may offset other increases and we end up with lower marketing margins.

4. *Marketing mix strategies.* The introduction of new products with advertising and promotion activities may increase demand. New shopping centres and more aggressive selling activities may increase competition and reduce margins.

5. *Administrative regulations.* Establishing standard regulation for quality may allow the consumer a better identification of the product, facilitate substitution effects and stimulate competition through prices. Similarly regulations of prices or buying or selling actions by government agencies can influence margins in various ways.

6. *Structural organizations.* Types of marketing channels (traditional, self-service), and market behaviour are other elements to consider. Monopoly and oligopoly formation may artificially increase prices and margins. However, some administrative controlled monopolies may have a social goal, with lower (or even negative) margins. As a rule, the more dynamic the structural organization and more competitive the markets, the lower the margins.

COMPARATIVE ANALYSIS BY PRODUCTS AND COUNTRIES

Marketing margins differ according to products, the market conditions they face, and the marketing services provided along the distribution channel. As we have seen, the degree of transformation will influence the methodology applied. Raw products not subject to substantial transformation from farmer to consumer (e.g. fruit and vegetables) facilitate the analysis of the farm–retail price spread. Special attention should be paid to time-lags among the different marketing stages.

Transformed products are more difficult to analyse, for example, wheat transformation into bread, grapes into wine, etc. Meat products are more complex – live animals compared with the carcass and then different cuts at the butcher's. A transformation coefficient has to be used to calculate the equivalence per unit of product at consumer level. This allows a comparison of one kilogram of bread and the kilogram of wheat necessary to produce it.

It is interesting to compare the situation in different countries in order to evaluate marketing efficiency. However, there will be problems (Hammond, 1990), because many of the factors we have considered above, e.g. local regulations, government interventions (subsidies, taxes), activities of private or official institutions, etc., will differ as well as marketing costs. In the next section we discuss changes which have been taking place in the components of marketing margins in developed countries – particularly the USA, for which the best data is available – and this is followed by a short section outlining a few aspects of marketing margins in less developed countries.

Table 8.4. Evolution of some significant indices in the USA (1982–1984 = 100) (source: USDA, 1992).

Year	Retail price	Farm value	Farm to retail spread	Farm value share of retail price (%)
1950	30	40	25	47
1990	134	106	154	30

Marketing margins in developed countries

The trend of marketing margins in developed countries is positive, with the farmer's share of consumer expenditure decreasing. Take for example two countries which differ markedly in size and economic organization. In 1950 the price spread from farmer to consumer in food products was 34% in Switzerland and 53% in the USA. In 1989, the figures were 51.5% in the first and 71% in the second. In both the increase has been significant, particularly in the USA (Senti, 1990).

In many countries price spreads have widened because farm prices have remained at approximately similar levels in real terms, but retail prices continue to increase, in spite of intense retail competition, due to higher processing and marketing charges. However, in countries like Japan, small margins at wholesale level led to economic difficulties (Tsukano, 1992). Sometimes fresh fruit and vegetables with great seasonality result in high marketing margins, due to the lack of transparency and poor competition, but exceptionally, in the Japanese market, processed food products' margins were in general greater than those for fresh products, reflecting the greater degree of value added.

Food expenditure in relative terms (as a percentage of personal disposable income) has been diminishing in all developed countries during the last decade. (In the USA, it was only 11.6% in 1991, according to the Economic Research Service of the USDA.)

For historical and comparative analysis we show in Table 8.4 and Fig. 8.4 the evolution of indices in the USA (1982–1984 = 100), and follow with a

Table 8.5. Evolution of marketing function components of consumer expenditure in the USA (billion dollars) (source: Dunham, 1992).

Expenditures and components	1981	1990
Expenditures at foodstores	194.0	276.2
Farm value	65.4	80.2
Marketing bill	128.6	196.0
Processing	60.1	87.4
Intercity transport	11.6	15.0
Wholesaling	17.7	28.5
Retailing	39.2	65.1

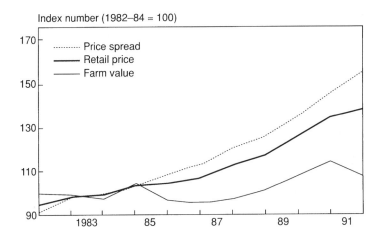

Fig. 8.4. Food price components in USA (source: USDA, various issues).

general discussion of the changing value of the components of marketing margins in the USA (see Table 8.5).

Food distribution (wholesaling, retailing and food services) has experienced a great transformation during the past quarter of a century. The size of store which provides the food we consume at home is much larger and the number of items handled has increased. This arrangement gives consumers greater choice, but it has caused most measures of efficiency to decline. The retail gross margin is up probably three to four percentage points. Measures of physical efficiency at the wholesale level also show decline. The number of employees in wholesaling has increased by 67% and in retailing by 102%.

The other part of food distribution relates to food eaten away from the home. This industry has grown from 32% of food expenditures in 1967 to over 45% in 1992 (USDA, 1994). The personal service nature of food service means that opportunities for automation are limited. The number of employees in food service has tripled during this period. While the so-called 'fast food' places have greatly increased in number, it is not as easy to get the very high sales volume needed for most efficient operation at each site. It is likely that a great part of the increased marketing margin is explained by increasing costs in distribution.

Within the general pattern of expanding marketing costs are three important exceptions – packaging, transportation and corporate profits. Each of these components have particular reasons for becoming a smaller part of food cost. It is likely that in each category, costs went up, but less than other marketing costs.

Better materials are available now and packaging processes are more automated. The issue of transport is of interest. It is likely that the 'typical' food item has more 'distance' in its history today than in 1968 in the USA.

Even with fuel cost increases, intercity transportation claims less of a share of food cost than many would expect. The low cost of this function may explain the very substantial increases in international movement of manufactured food products.

The slight decrease in corporate profits is also interesting. This has occurred during a time when the industrial structure has become substantially more concentrated at most levels. There has been a major consolidation of food manufacturing into multinational conglomerates. Concentration at a global level must have increased substantially.

In distribution, probably less change has taken place at the national or global level, but most communities have larger and fewer stores and fewer competing firms. There may be many contributing factors to explain the relative decline of corporate profits, but two observations are likely among the most important – the nature of competition among the multinational conglomerates and the tendency for distribution firms to compete on prices and costs. In the previous chapter, the implications for various kinds of market structure for agro-food marketing were considered in detail.

Economists associate highly concentrated markets with collusive behaviour. The tendency for competitors to 'cooperate' and behave like a monopoly is thought to increase as concentration increases. This argument leads to an expectation of output restriction and higher prices than would apply in a market characterized by a large number of small firms. The concept is usually applied most rigorously to commodities but frequently the implications are assumed to apply to differentiated products. The multinational conglomerates fit poorly into this scenario. The focus of their managerial strategy is competing in new product introductions. Their large size gives advantages in the development, physical distribution and introduction of new products to the public. This focus of activity is not conducive to collusion. An effective independent rivalry can be sustained at higher levels of concentration than is the case with commodities. It seems apparent that these firms are very large with substantial market power. Yet, it seems that there is an effective competitive interaction between and among them.

A specialized distribution system is one of the almost unique characteristics of the food industry. That condition leads to a pattern of economic behaviour which is less common in other industries and which influences the balance of market power in the distribution channel. The large food distribution chains (multiples) are major food manufacturers. In addition to the food they manufacture, they contract for manufacturing output from many other small food processors. Their brands are associated with the store – often called 'private' or 'own' label. They do not challenge the conglomerates in new products. Rather, they wait until successful products are identified. Then they produce copies of the successes and sell them at discounts of 20–30%. They control the logistics of these products from manufacture to shelf exposure. The efficiencies obtained gives them a significant cost advantage as compared with nationally advertised brands. They control the store merchandising activity and are able to favour their own products. Private label is profitable for the largest chains. It gives a 'competitive' flavour to the food market.

We should realize that farm value share varies among different foods. In general the more processed is the product, the smaller is the farm share. Foods derived from crops usually have lower farm value share than those derived from animal products, due to the lower incidence of farm inputs in the production process. Thus, in 1991 in the USA, farm value share for corn syrup was 4% versus 60% for eggs. When we try to understand the evolution of food retail prices, we should consider not only the farm prices, but also the charges for marketing services (assembling, storage, transportation and additional charges).

Another scenario of analysis relates to the prices of marketing inputs, which include wages and salaries, services and products that the industry acquired from outside. Most of the countries have indices to evaluate these items. In the USA the Economic Research Service (ERS) uses the Food Marketing Cost Index (FMCI) to try to evaluate changes in variable operating cost in wholesale, processing and retail activities. In fact the FMCI consists of hourly earning of workers and prices indices of different marketing inputs, weighted by the participation of each input in total operation cost. This index does not include changes in profits and productivity. In the USA during 1991, according to ERS, the largest component of the index was labour costs (45%), food containers and packaging materials (15%), transportation (11%), energy (8%) and some others (advertising, insurance, maintenance, etc.).

To evaluate profitability in the food industry, there are two financial ratios used in the USA: profit margin and return on stockholder equity. The profit margin (percentage of net income over sales) tries to explain the relative importance of profit. Return on stockholder compares the food industry with other industries in the country. In the USA in 1991, the after-tax profit margin of food and tobacco manufacturers averaged 4.9% of sales. Labour productivity in USA food manufacturing industries has seen some improvement over the years. However, food stores and eating places productivity has diminished during the last decade. Another important item to consider is food expenditure, usually reported by expenditure at grocery stores, eating places and institutions.

There is a general trend of increase of the marketing bill and maintenance of farm value, as mentioned earlier (Fig. 8.5). In developed countries, labour cost is usually the main component of the marketing bill. It is followed by packaging, transportation and others.

Marketing margins in developing countries

In Less Developed Countries (LDCs), it is now recognized that marketing systems need to be developed perhaps at even a higher rate than production, considering that some of the critical points of food problems are: (i) to serve the needs and interest of the farmers in order to stimulate production; and (ii) to supply food to the consumers at adequate prices, quantity and quality.

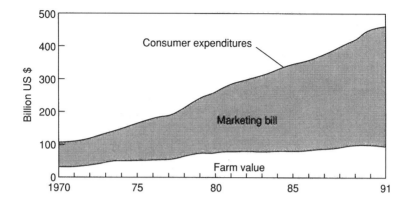

Fig. 8.5. Distribution of food expenditures in USA; the marketing bill was 78% of 1991 food expenditures (source: USDA, various issues).

In order to attain those goals, various measures should be taken, such as encouraging food production for commercial marketing, reducing losses in moving commodities along the marketing channel, increasing processing and preserving actions, and last, but not least, to supplement and improve distribution channels. It is sometimes argued that there is a waste in human resources along the marketing system of LDCs, with a large number of small retailers, market women, abundance of street markets, etc. However, the question is the lack of alternative jobs. A new marketing system should provide adequate opportunities for these people through the creation of additional manufacturers and services. For that, these LDCs need investment, either from national or foreign capital. Thus we may say that the price spread from producer to consumer is becoming more important for LDCs, with more demand for food supplies, rising incomes and migration of rural people to urban centres.

Marketing margins in developing countries may display special behaviour as a consequence of a different competitive situation. Some examples are referred to, but it would be a mistake to generalize from these.

Sometimes price fluctuations, irregularity of supply and lack of adequate variety and quality of products are the major problems. Very often marketing margins are quite high, due to high marketing costs. In India, the marketing margins in food grains reached 65% of the consumer price (Ghosh, 1988). In Egypt, the margin for red meat was 60% in spite of the low marketing services (Mohammed, 1991). But, as in other cases, strong competition among intermediaries can push down the marketing costs, with great varieties of service given to the consumer. For example, in the rice market, the farmers' share was around 72% of consumer price (Rahman *et al.*, 1991). The models of marketing margins behaviour in developing countries usually follow the general prototype. Thus, marketing margins for sunflower seed in Turkey were identified with an absolute and a percentage component (Taberler and Dawson, 1991).

Political power of inhabitants in the great urban areas can push food prices down. Under these circumstances food imports compete with domestic production and margins may be lower through a subsidy system or intervention of government agencies.

Evolution and trends

The general trend in most of the countries is an increase on the price spread from producer to consumer. The value added along the marketing process is increasing, with more service, higher quality, etc. Bargaining power has changed from farmers to processors, and from them to merchants. In these circumstances, distribution may get most of the increases in food margins. Only a competitive market may transfer those benefits to consumers through lower prices and higher quality and diversity of the products. Thus competition may reduce marketing margins. Even negative values are possible, either because of promotion programmes or because of an unexpected price evolution.

There is a relation between marketing margin behaviour and the market intervention system. In an interventionist economy, marketing margins are usually fixed at a maximum level by the authorities. Generally, food products are the ones under the most control, with governments trying to maintain consumer prices at low levels. In some cases the price spread is systematically negative or less than the costs, due to direct or indirect subsidies provided by the government.

There has been much debate over the rationale for these interventions. Arguments in favour of controlling prices artificially have been used in many countries, especially with weak economies and facing an uncertain developing process. In the short-term there may be an adequate economic policy when facing specific circumstances. However, it is difficult to control the whole of marketing margins, and frequently artificial measures fixing margins lead to another distortion in market behaviour: 'under the counter' trade, parallel channels, etc.

There are some other elements to consider in the political process of marketing intervention. There is a discriminatory situation against the food distribution sector, with a negative incentive for investment, innovation and modernization. However, political and economic circumstances may push the politicians to establish a rigid control of food marketing margins and we may find frequent examples of this in developing countries. In general terms, we have seen a market liberalization process during the decade of the 1990s. International trade, through the Uruguay Round Agreement in Marrakech, is starting a new era of liberalization. Domestic economies are in a process of transformation, especially those in Central and Eastern Europe.

Therefore we may appreciate a gradual trend of liberalization for food prices. It is not an easy process, and quite often the countries move through a gradual approach towards the more liberal market. For instance, in countries like Spain, the economy has been moving from a strong intervention system (1940–1960) towards a gradual liberalization until the integration into the European Union (1986). During a transitional period, government agencies controlled marketing margins, especially for food products. From fixed margins, the next step was to move into 'recommended margins' (on some significant items), and ended with a non-intervention system. In France, the Minister of Economics has been carrying out periodical enquiries since 1972, at retail level, in order to analyse the evolution of food marketing margins; shopping centres and independent retailers applied higher margins than hypermarkets, supermarkets or cash and carry.

The evolution is conditioned by macroeconomic factors, price policies, distribution channels and the nature of the food products (perishable or not). In the long-run, margins tend to move upward as costs change. While marketing cost may decrease as a result of innovation process or competition, in food margins the trend is to remain the same. In other words, wholesale and retail prices are more sensitive to increases than to decreases in food production costs, reflected through farmer and services prices. There are several reasons to explain such behaviour. Imperfect markets and poor competition may allow middlemen to capture the benefit of increasing food margins. On the other hand, consumers are in favour of price stability, with negative reactions to market prices at the retail stores, which in some ways are misleading. In a dynamic economy, the only way to avoid 'excessive marketing margins' is through competition, improving market transparency and creating the adequate atmosphere of concurrence.

CONCLUDING REMARKS

Marketing margins may be defined in different ways, either as the price difference between two marketing stages, or the cost of a collection of marketing services provided along the marketing channel. Measurement of marketing margins has been studied through a methodological process that includes a wide range of procedures: descriptive, analytical and dynamic according to the circumstances and restrictions. Frequent changes in quantity and quality of the products from farmer to consumer combine with interactions of market factors due to the political, economic and social events in the real world. For that reason it is important to analyse the significant components of marketing margins: input prices, innovation, marketing mix strategies, administrative regulation. Although no general rules may be given, it is useful to describe the special circumstances of food margins related to products and the comparative analysis of developed and less developed countries. The general trend is to increase the price spread from farmer to consumer due to the increasing value added along the marketing process and the lower bargaining power of producer organizations.

Distribution is in a much better position, and much more dynamic than farming, which may allow it to receive the greatest increase in marketing margin trends.

FURTHER READING

Gardner, B. (1975) The farm–retail price spread in a competitive food industry. *American Journal of Agricultural Economics* 57, 399–409.

Goodwin, J.W. (1994) *Agricultural Price Analysis and Forecasting.* John Wiley & Sons, New York.

Heien, D.M. (1980) Markup pricing in a dynamic model of the food industry. *American Journal of Agricultural Economics* 62(1), 10–18.

Tomek, W. and Robinson, K. (1982) *Agricultural Product Prices*, 2nd edn. Cornell University Press, Ithaca and London.

United States Department of Agriculture (USDA) (1972) *Farm–Retail Spreads for Food Products.* Miscellaneous Publication No. 741. USDA, Washington, DC.

REFERENCES

Briz, J. (1985) Consideraciones metodológicas en torno a los márgenes comerciales. *Información Comercial Española* 2009, 2779–3784.

Buzby, J.C., Jones, J.T. and Love, J.M. (1994) The farm–retail price spread: The case of postharvest pesticides in fresh grapefruit packinghouse. *Agribusiness* 10(6), 521–528.

Caldentey, P., Briz, J., Tifos, A. and Haro, T. (1993) *Marketing Agrario.* Editorial Agrícola Española, Madrid.

Dahl, D.C. and Hammond, J.W. (1977) *Market and Price Analysis.* McGraw-Hill, New York.

Dalrymple, D.C. (1961) *On the Nature of Marketing Margins.* Agricultural Economics Report 824. Michigan Agricultural Experimental Station, East Lansing, MI.

Dunham, D. (1992) *Food Cost Review, 1991.* Agricultural Economics Report No. 662. USDA, ERS, Washington, DC.

Elitzak, H. (1996) *Food Cost Review, 1995.* Agricultural Economics Report No. 729. USDA, ERS, Washington, DC.

Gardner, B. (1975) The farm–retail price spread in a competitive food industry. *American Journal of Agricultural Economics* 57, 399–409.

George, P.S. and King, G.A. (1971) *Consumer Demand for Food Commodities in the United States with Projections for 1980.* Monograph 26. Giannini Foundation, Berkley, CA.

Ghosh, M.M. (1988) Impact of regulation of markets on price-spread of important foodgrains: A case study in Bihar. *Indian Journal of Agricultural Marketing* 2(1), 60–64.

Hammond, J. (1990) *Marketing Costs and Efficiency for Agricultural Products: Conceptual Issues in Analysis and Measurement.* Staff Paper. University of Minnesota, St Paul, MN.

Heien, D. (1980) Markup pricing in a dynamic model of the food industry. *American Journal of Agricultural Economics* 62(1), 9–18.

Houck, J.P. (1966) A look at flexibilities and elasticities. *Journal of Farm Economics* 48(1), 225–232.

Mohammed, A.R. (1991) Estimation of marketing margin and costs for red meat in the A.R.E (Egypt). *Zagazig Journal of Agricultural Research* 14(1), 899–920.

Organization of Economic Cooperation and Development (OECD) (1978) *Agro-food Chains.* Ad Hoc Group-DAA/1580. OECD, Paris.

Rahman, M.L., Akbar, M.A. and Islam, M.Q. (1991) Food grain marketing in Bangladesh. *Bangladesh Journal of Agricultural Sciences* 18(2), 207–214.

Senti, R. (1990) Marketing spreads for food products in Switzerland during the 1980s. *Landwirtschaft-Schweiz* 10(3), 576–579.

Taberler, M. and Dawson, P. (1991) *An Econometric Analysis of Marketing Margins for Sunflower Seed in Turkey.* Discussion Paper DPI/91. Department of Agricultural Economics, University of Newcastle upon Tyne, Newcastle upon Tyne.

Tomek, W. and Robinson, K. (1982) *Agricultural Product Prices,* 2nd edn. Cornell University Press, Ithaca and London.

Tsukano, M. (1992) Present conditions and problems in the wholesale market. *Food Policy Study* 69, 57–99.

United States Department of Agriculture (USDA) *Agricultural Statistics.* Various issues. US Government Printing Office, Washington, DC.

United States Department of Agriculture (USDA), Economic Research Service (ERS) (1992) *Food Marketing Review, 1991.* Agriculture Economics Report No. 562. USDA, ERS, Washington, DC.

United States Department of Agriculture (USDA), Economic Research Service (ERS) (1994) *Food Marketing Review, 1992–93.* Agriculture Economics Report No. 678. USDA, ERS, Washington, DC.

9 Marketing Information and Support Systems

Gerhard Schiefer

Department of Agricultural Economics, University of Bonn, Meckenheimer Allee 174, 53115 Bonn, Germany

INTRODUCTION

Markets and market activities are key elements of market economies. They are determined by the activities of policy, business and consumers who are all in need of information about markets, i.e. in need of *market information*. Organizational structures which are designed to serve such needs through an ongoing routine of information collection and delivery are commonly characterized as *market information systems*.

In the agro-food sector, such market information systems have a long tradition in the monitoring of market activities and the analysis of market developments. In many countries, they are organized by well-established institutions with substantial public support like, for example, ISMEA, the Instituto per Studi, Richerche e Informazioni sul Mercato Agricolo (Italy), ZMP, the Zentrale Markt- und Preisberichtsstelle (Germany) or the USDA/AMS, the Agricultural Marketing Service in the USA Ministry of Agriculture.

The prime focus of these market information systems was, and very much still is, the support of market analysis for policy development. In the past, this limited focus was justified by the traditionally strong influence of policy regulations on markets in the agro-food sector and, to a lesser extent, by the limited capacity of traditional information collection and delivery technology to provide timely information appropriate for business and consumer decision support.

However, moves towards market deregulation and increasing competition in the saturated markets of the agro-food sector make the sector's industry increasingly dependent on its own marketing management activities and the availability of market information systems which could provide appropriate marketing decision support. Such systems, especially if combined with the utilization of modern information technology, are usually referred to as *marketing information systems* or MAIS (Schewe, 1976; King, 1977).

Traditionally, the provision of *marketing information* for companies in the agro-food sector is not limited to the institutionalized market information systems mentioned above but incorporates information from a variety of additional sources like, for example, information from national or international organizers of panels among households and food distributors or information collected at market places and through advertising activities.

Additionally, companies might derive marketing information from the internal analysis of their customers and competitors (*in-house* marketing information systems) or might even employ their own groups for general market analysis.

With the incorporation of new information technologies, these different information channels develop towards integrated information systems (sometimes and in the following referred to as *marketing support systems* or MASS) where the traditional distinction of channels loses much of its meaning. They are all based on information flows from the market to the firm (one-directional) or between the market and the firm in a bi-directional information exchange.

Presently, the organization of business-oriented MAIS and MASS has to deal with two major developments, the first relating to an ongoing change in business management concepts, and the second to the widespread implementation of computer-based information and communication technologies.

Modern management concepts employ a profound distinction between executive decisions on the selection and implementation of business activities and the rigid process orientation of operations management required by modern quality-oriented business approaches. Marketing activities are linked to both levels of management and require appropriate system support.

The new information and communication technologies reduce the problem of time and space and allow, in principle, the provision of up-to-date support for ongoing business activities at the place where support is needed.

The following sections deal with managerial requirements of the organization and the content of marketing information and support systems, the state of system availability, the technological development potential, and modern approaches for appropriate development activities. While most of the discussion is of relevance for all kinds of MAIS and MASS, the main focus will be on open systems, i.e. systems of which the use is not restricted to individual companies (in-house systems) but which provide information and support services for non-exclusive business groups of any kind in the agro-food sector. Special emphasis will be on activities of the institutionalized providers of market information systems in the agro-food sector to extend the focus of their systems beyond the traditional monitoring and analysis of market transactions towards the provision of comprehensive business marketing support.

MAIS AND MASS FOR MANAGEMENT SUPPORT IN MARKETING

Overview

The consideration of marketing aspects in management activities has changed during the last decade (Kotler, 1991). Initially, marketing activities were supposed to follow production and included of the line activities like distribution or the promotion and pricing of the sales products. Today's management view links marketing considerations to all steps of the design and production process and, in addition, sees them as an integral part of any strategic decision activities at the executive level of management. The following discussion of MAIS/MASS for management support will follow this view and outline the requirements for the organization of management-oriented MAIS/MASS on the executive and the process level of management.

Executive decision support

Since the early 1980s, there has been an extensive discussion on information requirements of executive management. According to pioneering work by Rockart and others (Rockart, 1979; Rockart and DeLong, 1988), executive information requirements could be summarized in the form of a few basic 'needs': (i) the need to remain generally informed on actual and expected developments; (ii) the need to 'drill down' from highly aggregated information to information details; (iii) the need for personalized analysis.

These needs demonstrate that an information system for executive management (*Executive Information/Support System* or EIS/ESS) (Rockart and DeLong, 1988) requires both aggregate information which can be grasped on a routine basis, especially if combined with exception reporting features, and detailed information for drill-down and personalized analysis through appropriate technical system features.

To meet the 'need to remain generally informed' is the most critical element in the design of information systems for executive management. Empirical studies suggest (Kelling, 1994) that most of the information needed in this respect refers to market information which should be served by a MAIS/MASS. The difficulty, however, is the type of market information that is required which differs from the traditional transaction-oriented monitoring information. It is sometimes referred to as 'soft' information and includes impressions, vague information about development tendencies, changes in the political environment and its potential marketing consequences, etc.

Present approaches to deal with this type of information are summarized under the heading *environmental scanning*. The literature on environmental scanning is fast growing (for an overview see Choo and Auster, 1993), including reports on experiences from various sectors of industry (see, for example, Jain, 1984). Environmental scanning usually requires an organizational structure which goes beyond traditional data collection schemes and might even include expert workshops which support the analysis of developments on a routine basis. The use of such information gathering elements is not unique to environmental scanning systems. What is new, however, is their integration into an organized information collection routine as part of an institutionalized information system for executive management.

Executive management information needs focus on its key decision responsibilities which include the identification and selection of strategic alternatives. In a further move towards the provision of executive support (Kelling, 1994), the respective information could be directly linked to the principal *decision support systems* or DSS (Sprague and Carlson, 1982) used in strategic planning. A case in point is portfolio matrix analysis (Hax and Majluf, 1984), one of the central approaches for strategic decision support. It requires information on market growth, market share or on the 'attractiveness' of a market, a measure derived from a variety of external factors which should capture the market situation and might include factors as different as policy developments, competition, market volume, technology, etc.

The literature does not yet explicitly report on experiences with EIS applications or environmental scanning routines as part of MAIS/MASS developments for the agro-food sector. However, there is no apparent reason why experiences from other industries could not be adopted for the sector. This argument is supported by experiments with early EIS prototype applications for the meat industry (Kelling, 1994) and agricultural trade (Kuron, 1993).

Marketing function in modern quality-oriented management approaches

Today, the food industry in most Western countries, as with many other industries, has to deal with saturated markets with little or no growth and in which companies have to face increasing competition. This situation forces industry to focus production and services more clearly than before on meeting the needs and expectations of their customers. This customer orientation is expressed in the modern view on quality which might differ from and reach beyond traditional product-related and measurable quality concepts and include non-measurable elements like taste, health considerations, environment-consciousness of production or farm animal treatment in meat production. The customer orientation of modern quality management links it closely to marketing management activities and, in turn, to the organization of MAIS/MASS.

Traditionally, an increase in quality is met by stricter product controls and the elimination of products which do not meet the higher quality requirements. In today's market situation, this approach is no longer an appropriate alternative as it increases costs and does not capture all of the food quality characteristics that might be required by consumers.

As an alternative, modern quality management approaches (Brocka and Brocka, 1992) focus on *processes* and attempt to reach high quality in products through the identification of quality processes which, if appropriately controlled, deliver quality in products and, in principle, eliminate the need for product control and product selection. In this concept, product control is replaced by process control and quality through product selection by quality through process improvement. It is widely argued that this switch should allow industry to increase quality with no (or at least lower) increase in costs (Taguchi, 1986).

To assure quality production without product control, the International Standardization Organization (ISO) has specified in its standard series *ISO 9000* (Johnson, 1993) recommendations for process documentation, process control, and (internal and external) process management audits which should assure the elimination of divergences between actual and documented process activity, i.e. between actual and proclaimed product quality. The specified process controls are direct links to modern MAIS/MASS requirements as their focus is on the potential consequences of process failures to customers and on customer evaluation and customer reaction. This is exemplified in the Failure Mode and Effects Analysis (FMEA), a tool for process analysis and the identification of appropriate process controls (Oakland, 1989).

Quality management approaches for process improvement as, for example, the approach of *Total Quality Management* or TQM (Oakland, 1989), depend on information about customers' expectations and product evaluations, key elements of a Quality Function Deployment (QFD) analysis used in TQM. Modern MAIS/MASS have to deliver this information. Furthermore, the need to inform consumers on process-related quality characteristics of food products requires, in principle, a MAIS/MASS organization with a two-way communication capability.

The case of agro-food production chains

In today's quality management view it is assumed that consumer requirements are communicated within production chains through close customer–supplier relationships. Similarly, products and their characteristics are assumed to be traceable from the final customer back to the initial producer.

The food industry is no exception. On the contrary, requirements on product safety, which are characterized by moves in the European Union to require the implementation of the HACCP concept (Hazard and Critical Control Point Analysis) (Schothorst and Jongeneel, 1993) and on quality assurance throughout agro-food production chains, stress the need for chain encompassing information flows.

The weak point in the discussion is the ISO 9000 series of standards with its focus on individual firms and the support of quality assurance activities with regard to firm-specific quality definitions which might vary between different firms within a production chain. The efficient delivery of consumer-oriented quality in food production requires, therefore, not only a quality assurance initiative as specified by the ISO 9000 standards but, in addition, a coordinated quality specification throughout a production chain.

Such a coordination provides the basis for a similarly coordinated information flow between the consumer and each member of the production chain, i.e. a system of interlinked MAIS/MASS within a production chain.

Requirements on transaction support

The principal market-related transaction activities include the ordering (purchasing) of material or goods from suppliers and the sale of goods to customers. From an organizational point of view, these transactions can be linked to three principal types of situations: (i) an 'order situation' where goods are purchased from one or several selected suppliers; (ii) a 'sales situation' where goods are sold to one or several interested customers; or (iii) a 'market situation' where goods are purchased and sold among a number of market participants through more or less elaborated bidding or auctioning systems.

Traditionally, decisions on transactions are based on the exchange of information about the goods (prices, etc.) and on their inspection. This combination is especially apparent in the traditional organization of agricultural markets and auctions where commodities are sold and bought at some kind of a market place. However, efficiency requirements have led to attempts to eliminate the need for the inspection of goods and to base transaction decisions on information only. This separation of goods and information requires an identification of all product characteristics which might be relevant for transaction decisions and to capture them in appropriate information items.

The availability of comprehensive product information is, however, not sufficient to replace the traditional organizational structures (like, for example, the market places) for the support of transaction decisions. They provide *transaction management systems* which define the transaction rules (i.e. rules regarding market access, pricing, bidding, quality guarantees, etc.) and offer supporting organizational and other services.

Any system which wants to utilize fully the potential advantage of separating goods and information and to eliminate the need for goods and transaction participants to meet, requires, therefore: (i) the identification of appropriate exchanges of product information; (ii) the utilization of modern developments in information and communication technology for efficient information exchange; and (iii) the establishment of transaction management systems which provide appropriate organizational, technical, and administrative support for the information-based transaction activities.

This integration of marketing information (product information) and marketing support services develops out of efficiency considerations and demonstrates an evolving development path from MAIS to comprehensive MASS.

Situation in the agro-food sector

The separation of goods and information about their product characteristics has met with some reluctance in the agro-food sector, especially with regard to the marketing of primary products in the early stages of the agro-food production chains. These products were considered to be of insufficient homogeneity, which distinguished the marketing situation from situations in other sectors despite early studies (Debatisse *et al.*, 1984) which arrived at different conclusions and recommended MASS developments for a variety of primary products.

Efforts to facilitate the separation of goods and information have resulted in the specification of quality grades for various primary products which, in principle, facilitate the communication of product characteristics and have supported the development of transaction systems which do not depend on inspections by individual buyers. Still, such systems scarcely meet the marketing needs of farms at the first stage of the production chains. This is partly due to difficulties in the identification of product characteristics (e.g. meat quality characteristics based on the inspection of live animals), partly to the organization of grading systems which require off-farm support by grading organizations or depend on grading activities of potential buyers like traders or the processing industry.

Actual developments in quality assurance open, in principle, the way for a more widespread change to (purely) information-based transaction activities. Appropriate process control allows the consideration of a more comprehensive list of quality characteristics including those which might not be measurable with the product and improves assurance in the production of a pre-defined quality which reduces needs for inspection.

However, the quality-induced tendency towards the establishment of integrated production chains might reduce the interest in open-transaction-oriented MASS developments and support chain internal development activities.

Consequences for MAIS/MASS organization

The analysis of management support requirements in marketing revealed support potential on the executive level with its focus on strategic decision activities and on the operational level of management with its focus on the management of processes and marketing transactions. Despite the differences in management orientation, the interest in information delivery points to a similar direction. With regard to sales activities, the principal interest on the executive and on the process level of management is in information about the 'market potential' in general, for the individual firm, and for the specific product.

Differences in interest relate to the level of aggregation and specification and to the speed and frequency of information delivery. The activities at lower levels of management tend to require information which is less aggregated, more specific, more time-sensitive, and more frequent and vice versa for activities at higher levels.

Information links with management activities

Requirements on information delivery formulate two levels of linkage to management activities. The first level asks for a link to management personnel and requires appropriate procedures for information presentation like the 'drill-down' or 'exception reporting' system capabilities discussed for EIS or the design of user-oriented presentation interfaces.

The second level asks for the linkage of information to the management and decision support tools which are part of management's decision and transaction activities. This level links MAIS with decision support systems and with transaction support systems. In this concept, it is a question of definition where one ends and the other begins. As it is common to see market transaction activities as part of marketing science but not, for example, decision support tools for strategic planning or quality management, the definition of MASS follows this distinction if not specified otherwise.

Information delivery

The general interest in information about 'market potentials' translates into information about customers, competitors, and structural market developments.

Information about customers supports the identification of a general market potential and complements traditional market analysis based on econometric methods of analysis. It might include information about customers': (i) purchase habits (prices, quantities); (ii) purchase potential (as determined by customers' production programmes or (consumers') living standards); (iii) expectations and needs; or (iv) evaluation of products (product images).

Information about competitors could provide a better understanding of the attainable market potential. It might include information about: (i) production and marketing efficiencies (benchmarks) (see, for example, Oakland, 1989); (ii) strengths and weaknesses of major competitors; or (iii) market share and price policy.

These aspects, together with the principal system requirements formulated for the executive and process level of management, constitute the principal design concept for a baseline model for MAIS/MASS developments.

However, the further specification of the model towards a MAIS/MASS for implementation requires additional conceptional and technological considerations which are discussed in the following sections. The discussion is complemented with a description of examples of development activities in the agro-food sector which demonstrate ongoing development alternatives within the general MAIS/MASS framework.

DEVELOPMENT CONCEPTS

Critical success factors in development

The further specification of the baseline model for MAIS/MASS developments has to deal with a number of critical success factors which are independent of the level of management the systems are supposed to support.

They include: (i) the identification of management information and support needs beyond the general principles discussed above; (ii) the identification of the data which relate to the needs; (iii) the efficient organization of the collection, communication, storage and processing of the data; and (iv) the transformation of the data into management information. The organization and management of these factors is usually summarized as *Information Management* (IM), a new management function which assumes an increasingly central role in business management.

System analysis and design

Traditionally, the design of information systems builds on experts' knowledge and their perception of needs and appropriate information deliveries. This approach is supported by a wide variety of logical system analysis and design methodologies and techniques (Senn, 1989). However, increased requirements on the user orientation of systems has led to a reorientation of system development towards participative approaches which incorporate users in the analysis of information needs and the system design in a stepwise system improvement process which involves the development, testing, and use of system prototypes of various levels of sophistication (Boar, 1984).

The most well-known interview-based approach for user involvement in the analysis of information needs is the *Critical Success Factors* (CSF) analysis proposed by Rockart (1979). It employs a multi-step approach to specify indirectly information needs from critical success factors identified by the target group.

In system developments for sector use, user participation supports the identification of an appropriate compromise in the differentiation of user groups which is economically feasible (i.e. would ask for as little differentiation as possible) but still finds acceptance with users (i.e. would ask for as much differentiation as possible).

Experiences from the agro-food sector

There is little literature available on the systematic use of participative development concepts for the design of MAIS/MASS in the agro-food sector. An exception is the work by Ditz (1995) who designed and developed a MAIS for tree growers. It started from a baseline model derived from experiences in other sectors and employed a multi-step participative prototyping approach which involved extended field trials and finally led to the implementation of the MAIS for routine business use. Other reported developments concern prototype applications which remained at an experimental stage. They include the application of the critical success factor analysis in prototype information system developments for agriculture (Huirne *et al.*, 1993), the meat industry (Kelling, 1994), and agricultural trade (Kuron, 1993; Schulze-Duello, 1995).

TECHNOLOGY

Technology alternatives for MAIS/MASS developments

Electronic information and communication technology assumes a central role for the development of modern MAIS/MASS. At present, the following technologies are of foremost interest to service providers: (i) *telefax* as a telephone based two-way (1:1) communication service; (ii) *videotext* as a television based one-way (1:*n*) information distribution service; (iii) *videotex* as an online two-way (*n*:*n*) information and communication service based on telephone communication lines and computer networks; (iv) *Internet* as an online two-way (*n*:*n*) information and communication service based on computer networks; and (v) *Integrated Services Digital Networks (ISDN) online services* as general two-way (*n*:*n*) information and communication services based on telephone communication lines and computer networks.

Their application potential is discussed in the following sections.

Telefax

Telefax is a communication service primarily based on the transmission of graphics images over telephone lines. Traditionally, the images are scanned from paper displays at the source and are again displayed on paper at the point of destination. However, more recent alternatives communicate the images directly between computers linked by telephone lines. They have provided the basis for some MAIS developments.

Use in MAIS/MASS

Telefax has become popular with providers of market information for both information collection and the dissemination of information to service subscribers. The transmission of information to the point of destination depends upon its availability and the use of readily available telephone lines gives it some comparative advantage over other telecommunication alternatives in situations where the information requirement is well-defined and does not need further processing.

Videotext services

Videotext services utilize the fact that the transmission of television screen images is not based on a continuous signal transmission but leaves some space between consecutive screens. In videotext, this intermediate space is used to transmit screen images not detected by the standard television but by a specific videotext 'decoder'. Television sets which host both types of decoders can switch between them and display either the standard television or the videotext screen images.

In contrast to television displays, videotext screen images are not designed to display dynamic movements but static information screens. This allows different information screen images to be put in the television inter-transmission space and their identification by codes that can be detected by the decoder to select any one of them for display. Initially, the limited space between television screen image transmissions left only room for a small number of alternative videotext screen images that could be utilized in the service. However, today's decoders allow the capture of individual screen images and their storage in memory for continuous display. This opens the possibility of putting a larger number of television inter-transmission spaces between consecutive transmissions of a certain videotext screen image. The capacity of such systems can reach 15,000 pages of screen information.

Newest additions to the technology attach an individual identification code to decoders which allows them to capture certain, not generally identifiable, information screens and to automatically transfer them to printers or tape recorders upon receipt. This technology is sometimes referred to as 'videoprint'.

Videotext services require no cable infrastructure and can be utilized wherever television programmes are available. The videoprint addition should make it an attractive alternative for one-way MAIS developments.

Use in MAIS/MASS

Despite its attractiveness for information dissemination in regions with poor communication infrastructure, videotext has been scarcely used for MAIS/MASS. One of the few exceptions is the British Milk Marketing Board which had switched information services from videotex (see below) to videotext upon the availability of decoders with personal identification codes.

Videotex services

Videotex Services describe online computer services which are easy to use by non-experts, readily accessible, and affordable to a broader public.

This general description refers to a variety of services which employ different technologies (for overviews see Schiefer, 1988; Harkin, 1990). Originally, the description excluded online services which required computers as access terminals and did not offer a more affordable alternative like television sets or special purpose terminals. However, the widespread availability of microcomputers has extended the videotex philosophy to include online services solely based on microcomputer access. Furthermore, traditional videotex services increasingly promote microcomputer access and deviate from the traditional videotex concepts discussed below.

Traditional videotex technology has its roots in the 'Prestel system', a British development of the 1970s, a time when microcomputers were scarcely in use. The development was primarily supported in Europe where different videotex systems emerged during the late 1970s and early 1980s. They had in common: (i) the implementation of an easy-to-use full-screen and graphics oriented communication system; (ii) the possibility to use television sets or specific but readily available user terminals (as the 'minitel' in France); (iii) access through ordinary telephone lines; (iv) and public support through the provision of subsidized connection charges or subsidized information services.

The graphics orientation has been based on certain standards ('Conférence Européenne des Administrations des Postes et des Télécommunications (CEPT) standards' for the coordination of European postal and telecommunication services) that evolved in Europe. However, the use of different levels of CEPT standards in Europe has confined the services primarily to their host countries (Schiefer, 1988). The possibility of using television sets or minitel terminals was intended to facilitate service access and, in addition, to implement a graphics communication interface at a time when the graphics capabilities of microcomputers were still rather poor. The access through telephone lines ensured easy service access even in remote and rural regions. In addition, countries subsidized service access through special access rates, e.g. by confining telephone rates to those of local calls.

The graphics orientation combined with the simplicity of access through telephone lines was a major breakthrough in online information service availability. However, it also proved to be one of the services' weaknesses. The slow communication speed of telephone line communication together with the high communication requirements of graphics information made service access rather slow. Furthermore, the graphics orientation of the CEPT standard aggravated information exchange with character-oriented computer programs.

Actual developments in videotex services adjust them to the new developments in telephone communication technology and desktop computer performance and availability. They include high speed access through analogous and digital telephone networks and the transfer of graphics interfaces to microcomputer terminals.

Use in MAIS/MASS

Videotex systems have been the prime choice in most European countries for the dissemination of market information and the provision of marketing support in the agro-food sector. Most examples presented later in this section are based on this technology.

Internet

The newest addition to the communication networks suitable for MAIS/ MASS is the 'Internet'. It consists of a worldwide decentralized computer network based on a common communication protocol, TCP/IP (Krol, 1992). Access to the network can be organized by (i) integrating an individual computer into the international Internet network, or (ii) connecting an individual computer as terminal to the Internet computer of an Internet access service provider. Most users rely on the second alternative where access usually can be organized through a telephone connection. It has become common in the USA, the home country of the Internet, that local access providers offer both local information services on their access computers and access to the Internet network.

The principal advantage of the Internet is its utilization of a worldwide common communication standard and the concentration on a computer-based digital communication technology which avoids technology breaks. Its principal disadvantage is the dependency on computers for information access, the (still) limited availability of regional access points, and the lack of a central service guide.

To offset the lack of a central service guide, the Internet community has developed various supporting search structures. The most prominent ones are the 'gopher' and the 'world wide web (www)' (Manger, 1995).

The gopher service consists of screen menus provided on Internet computers which guide users to locate information in the Internet and to automatically connect to the host computers. There is no 'central' Internet gopher menu and each Internet computer can build its own one. However, they usually include the option to connect to other gopher menus on other Internet computers, thus facilitating the travel through 'gopher space'. The www follows the same organizational principle but is further developed. Its menus are built on the hypertext technology and it supports the communication of all types of information including graphics or sound.

Use in MAIS/MASS

The international network of the Internet is of interest in MAIS/MASS where the access to internationally available information is of interest. Some services for the agro-food sector are already available (see, for example, Braun, 1994). A case in point are USA commodity prices. A comprehensive MAIS/MASS for horticulture is presently being developed within the Internet framework in Germany. Furthermore, the traditional videotex systems increasingly provide links to the Internet or connect their computers directly to the Internet computer network. However, the Internet is still in its infancy and the service level for users is still below the standards implemented in the videotex systems.

ISDN communication systems

Actual developments in telephone networks and their integration into networks for digital communication, commonly referred to as ISDN, have supported the emergence of special purpose information and communication services which act outside the large-scale and general-purpose-oriented videotex online services.

The speed and reliability of digital data communication in ISDN supports the communication needs of online services, the ISDN infrastructure facilitates the establishment of readily accessible services in regions with an ISDN telephone network. Some MAIS developments are known from regional extension services with no professional interest in the utilization of an international information network as provided by the Internet.

DEVELOPMENTS IN AGRO-FOOD

Directions

Actual developments in MAIS/MASS for the agro-food sector involve all critical success factors of system development. They focus on improvements in traditional services of established market information organizations and on the design of new services in and outside these organizations where the emphasis is on services for increased marketing transaction support.

New services relating to the management information needs, discussed for the executive and process level of management in the section on MAIS and MASS for management support in marketing, are not yet reported in international accessible literature. While some of the basic requirements on system capabilities formulated for the design of EIS might be elements of actual developments and have been implemented in the prototype systems listed above, deficiencies include approaches for environmental scanning and for quality-oriented process management support.

Improvements in traditional services

Established market information organizations follow a stepwise approach to adapt to the new situation in information and communication technology.

The focus during the first phase is on the utilization of the new technologies for efficiency improvements in traditional market information services. Such improvements include, for example, the integration of telefax technology into data collection or the provision of market information through videotex or Internet online services.

The second phase involves efforts to utilize the potential of the new technologies for quality improvements in information delivery. Such improvements could focus on data collection or the improved linkage of data to users' individual information needs.

An example of efforts to improve data collection is reported in Irps and Christiansen (1993) who describe a national pilot project to improve information on market transactions, especially on market prices of agricultural products. If prices are influenced by market transactions, price information for marketing support must represent the most recent developments. This leaves no room for detailed market analysis. Traditional price information services assure this accuracy through data collection schemes which collect, where possible, prices at certain concentration points in the market. Where these do not exist, the information is derived from market experts, trade or the processing industry. However, those data do not always adequately reflect the outcome of the price negotiations between farms and their trading partners. The project utilizes new technology (including videotex technology) to collect directly the data from a large number of farms who monitor and report all their transactions as they take place.

Marketing transaction support in purchases and sales

Developments to improve marketing transaction support in purchases and sales which go beyond the traditional recording of transactions are all based on the utilization of some kind of electronic bulletin boards. They require a two-way interactive service technology as utilized in online services on, for example, Internet or videotex systems.

The development potential and direction is best illustrated by listing some of the examples that are known to exist, at least in some form of prototype application. They are scarcely recorded in scientific literature but in project reports and project leaflets.

Example 1: Public announcement of trading interest

Such systems include bulletin boards which allow the listing of purchase or sales interests. Any resulting further trading activity will take place outside the system. Additional system support may include the ordered display of bulletins or the selection of individual bulletins according to specified interests.

Implementations are known from a variety of videotex services and for different commodities (e.g. agricultural machinery).

Example 2: Aggregation of trading interest

Trading organizations (e.g. marketing cooperatives) may use electronic communication systems to collect commodity trading interests from trading partners which they represent in the market and use the combined sales or purchase interest to negotiate a joint contract with others.

Implementations are known from a variety of videotex services and for different commodities (e.g. vegetables).

Example 3: Basic sales/ordering system

In this application, products may not only be offered on electronic bulletin boards but sold through system order forms on screen pages. For payments, different alternatives are in use. They include payment through videotex page charges, credit cards or through billing activities outside the system.

Successful system implementations are known for wine sales through videotex and Internet systems.

Example 4: Advanced sales/ordering system

A more advanced computer support is utilized in systems which basically build on order lists with products from many different suppliers but provide support in supplier selection. This could be achieved by including in the order lists the price of the cheapest supplier for each individual product and by automatically adjusting the lists if price changes occur. This principal approach could similarly be applied to sales activities and the selection of the best price offer from potential customers.

Order systems of this type are known to exist as prototype videotex applications for breweries and agriculture.

Example 5: Basic market-place system

If a system incorporates different bulletin boards for the announcement of purchase and sales interests, it might provide support which automatically connects matching trade interests.

Such applications are not known for the agro-food sector but are common in other sectors as, for example, the transportation sector.

Electronic market-places

The organization of electronic market-places is the most sophisticated front-end application of a MASS. They are based on a computer-supported communication system and include an electronic operator who defines and enforces market rules. Such market places allow (Mueller, 1984) traders to search for others who may want to take an opposite market position and to convey to them their bids or offers which might be followed by a negotiation process supported by the system.

There is a variety of alternatives in the organization of electronic market places which have been discussed in literature and which differ substantially in the rules they employ. Mueller (1984) describes seven examples (of which three were operational at the time of publication) from Europe and the USA which dealt with cotton, eggs, cattle, pigs, piglets, lambs and meat, i.e. a wide variety of commodities. The systems employed different pricing mechanisms including English (ascending bid) auction, Dutch (descending price) auction and double auction as the most elaborate one. In double auctions, purchasers stepwise increase their bids, whereas sellers decrease their price requests until offers and requests meet. In case the adjustment process leads to offers higher than the requests, the system must provide rules on which price to choose for the transaction, the offered price, the requested price or some price inbetween.

This example provides some insight into the need for rules in such systems and the influence of the system organizer who not only chooses the principal pricing mechanism but provides the rules for its utilization. Pricing mechanisms and their consequences for the development of markets have been discussed in, for example, Smith (1964), Frahm and Schrader (1970), Coppinger *et al.* (1980) and Schrader and Henderson (1980).

Experiences with the implementation of electronic markets revealed initial organizational difficulties and high investment costs (Mueller, 1984) which were not offset by the perceived system advantages and, in turn, prevented the systems' widespread acceptance. Part of the reason for the limited acceptance of electronic markets in early application experiments might also have been due to (i) deficiencies in technology like the initially slow speed of videotex services or (ii) requirements on changes in established trading habits and trading rules which were not due to requirements set by technology or by potential users, but by system organizers – for example, the requirement in the proposed USA wholesale electronic meat market for public disclosure of private firm price and quantity information.

A more recent and promising development concerns efforts to introduce electronic marketing at flower auctions in The Netherlands. In these flower auctions, each sale's unit of flowers is shortly presented in the auction hall and immediately auctioned against a clock which represents descending prices. The electronic system displays system screens which include both the presentation of the sale's unit of flowers through video pictures and the clock against which to auction. To get potential users used to the system, the system's auction screens will be displayed in auction halls in support of the traditional auctioning activity.

The experimental system has a good chance of success as: (i) it models the system's handling after an accepted system based on traditional technology, utilizing the results of past trial and error improvement processes; (ii) it avoids trade-offs in system capability, i.e. of system advantages against system deficiencies in relation to an established system alternative; and (iii) it provides support for the gradual adjustment of potential users.

However, irrespective of past experiences, general efficiency requirements support a general development path towards more well-defined trading structures and the further integration of modern information and communication technology into all areas of agro-food management activity. This, in turn, will facilitate electronic market transaction support and reduce the need for special purpose investments in information and communication technology.

Organizational consequences

The new technologies facilitate and lower costs for the market entry of service providers, the distribution of information and the utilization of services by target groups.

This should, in principle, open the way for a widespread improvement of the information situation in the agro-food sector. However, it is not yet clear what consequences it could have for the future role and establishment of public (open) or limited access (special purpose) information providers. The emerging information service market improves the competitiveness of newly established service providers but also facilitates a forceful market penetration by established institutions.

In this situation, the success of a service will much more than before depend on its quality and customer orientation and follow rules that are known from competitive product markets. A 'mass information market' which covers basic information needs at low costs might be complemented by special purpose ('niche') information markets with higher priced but more focused information services. One could speculate that, in the longer run, the traditional public market information institutions might focus on the first, and specialized private information providers on the second market. Furthermore, the management support orientation of advanced MASS might support a tendency to link the specialized information providers to groups of businesses or individual agro-food production chains.

CONCLUDING REMARKS

MAIS and MASS are supposed to support marketing activities, an objective which requires more than traditional market information systems are designed to deliver. The target groups, however, are more clearly defined than users of traditional market information systems and include people on executive and operational levels (process levels) of management with marketing-related responsibilities.

The target group orientation relates MAIS/MASS to the strategic and process-oriented management concepts and allows the design of baseline models for management support requirements which involve information deliveries and marketing transaction support.

However, the further specification of a MAIS/MASS for implementation requires the incorporation of potential users in participative approaches of system design and system development which might build, if available, on system design elements from established marketing support systems used in comparable situations of support needs (analogies).

Actual developments of MAIS/MASS applications for groups of potential users in the agro-food sector seldom employ rigid approaches to management orientation and management participation, prerequisites for system developments that are supposed to deliver appropriate support and to find acceptance with target groups. The main focus of developments is on the integration of new technologies in established information and support systems and the improvement of the services linked to them.

However, the further link of MAIS/MASS with the marketing support needs of management requires studies on the linkage of MAIS/MASS with management decisions and activities at different levels of management as well as studies on the suitability and need for the integration of different developments paths into an infrastructure of interlinked MAIS/MASS for comprehensive marketing support.

FURTHER READING

College of Agriculture and Life Sciences (ed.) (1980) *Market Information and Price Reporting in the Food and Agricultural Sector.* University of Wisconsin-Madison, Madison, WI.

Schewe, C.D. (ed.) (1976) *Marketing Information Systems.* American Marketing Association, Chicago, IL.

Watson, H.J., Rainer, R.K. and Houdeshel, G. (eds) (1992) *Executive Information Systems.* Wiley, New York.

REFERENCES

Boar, B.H. (1984) *Application Prototyping.* Wiley, New York.

Braun, E. (1994) *The Internet Directory.* Fawcett Columbine, New York.

Brocka, B. and Brocka, M.S. (1992) *Quality Management: Implementing the Best Ideas of the Masters.* Irwin, Homewood, CA.

Choo, Ch.W. and Auster, E. (1993) Environmental scanning: acquisition and use of information by managers. In: Williams, M.E. (ed.), *Annual Review of Information Science and Technology*, Vol. 28. Learned Information, Medford, OR, pp. 279–314.

Coppinger, V.M., Smith, V.L. and Titus, J.A. (1980) Incentives and behavior in English, Dutch and sealed-bid auctions. *Economic Inquiry* 18, 1–22.

Debatisse, M.L., Deshayes, G. and Henry, R. (1984) *Study on the Opportunities for Coordination of Information and Transaction for Agricultural Markets in the European Economic Community.* Unpublished report. Institut de Gestion International Agro-Alimentaire (IGIA), Cergy Pontoise.

Ditz, S. (1995) Concept of a market information system to support pricing in tree growing companies (in German). PhD Thesis, University of Bonn, Bonn.

Frahm, D.G. and Schrader, L.F. (1970) An experimental comparison of pricing in two auction systems. *American Journal of Agricultural Economics* 52, 528–534.

Harkin, M. (1990) *Information Technology in Agriculture, Food, and Rural Development.* ECSC/EEC/EAEC, Brussels.

Hax, A.C. and Majluf, N.S. (1984) *Strategic Management. An Integrative Perspective.* Prentice-Hall, Englewood Cliffs, NJ.

Huirne, R., Dijkhuizen, A., King, R. and Harsh, S. (1993) Goals, critical success factors, and information system needs on swine and dairy farms identified by workshops. *German Journal for Agricultural Informatics (Zeitschrift fuer Agrarinformatik)* 1, 61–65.

Irps, B. and Christiansen, M. (1993) Federal pilot project of market information systems (in German with English summary). *German Journal for Agricultural Informatics (Zeitschrift fuer Agrarinformatik)* 2, 50–52.

Jain, S.C. (1984) Environmental scanning in US corporations. *Long Range Planning* 17, 117–128.

Johnson, P.L. (1993) *ISO 9000 – Meeting the International Standards.* McGraw-Hill, New York.

Kelling, A. (1994) *Development and Design of an ESS-Concept: The Example of the Meat Industry* (in German). Wehle, Bonn.

King, W.R. (1977) *Marketing Management Information Systems.* Mason and Charter, New York.

Kotler, Ph. (1991) *Marketing-Management: Analysis, Planning, Implementation, and Control.* Prentice-Hall, Englewood Cliffs, NJ.

Krol, E. (1992) *The Whole Internet User's Guide and Catalog.* O'Reilly and Associates, Sebastopol.

Kuron, U.H. (1993) *Management Support System for Agricultural Commodity Trading: Analysis and Concept for an Integrated MIS* (in German). Wehle, Bonn.

Manger, J.J. (1995) *The World Wide Web, Mosaic and More.* McGraw-Hill, London.

Mueller, R.A.E. (1984) *What Future for Electronic Marketing?* Paper presented at the Annual Conference of the Agricultural Economics Society, Cambridge.

Oakland, J.S. (1989) *Total Quality Management.* Butterworth-Heinemann, Oxford.

Rockart, J.F. (1979) Chief executives define their own data needs. *Harvard Business Review* 57, 81–93.

Rockart, J.F. and DeLong, D.W. (1988) *Executive Support Systems.* Irwin, Homewood, CA.

Schewe, C.D. (ed.) (1976) *Marketing Information Systems: Selected Readings.* American Marketing Association, Chicago, IL.

Schiefer, G. (ed.) (1988) *Videotex, Information, and Communication in European Agriculture.* Vauk, Kiel.

Schothorst, M. van and Jongeneel, S. (1993) Line Monitoring, HACCP and Food Safety. *Food Control* 5, 107–110.

Schrader, L.F. and Henderson, D.R. (1980) *Auction Pricing: Options and Implications.* Working Paper. Purdue University, West Lafayette, IN.

Schulze-Duello, M. (1995) *Logistic Controlling in Agricultural Trade: Analysis and Design of a Management Support System* (in German). Wehle, Bonn.

Senn, J.A. (1989) *Analysis and Design of Information Systems.* McGraw-Hill, New York.

Smith, V.L. (1964) Effect of market organization on competitive equilibrium. *Quarterly Journal of Economics* 78, 181–201.

Sprague, R.H. and Carlson, E.D. (1982) *Building Effective Decision Support Systems.* Prentice-Hall, Englewood Cliffs, NJ.

Taguchi, G. (1986) *Introduction to Quality Engineering.* Asian Productivity Organization, Tokyo.

10 Consumer Behaviour

Reimar von Alvensleben

Lehrstuhl für Agrarmarketing, Institut für Agrarökonomie, Universität Kiel, Olshausenstraße 40, 24098 Kiel, Germany

CONSUMER BEHAVIOUR AND THE MICROECONOMIC THEORY OF DEMAND

The theory of consumer behaviour is a complex, multidisciplinary approach with contributions of different social sciences such as economics (micro-economic theory of the demand), psychology (motives, attitudes, perception, learning), sociology (consumer socialization, reference groups), anthropology (culture, tradition), geography (regional factors), and nutritional sciences and medicine (nutritional needs, physiological regulation, sensory factors, etc.).

The basic model of microeconomic theory, as outlined in Chapter 3, explains demand as a function of product prices, consumer income and preferences. Since preference changes are difficult to measure, the main focus of microeconomic demand theory is the analysis of the prospective effects of income and price changes on demand, under given preferences. The microeconomic theory does not explain how consumer preferences are determined (why do they change?); does not explain how consumer decisions evolve (decision is a process rather than a discrete act); and does not take into account that preferences may be influenced by income and prices, too.

With rising consumer income, the relative influence of prices and income on food demand is decreasing while the influence of preferences is increasing (Table 10.1). For example, in their study of the causes of changing patterns of food product consumption in the UK, Ritson and Hutchins (1991) show that, during the 1960s and 1970s, most of the changes in patterns of food consumption were caused by changes in prices and incomes but, since 1980, changes in tastes and preferences have dominated. Similar developments have been observed in other industrialized countries (von Alvensleben, 1989). In affluent societies, future changes in food demand will be more and more caused by preference changes rather than by price and income changes. To influence consumer preferences is the major aim of the

Table 10.1. The change of demand-determining factors in economics with rising consumer income.

	Relative influence on the food demand	
Demand-determining factor	Low income	High income
Income	Strong	Small
Prices/Price relations	Strong	Small
Consumer preferences (motives, attitudes)	Small	Strong
Population	Same influence	Same influence

marketing efforts of private food firms.

Consequently, the survival of a private firm in a competitive system requires more knowledge of consumer behaviour: why they buy, how products are perceived, how the buying decision is taken, where they buy, what they buy and, last but not least, how buying decisions can be influenced by marketing measures (Foxall, 1988). The answers to these questions can be given only by a consumer theory which goes far beyond the microeconomic theory of demand.

SOME BASIC VARIABLES AND THEIR INTERRELATIONS

The basic forces of consumer behaviour are emotions, motives and attitudes (Kroeber-Riel, 1992). The definitions are given below:

1. *Emotion.* Internal tension, which may be felt as pleasant or unpleasant, and may be more or less conscious to the consumer. Examples of verbal expression of an emotion are 'I am concerned about my health' and 'I'm not feeling very well'.
2. *Motive.* Internal tensions combined with a certain activity as objective (activity oriented). An example for a verbal expression of motive is 'I want to take care of my health'.
3. *Attitude.* Willingness or predisposition of the consumer to react positively or negatively to a stimulus pattern of a product offer: the consumer's evaluation or image of a product (object oriented). An example for a verbal expression of an attitude would be 'Fruits are good for my health'.

Without emotional basis, there is no motive. Without motivation there is no attitude towards a product. The stronger the emotion the stronger the motive, the more positive (negative) is the attitude towards the product and the higher (lower) is the probability of purchase:

Emotion → Motive → Attitude → Behaviour (Purchase)

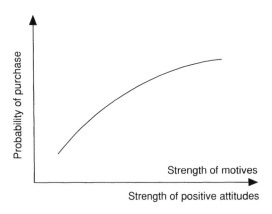

Fig. 10.1. The relation between attitudes and behaviour.

Regarding the consumption of fruits, the demand evolvement may be described as follows: the stronger the health concern, the stronger the health motive in nutrition and the more positive the health image of fruit, the higher is the probability of purchase (Fig. 10.1).

In consumer research, motives and attitudes are usually measured by survey methods, using rating scales as described in Chapter 11. Emotional reactions to marketing stimuli are often determined by measuring certain psycho-biological indicators, such as electro-dermal reaction (EDR) or blood pressure. Figure 10.2 shows the linkages between some variables influencing food demand. The model may be interpreted as an extension of the microeconomic demand model (von Alvensleben, 1989).

Consumer behaviour is determined by motives and attitudes. However, the relations between motives/attitudes and consumer behaviour are not unilateral. Consumption leads to experience with the product, and vice versa this affects attitudes:

Motives/attitudes ↔ Behaviour

The major motives for food demand are nutritional needs, health, enjoyment (taste, diversity, social events), convenience, safety, transparency, compliance with the norms of a reference group, prestige, and environmental/political motives. This list may be extended, reduced or modified depending on the research object and purpose. The different motives may be ranked into an order according to Maslow's hierarchy of needs (Foxall, 1988).

The motives depend on some consumer-related variables, which may interfere with each other. These variables are: (i) the general norms and values of the society and of the family/reference group; and (ii) the socio-economic situation of the consumer. For example, the age of the consumer

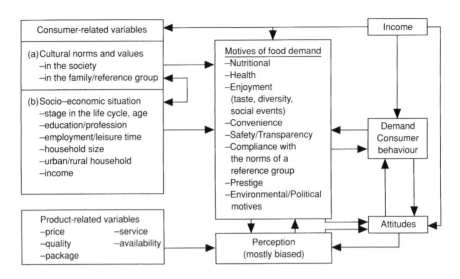

Fig. 10.2. Some variables influencing food demand.

may influence the motive (e.g. the health motive) and this affects demand. In many market research projects the socio-economic variables are used as independent variables to explain consumer behaviour. In other research projects the influence of a general value modification within the society on consumer behaviour has been investigated. Furthermore, the motives can be affected by the product presentation rather than the perceived information on the products. For instance food scandals with a broad media emphasis may create discredit to the food supply and may increase the desire for more food safety and more transparency in production and distribution.

The attitudes towards a product are not only determined by the motives and the consumption experience, but also by the consumer's *perception* of the product and its properties. Perception is most likely to be distorted. The perceived world and the real world mostly do not correspond with each other. Attitudes and the perceived product properties are linked variables:

Perceived product properties ↔ Attitudes

The more there is a positive (negative) attitude towards a product, the consumer prefers the selective perception of positive (negative) properties of the product – leading to a stabilization of the attitude towards the product.

The effect of the income on demand may be modelled in two approaches: (i) as a demand restriction – according to the microeconomic theory of the household; and (ii) as a factor influencing attitudes, motives and other consumer-related variables, such as values. Within the context of the microeconomic theory the latter case means that income has an influence on the preference function of the consumer.

MAJOR MOTIVES OF FOOD DEMAND

The following section will give a brief survey of the major motives of food demand, which are listed in Fig. 10.2.

Nutritional needs

Every consumer has certain requirements for energy and nutrients such as fat, protein, carbohydrates, minerals, vitamins, etc. depending on his or her age, sex, weight, working conditions, temperament and the climate. However, many people are consuming more rather than less food than they need.

In rich societies the following development has occurred: the mean food calorie consumption increased, while the calorie requirements decreased – resulting in a widespread overnutrition. Many people suffer from overweight and health problems caused by a surplus of food energy. The modern consumer did not adapt to the situation of plentiful food. In contrast, the share of people with malnutrition is estimated to be 15–40% of the population in the low-income countries. In these segments of the population the available income is restricting the demand to satisfy the nutritional needs.

Health motive

In high-income societies the food energy surplus has led to a widespread desire to reduce overweight, to control the calorie intake and to pay more attention to healthy nutrition. The consequences are an increasing demand for 'health foods', calorie-reduced food and dietetic food. On the other hand many consumers avoid food containing cholesterol. This is probably the major reason for the decline in egg consumption. The health motive differs significantly between the sexes and the age groups. Women between 30 and 40 are much more concerned in controlling their calorie intake than men and other female age groups. At advanced age the consumer has to cope with growing health problems. This leads to more health consciousness and demand for a healthy nutrition among the older people.

Desire to enjoy food

In every society we find a basic desire to enjoy food. Food consumption means much more than the intake of nutrients according to the physiological requirements. People prefer to consume tasteful food and diversified food. Furthermore, people like to purchase, prepare and eat food eventfully (e.g. at barbecue parties, in speciality restaurants, purchase direct at a farm). With the increasing saturation of consumers, the demand for quality, diversity, specialities and products which offer more than just food, is going up, while the demand for inferior qualities and everyday products is going down. The desire to enjoy food may cause some conflicts with the health motive. This may offer good market prospects for products providing a solution to this conflict.

Convenience motive

People like to avoid effort and inconvenience in buying, preparing and consuming food. Products which offer solutions are called convenience goods. Examples are canned or deep frozen foods, ready-to-cook soups, potato mash, ready dishes and fast food. Most of our products have convenience advantages or disadvantages. For instance, some consumers prefer mandarins as easy peelers to oranges for convenience reasons. The decline in the consumption of fresh potatoes is partly due to the inconvenience of their preparation. Many children prefer margarine because it can be spread on bread more easily than butter. The convenience motive has gained increasing importance due to changing attitudes towards housework, better education of women and changing roles of housewives/men, leading to an increasing participation of women in out-of-home employment combined with a rising income. The consequence is an increasing share of services, which are consumed together with food, and the fast development of time-saving shopping systems and habits.

Safety motive

Especially in affluent societies we observe a growing concern about residues in food. Confidence in the food supply has deteriorated. On the other hand – according to most nutritional scientists – food safety has never been at such a high level as today. Obviously many consumers are subject to perception distortions caused by numerous food scandals with extensive media emphasis. This development has been accompanied by an increasing alienation of the consumer in the mass distribution system, leading to more distrust. Furthermore, a general deterioration of the confidence in authorities and institutions in our society can be observed, which includes food inspection and food science. The result is a growing demand for controlled food, health food and organic food, which promise more food safety, and a desire for more transparency and personal atmosphere in the production and distribution system. In some segments of the population we find a growing interest in buying direct from the farmer, for home garden production and baking in the home. The wish for more confidence is also supporting brand and store/seller loyalty.

Compliance with the norms of a reference group

In general, the consumer likes to live in compliance with the norms of his reference groups. Examples are religious motives for nutrition habits, conventions and fashions. However, we will also find the motive to escape the conformist pressure of the society. Every society has its drop-outs, subcultures and snobs. In the case of a snob effect, a good is preferred because it is consumed by only a few people. In some cases the nonconformist behaviour of a subculture may have a pioneer function for the change of consumer attitudes of the majority. In the early stages of the development of the organic food market, consumers belonged to a small, not very well accepted subculture in society. More recently, organic food has gained a positive image within the majority of the society.

Prestige motive

The wish for compliance with the norms of a reference group is often combined with the prestige motive. The consumer is buying certain goods to seek the recognition of his reference group. People with low self-confidence are tending more often to prestige consumption than people with high self-confidence, who depend less on the opinion of other people. Prestige consumption is widespread in lower classes and in upper classes as well. Veblen (1899) criticized a typical upper class behaviour: a good may be preferred more, the higher the price non-buyers assume to apply. The prestige motive is important in the market for cars, clothes and travelling. On the food market, certain products, such as champagne, caviar and others have a prestige value. The decline of butter consumption is perhaps partly due to a loss of its prestige value compared to margarine.

Environmental/political motives

The demand for certain goods may be motivated by political issues, e.g. the demand for products which are supposed to preserve the environment and resources (organic food, reusable bottles, etc.). Another important factor is the widespread consumer preference for products grown in the local region or country, which may be supported by special campaigns ('Buy British') or by labels of origin. The driving force of these preferences is either patriotism or the belief that these products are fresher or cheaper or have saved resources. Furthermore, certain products have been subject to political boycotts ('No fruit from South Africa', or 'No bananas from multinationals'). Others have been promoted with political reasons ('Buy coffee from Nicaragua', or the 'Fair Trade' campaign in Germany).

BASIC TRENDS IN FOOD DEMAND

Table 10.2 shows in a schematic way the long-term change and differentiation of consumer preferences for food in a growing economy. The development may be divided into three phases:

Table 10.2. Change and differentiation of consumer preferences for food in a growing economy.

Influence of income and prices		
Strong (1st phase)	Decreasing (2nd phase)	Small (3rd phase)
Get enough food Eat more	(Health trend) Eat healthy Less calories More vitamins (Diversification trend) Eat better and more diversified Enjoy food (Convenience trend) Eat, buy and prepare food with more convenience	Concern about residuals, etc. Concern about the environment Eat, buy and prepare food more eventfully More transparency Less anonymous mass consumption Back to nature

1. *Phase 1.* The situation is characterized by a general food scarcity. For this reason food demand is dominated by the nutritional need of getting enough food. The income and price elasticities of demand are high. The development of the per capita demand for food depends very much on income development.

2. *Phase 2.* With growing income the basic physiological needs are satisfied resulting in a decline of the income and price elasticities of demand for basic food. Other motives behind food demand are gaining relative importance: the health trend, the desire to enjoy food, the diversification trend and the convenience trend.

3. *Phase 3.* The income and price elasticities of food demand are very small. The main trends of phase 2 prevail; however, they are differentiated and partly reversed. A growing concern about residues is promoting the demand for more food safety. The growing concern about the environment is stimulating the search for problem solutions, which preserve resources. These trends are partly accompanied by a nostalgic move 'back to nature' and by the wish for more transparency and less anonymous mass consumption. They are contrasted by an increasing hedonism, the desire to eat, buy and prepare food more eventfully. For many people food consumption is becoming part of an 'adventure seeking behaviour'.

CONSUMER TYPOLOGIES

When discussing basic motives and trends of food demand, we have to keep in mind that all consumers are different. The so-called 'average consumer' is a small minority of the society, only. For this reason, it may be useful to

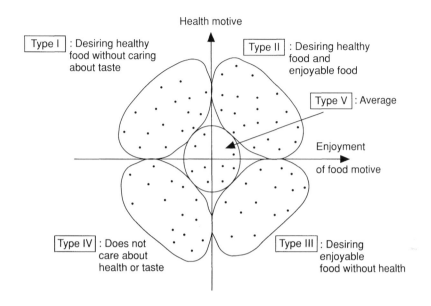

Health motive

Type I : Desiring healthy food without caring about taste

Type II : Desiring healthy food and enjoyable food

Type V : Average

Enjoyment of food motive

Type IV : Does not care about health or taste

Type III : Desiring enjoyable food without health

Fig. 10.3. A consumer typology for food.

divide the total population into different consumer types or market segments, each representing similar motives and behaviour.

To illustrate the problem the author has constructed Fig. 10.3. Assume we have measured the intensity of two motives in a consumer survey, the health motive and the desire to enjoy food, and we draw the results in a bidimensional system of coordinates. Each point represents a combination of motives for consumers. In the next step we can put all the consumers with similar motive combinations into different groups, for instance: Type I desires healthy food without caring about taste; Type II favours healthy *and* enjoyable food; Type III desires enjoyable food without caring about health; Type IV does not care about health *or* taste; and Type V is an average consumer.

In reality we have to consider more than two motives to delineate different consumer types. Thus, the problem becomes more complex. In market research some multivariate methods, especially cluster analysis (see Chapter 12), have been developed to solve such a problem.

The construction of a consumer typology has mainly two purposes:

1. It provides a better understanding of the market and a better basis for evaluation of possible trends; for instance if it is possible to find a group of opinion leaders or so-called trend setters, the motives and behaviour of such a consumer type may be an early indicator of the behaviour of the majority in the future.

2. It provides the basis for market segmentation strategies. Marketing measures, which are designed for different target groups, will satisfy the consumer wishes better than a marketing policy designed for an average consumer.

PRODUCT PERCEPTION

As mentioned above the consumer's attitudes towards a product depend heavily on his perception of the product (von Alvensleben and Meier, 1990; Kroeber-Riel, 1992). Usually perception is distorted. The perceived world and the real world are mostly not identical. How does this occur?

The consumer is subject to a large number of information stimuli every day. Only a very small proportion of them can be processed. The information overload and convenience aspiration leads to strong selectivity of information stimuli. In addition, the perception is subjective: every consumer may select, decode and interpret the information stimuli in a different way. We all perceive first those stimuli which correspond to our wishes and needs. Irrelevant stimuli are neglected, pleasant stimuli are preferred and unpleasant avoided (perceptual defence).

The selective and subjective perception has significant consequences for market research and marketing. Not the objective but the perceived product offer is determining consumer behaviour. It is not sufficient to make a good product. It is also necessary to take care that the consumers perceive the superior quality of the offer.

Information used for the perception of products

The perception of a product is the result of: (i) the actual product information consisting of the direct product information and the product environment information which are processed together; with (ii) the stored product image, using information processing programmes, which may be of very simple or complex nature. Figure 10.4 provides an overview.

The *actual information* is a complex pattern of stimuli, which can be divided into the direct product information and the information about the product environment.

The *direct product information* includes all properties of the product, the physical, technical and sensory properties as well as the price, the brand, the variety or the origin. Some of these product properties cannot be perceived directly at the point of sale, e.g. the taste and other internal quality criteria. In these cases, the consumer has to deduce the non-perceptible from the perceptible properties using his past experience with the product.

In the case of *product environment information*, the consumer perception of a product is influenced by the environment in which the product is offered, such as the type and atmosphere of the store, the properties of the vendor, the assortment and the information, which together are supplied

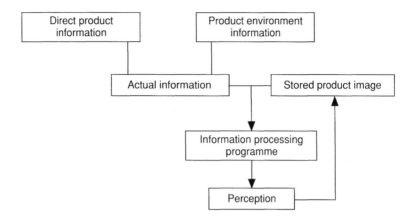

Fig. 10.4. The perception of products.

with the product. For example, an apple sold in the supermarket may be perceived differently from the same apple offered by a grower in his farm store.

In the case of *stored product image*, usually the consumer has already had some experience of the product, which is stored in his memory and can be made available for the further perception process. Especially in the case of low involvement products, which are bought largely in a habitual way, the product image is playing the dominant role in consumer perception. In most cases the consumer is applying some simplified information processing programmes, which are using the actual information and the stored product image for the further perception process and which may lead to perception distortions.

Information processing programmes and perception distortions

The information processing programmes can be described as follows. The consumer is deducing: (i) the total product quality from one or a few product properties (= use of *information chunks* or *key information*); (ii) a certain product property from another product property (if distorted: *irradiance*); (iii) a certain product property from his stored evaluation of the total product (if distorted: *halo-effect*); and (iv) a total product quality or certain product qualities from the image of a similar product (*image transfer*).

The use of information chunks or key information

If the consumer is deducing the total product quality from one or a few product properties, he is simplifying the total perception process. The information about single product properties (= information chunk) has the function of key information, which saves the consumer a further search and processing of information. Widely used key characteristics are the price, the brand name, the advice of experts or friends, or in the case of fresh agricultural products, the appearance. The use of key information may be efficient behaviour, if the consumer has to decide under time pressure. However, it bears the risk of perception distortions. Very often consumers associate a high price with a high quality of the product. Fresh products are 'bought with the eye', the consumer is deducing the total quality from the external quality. The introduction of product brands can only be regarded as successful when the brand name has the function of a key information to the consumer.

Irradiance

The deduction of a product property from another may be based either on logical relations (for example, 'a rotten apple tastes bad') or on illogical ones. The latter case is a perception distortion, which has been investigated by psychologists under the term 'irradiance'. Examples are numerous: the kind of package influences the perception of the freshness of a bread, or the softness of margarine is deduced from its colour. In many cases the appearance and the colour of food and drinks have a significant impact on the taste perception.

Halo-effects

If the product has a stable image, which is stored in the memory of the consumer, he tends to deduce the single product properties from the stored image. The result is an interdependency between the image and the perception of the product properties – the so-called halo-effect.

Positive (negative) image $\rightarrow$ Selective perception of positive
                              $\leftarrow$ (negative) properties

If the product image is positive, the consumer tends towards a selective perception of the positive product properties. If the product image is negative, the same process will take place with negative expression. In both cases the perception is distorted and the distorted positive or negative image will be stabilized. A possible explanation for this behaviour is given by the theory of cognitive dissonance. The consumer tends to avoid cognitive dissonance by a selective perception of information, which leads to cognitive consistency.

For example, the same smiling of a man may be regarded as friendly by people with a positive attitude and as ironic or sarcastic by people with a negative attitude towards this man. The demand for organic food seems to benefit from positive halo-effects, while conventional food is very often underlying a negative halo-effect. Even the perception of taste of food products may be distorted by a halo-effect. This has been shown in experiments with branded turkey meat (Makens, 1965) or glasshouse tomatoes (von Alvensleben and Meier, 1990).

Image transfer

A positive or negative image of a product may affect the image of similar products. Many consumers tend to generalize their experience with a product to avoid the search for and processing of information. This behaviour may lead to perception distortions, which are positive or negative: food scandals caused by single firms often have negative effects on the total food sector. On the other hand, private companies – having established a strong product brand – try to transfer the positive image of the brand to other products and, in particular, new products of their company.

CONSUMER DECISIONS

The decision process

As already mentioned, decision is a process rather than a single act. To understand the consumer decision process, it is common to use phase models that subdivide the process in different stages, for instance:

1. Problem recognition. What happens to initiate the process?
2. Search. What sources of information are used to help arrive at a decision and what is the relative influence of each?
3. Alternative evaluation. What criteria are used by consumers to assess alternatives? What is the status of purchase intention?
4. Choice. What selection is made from among the available alternatives?
5. Outcomes. Is the choice followed by satisfaction or by doubt that a correct decision was made? (Engel *et al.*, 1993).

In this model feedbacks to prior stages are possible in each phase. Single phases may be omitted. The process may be conscious or unconscious to the consumer and may last different times – from seconds to years. When the consumer is moving through all phases thoroughly and consciously, the process is often referred to as an extended problem solving.

Types of consumer decisions

Decision processes may be classified in different ways. The classification of decisions into high and low involvement products is very useful (Engel *et al.*, 1993).

High involvement products bear a high risk of a wrong decision because of, for example, a high price, high importance for the self-image (e.g. certain clothing, jewellery, cosmetic items), or strong influence of outside reference groups, etc. This leads to extended problem solving: active search and use of information, careful processing of information, weighing and evaluating of many product attributes in a complex manner. The outcomes are beliefs about various alternatives, evaluation of the pros and cons (attitudes) and purchase intention.

Low involvement products bear a low risk of a wrong decision because of, for example, low price, product alternatives are not very different, low importance for the self-image, costs of extended information search outweigh the expected benefits, etc. This leads to simple problem solving, e.g. when passive learning and information processing, information is stored in the form of an image, limited number of evaluation criteria, choice is made on the basis of existing information. Beliefs, attitudes and intentions are the *outcome* and not the cause of the purchase.

Another common classification uses the degree of cognitive control as a criterion for the consumer decisions (Kroeber-Riel, 1992). The result is four types. Decisions with strong cognitive control: (i) extensive decisions; and (ii) limited (simplified) decisions. Decisions with weak cognitive control: (iii) habitual (routine) decisions; and (iv) impulsive decisions.

The differences between these types are fluent. The different decision types may be ordered on a continuum between two poles of very strong and very weak cognitive control:

Strong cognitive control ⟷ Weak cognitive control

Extensive decisions are characterized by an active search and use of information and a complex evaluation of the alternatives. Examples are the purchase of a house or the search for a restaurant to celebrate a wedding.

In the case of *limited decisions* the search is discontinued, when a certain satisfaction level is reached. For example, after having found a good wine (or in a nice restaurant) the decision is made without checking further alternatives. In the case of repeated buying experience and satisfaction lead to simplified decisions by limiting the number of alternatives and/or decision criteria. For example, choice of fruit is based on external quality and price. Similarly, after having tried several wines, a loyalty to a special type is developed and the incentive to search for better alternatives declines. This represents the transition to habitual buying.

In the case of *habitual decisions* we have no search for information and no evaluation of alternatives. The decision relies on past experiences based on a former extended or limited decision or a former impulse or curiosity purchase. The consequence is a rather stable pattern of consumer behaviour, which is based not only on own experience but on the experience of reference persons and groups, too. Examples are many of our daily purchase decisions for food.

Impulse buying is a quick reaction to a buying stimulus – without former search of information and evaluation of alternatives. In the literature four types of impulse buying are found: (i) pure impulse buying, a stimulus leads to deviation from 'normal' buying; (ii) memory impulse buying, a stimulus activates the memory and reminds you to buy a certain product; (iii) suggestive impulse buying, the buyer is reacting to a suggestion of the seller; and (iv) planned impulse buying, the buyer enters the store with a general intention to buy – the decision is made on the basis of the offer and information in the store, for instance, the purchase of special offers.

The different types of impulse buyings have gained increasing importance in high-income countries supported by the introduction of the self-service distribution system. Purchase decisions are more and more taken in the store and not at home.

The purchase of food can be classified as predominantly habitual and impulse buying with a small proportion of limited decisions. Extended decisions seem to be rare.

Finally, it may be interesting to compare the situation in food markets with the assumptions of the microeconomic theory of demand. This theory assumes the ideal of an extended decision process with a strong cognitive control. In reality most of our purchase decisions for food are underlying a rather weak cognitive control – a fact which has to be considered in market research and marketing.

CONCLUDING REMARKS

The theory of consumer behaviour is a complex multidisciplinary approach which goes far beyond the classical economic demand analysis, which explains the food demand mainly as a function of income and prices.

With rising consumer income the relative influence of prices and income on the food demand is decreasing, while the relative influence of preferences is increasing. In affluent societies, future changes in the demand for food will be caused mainly by preference changes rather than by changes in prices and incomes.

The better understanding of consumer behaviour, the evolvement of attitudes, the perception and the decision process, is supplying the basis for market research and marketing measures such as product development, pricing and promotion as discussed in Part V of this book.

FURTHER READING

Engel, J.F., Blackwell, R.D. and Miniard, P.W. (1993) *Consumer Behavior: The 25th Anniversary Edition.* Dryden Press, Hinsdale, IL.
Foxall, G.R. (1988) *Consumer Behavior. A Practical Guide.* Routledge, London.
Howard, J.A. and Sheth, J.N. (1969) *A Theory of Buyer Behavior.* Wiley, New York.
Kroeber-Riel, W. (1992) *Konsumentenverhalten*, 5 Auflage. Franz Vahlen, München.

REFERENCES

v. Alvensleben, R. (1989) Sozialwissenschaftliche methoden der agrarmarktforschung. In: *Neuere Forschungskonzepte und -methoden in den Wirtschafts- und Sozialwissenschaften des Landbaus.* Schriften der Gesellschaft für Wirtschafts- und Sozialwissenschaften des landbaus e.V. Band 25. Landwirtschaftsverlag, Münster-Hiltrup, pp. 239–246.
v. Alvensleben, R. and Meier, Th. (1990) The influence of origin and variety on consumer perception. Some psychological factors causing perception distortions. *Acta Horticulturae* 259, 151–162.
Engel, J.F., Blackwell, R.D. and Miniard, P.W. (1993) *Consumer Behavior: The 25th Anniversary Edition.* Dryden Press, Fort Worth, TX.

Foxall, G.R. (1988) *Consumer Behavior. A Practical Guide.* Routledge, London.

Kroeber-Riel, W. (1992) *Konsumentenverhalten.* 5 Auflage. Franz Vahlen, München.

Makens, J.C. (1965) Effect of turkey brand preferences upon consumer's perceived taste of turkey meat. *Journal of Applied Psychology* 49, 261–263.

Ritson, C. and Hutchins, R. (1991) The consumption revolution. In: Slater, J.M. (ed.), *Fifty Years of National Food Survey 1940–1990.* Ministry of Agriculture, Fisheries and Food, HMSO, London, pp. 35–46.

Veblen, Th. (1899) *The Theory of the Leisure Class. An Economic Study of the Evolution of Institutions.* New York.

11 Marketing Research

Leslie Gofton

Department of Social Policy, Faculty of Law, Environment and Social Sciences, University of Newcastle upon Tyne, Newcastle upon Tyne NE1 7RU, UK

INTRODUCTION

Marketing Research (MR) is becoming more and more important to researchers and practitioners in the field of agro-food marketing. Recent interest in food and drink (Sanderson and Winkler, 1983; Beardsworth and Keil, 1990), sparked by health concerns, has revealed the paucity of academic knowledge about food use and focused attention on the nature of marketing research data related to food and agricultural products. Differences arising from the nature of marketing research processes are apparent. While the application of marketing research techniques to food and farming gathers distinctive kinds of information, changes in the nature of markets for food and agricultural products and the way in which they are modelled also produces new types of measures, as a result of the techniques employed and information gathered (for instance, quality or risk indicators) (Perloff and Caswell, 1993).

While recognizing the common ground with, for example, standard social research data and methods, it is essential to understand the ways in which the origins, purpose and nature of marketing research practices imbue the data produced with distinctive characteristics.

THE NATURE AND ORIGINS OF MARKETING RESEARCH

According to the American Marketing Association definition:

> Marketing research is the function that links an organization to its market through information. This information is used to identify and define marketing opportunities and problems; generate, refine and evaluate marketing actions; monitor marketing performance and improve understanding of marketing as a process ... Marketing research specifies the information required to address

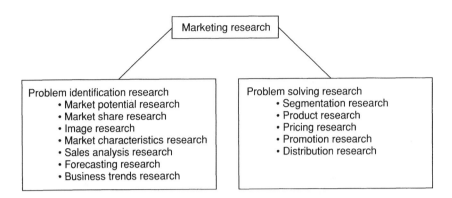

Fig. 11.1. The functions of marketing research (source: Malhotra, 1993).

these issues; designs the method for collecting information; manages and implements the data collection process; interprets the results and communicates their findings and implications.

Essentially, marketing research is an aid to decision-making, an '… information input to decisions, not simply the evaluation of decisions that have been made' (Aaker and Day, 1991). Since marketing activities are very varied and the needs of decision-makers vary greatly, this involves a wide range of decisions and consequently a variety of data and methodologies.

Marketing research may be required to describe a market (gathering and presenting data which are 'market facts'), to diagnose ('what if?' the marketing mix is changed), to identify the factors which are relevant to a particular state of a market or consumers' decision-making; or to predict outcomes of various marketing mix combinations. The range of decision-making situations described by Malhotra (1993) is shown in Fig. 11.1.

Marketing research, then, is only undertaken to provide information which can be used to inform commercial decision-making. As we shall see, this dictates what is gathered, how, the analysis to which it is typically subjected and the forms in which it is presented to end users.

THE GROWING IMPORTANCE OF MARKET RESEARCH INFORMATION IN THE INFORMATION AGE

Theorists such as Peters (1987) and Drucker (1985, 1990) see modern business as serving increasingly fragmented markets, with customers able to pick and choose from ever more varied and diverse product assortments. Failure rate amongst new products is very high and growing. Competitive, unstable and rapidly changing markets have increased business's need for information, to monitor and 'target' their segment of the market by identifying needs, formulating products and communicating with consumers. This is particularly the case in relation to food markets. Consumers now are

Table 11.1. Information needs of large organizations.

	Recurrent	Monitoring	Requested
Directors	Regional economic data Competitor prices/promotions Regional market share	Competitor product changes Customer acquisitions New entrants	Customer profiles Customer needs/satisfactions
Sales managers	Product margins Costs per call per customer Share by salesperson	Regional economic changes New competitive activities	Contribution per customer Salesforce effectiveness vs competitors
Brand managers	Brand share Customer satisfaction Feature preferences	Competitor activities Technology changes Government regulations	Test of new formulation Price elasticity data
Advertising managers	Advertising awareness Media habits of target audience	Media rates Ad themes of competitors Media effectiveness studies	New commercial theme test Communications impact of competitor's ad
Public relations manager	Key public attitudes towards the firm Company plans that affect the public	Legislative activities Trade and popular publications	Impact on buyers' attitudes of strike by the union Impact on the firm of other industries' responses to safety problems
Sales and marketing	Net contribution by product line Market share by product line Customer satisfaction levels	New competitors Developments in related markets Customer satisfaction levels New product launches by competitors	Impact on related products of dropping one product line Price and advertising elasticities across products

more discriminating and informed, demanding changes in production, processing, distribution and regulation.

The reasons for this activism are various. From the demographic standpoint, population growth is slowing, while there is more longevity, coupled with slowing demand for food. Other factors include profound economic changes, changes in the value of time and a changing significance for information. Overall, there is a much stronger quest for control amongst the population.

The marketing implications of the changes in social awareness and food preferences are diverse. These include: (i) value-added services in the marketing of new food products – basic foods are performing worst; (ii) convenience; (iii) nutrition education based on avoidance – this must be taken into account in marketing many types of products; (iv) new definitions of quality and safety; (v) high quality regulation and enforcement; (vi) environmental responsibility; and (vii) more international trade (see Perloff and Caswell, 1993; Kinsey, 1993).

Competition has created enormous pressures to cut margins and make every decision count. Information reduces the risks involved. Consequently, commercial success requires a marketing orientation (see Chapter 2).

Large organizations in the food and agriculture sector contain many decision-makers with widely varying types and levels of expertise, so that marketing research data must be rendered into many different forms (see Table 11.1).

The range, quality and amount of MR information has increased enormously with the electronic storage, retrieval and manipulation of data on stocks, purchases and a whole range of information about customers.

Precision marketing

Precision marketing encompasses a variety of types, such as 'Database', 'Relationship' and 'Niche' marketing. It is a form of direct marketing which uses information to precisely identify and meet consumer needs over time, using this information to identify market segments by combination (with, for example, census information) or projection (making inferences, for example, about attitudes, opinions or interests based on information such as place of residence or first name). The latest variant of this involves 'Data Mixing' and 'Data Warehousing', in which information relevant to marketers' concerns can be acquired 'off the shelf' from companies exploiting the latest developments in parallel processing to analyse the huge amounts of data yielded by Electronic Point Of Sale (EPOS) accounting systems. Markets are subjected to *psychographic segmentation* according to differences between consumers' lifestyles, using socio-demographic, behavioural and attitudinal data. *Strategic Analysis and Corporate Strategy* must take into account not simply 'market facts' but the macroeconomic environment, political factors, social and cultural changes. The idea that business simply means 'selling' is, as other chapters in this book also maintain, oversimplistic and dangerously misleading. Businessmen pursue different objectives at different times. Long-term objectives (for instance, market dominance) may very well necessitate short-term strategies which do not involve profit (such as loss leader price cutting, improving public relations, developing consumer awareness or repositioning a product).

Good decisions depend on comprehensible and actionable information about these markets being accessible to decision-makers – from sales personnel dealing face to face with retailers or customers, through research/development and production staff concerned with product form or uses or satisfactions they must deliver, to marketing staff concerned to identify customer needs and wants and formulate and communicate to those customers.

Management Information Systems (discussed in detail in Chapter 9) accomplish this purpose. Market information – about sales, customers, competitor activity, trends in the market, financial movements and government policies – can be accessed and used, perhaps incorporated into databases and software which enables non-specialists to perform complex projections and analyses, as well as monitoring their own and competitors' performances.

MR data, then, are produced in a variety of forms, for members of business organizations making decisions about their operations. This information flow has become more important with the rise of the marketing concept within the agro-food sector, as well as in business generally.

METHODOLOGICAL ELEMENTS

The research setting

MR is conducted, very often but not always, in response to the client's particular need for information, to solve a problem or to inform strategic decision-making. Thus, for example, Birdseye, the largest producer of fish products for the UK market, carries out market studies which reflect very specific concerns and not simply 'the market for fish'. The company focuses upon consumer preferences in mass markets (e.g. for frozen products based on white fish) and is not interested, for instance, in markets for fresh fish or little-used species (Gofton and Marshall, 1992).

Although market research is almost invariably initiated by a client problem, the researcher must translate the problem of the client into a meaningful research programme. Clients often do not have a clear idea of what kinds of information they need, nor what information it is appropriate or feasible to gather. Marketing research problems (as perceived by the client) and methods, concepts and programmes (as conceived by researchers) are then in a dynamic, interactive relation. Thus, research problems are, effectively, negotiated.

Resource constraints

Primary constraints are time (decisions generally have clear time limits) and finance. The numbers of personnel involved, the scale, the size of the sample, the number of stages, the choice of methodology and, in some cases, the timescale itself are, of course, derivable from the research appropriation. This, in turn, will be determined by the value of the information, or rather the potential profit which would be generated by the decision(s) which it informs. A Bayesian or rational decision analysis may be carried out to indicate whether or not it is cost-effective to conduct the research. Negotiation, from the researcher's viewpoint, involves a claim to inform a wider range of more important and valuable decision areas and a necessity to increase the range of information to be gathered (and resource allocation).

Marketing research instruments, ideally, relate only to: (i) the actual needs and capacities of the decision-maker; and (ii) the value of the decision. Surplus information involves unnecessary time and cost. Information should be relevant, timely and also 'actionable' – and hence, completely case-specific. By definition, primary market research should not produce generalities. Where research is carried out by an agency in order to publish a report, what information is gathered and included in the report is based on assumptions about what non-specialist decision-makers would be able to use. The client here is the purchaser or the market itself. Thus, for example, customer profiles are provided in terms of 'actionable' characteristics, rather than, say, age, occupation, educational level or registrar general's classification of social class. The popularity of 'psychographics', which is the language within which these profiles are couched, resides in the fact that it promises to describe key aspects of the ways in which different groups of existing customers think, feel and act, what kinds of activities they enjoy, what press and television they see – a description of their 'lifestyle' which should enable the marketer to see where the product could be fitted in, how it should, *ergo*, be 'conceptualized' and, consequently, how this 'product concept' can best be communicated to the consumer.

The market for research

Market research companies, from the largest (e.g. Nielsen, NOP, BMRB, CACI, Marketing Intelligence) to small, local or regional businesses, and even some college and university departments and business schools, are in the business of selling their services to food industry clients. Clients for market research information about agriculture and the food industry obviously have differing needs and differing capacities to make sense of potentially complex data. Market research data is sold on the basis that it combines up-to-date information which is relevant and understandable with scientific probity.

Marketing research, then, is produced in response to the practical needs of non-specialist businessmen by researchers who are competing with each other to sell their information's capacity to solve problems. Information must be understandable and relevant, but also trustworthy – users must be assured that it provides an accurate account of the phenomenon to which it refers.

Much of the information which is on sale in the form of marketing reports, reviews or studies has been derived from data gathered by either official sources (for example, in the UK, Office of Population Census Statistics, National Food Survey, Household Expenditure Survey) or trade associations – that is to say, it is largely secondary data gathered, at least in part, for general record keeping purposes – and then subjected to various kinds of analysis. Analysis can be tailored to the specific needs of a client – for instance, buying an ACORN (A Classification Of Residential Neighbourhoods) report on a particular area gives a street by street breakdown of the housing/lifestyle in a locality – but the data on population, rates of car ownership, employment statistics and consumption patterns are those collected by the government.

Primary marketing data are gathered, in the vast majority of cases, for the specific needs of particular clients and will only become available to the general public when the information it contains no longer has particular commercial value or strategic sensitivity.

The final product, the findings and the report are fitted to the client's organizational needs, information which can be understood by those who have to make the decisions and is relevant to their problems. There would simply be no commercial sense in producing data which were too complex to be grasped or easily absorbed by the client, or buttressed by 'superfluous' information.

What information is gathered has a precise cost. That should always relate to the financial value of the decisions it informs. The best market research has very limited application; like a tailored suit, it should fit the client, but no-one else, perfectly. As a consequence, the more precisely marketing research is targeted at the requirements of a particular client, the less generalizable are its findings. The better marketing research is, the less use it is likely to be to anyone apart from the client.

Market researchers commonly distinguish three main groups of data types: (i) secondary research; (ii) qualitative research; and (iii) quantitative research.

SECONDARY RESEARCH

Secondary research, sometimes called 'desk research' or 'background research', is mostly concerned with the use of secondary data (i.e. information which has not been gathered specifically for this research project). This includes company reports, government statistics, newspaper and journal articles and reports produced by commercial market research agencies. Most secondary data are quantitative, gathered by government or industry in accounting and record keeping. What data are available varies between countries; some are well provided for, in others the range and quantity of the data are very limited.

Secondary data are usually gathered at the beginning, exploratory, stage of the research process and involve two main sources.

Internal data sources

Internal data sources might produce: (i) sales records; (ii) delivery and stock records; (iii) prices and quotations; (iv) sales promotions; (v) advertising – media, messages budget; (vi) salesman's call reports and estimates of effectiveness; (vii) past studies on marketing effectiveness; and (viii) public relations strategy and expenditure.

Internal accounting and control systems provide data on inputs and outcomes. The principal virtues of these data are ready availability, continuing accessibility and relevance to the organization's situation. The main limitations arise from the fact that on many occasions systems are: (i) designed to satisfy many different information needs, consequently, reporting formats are frequently rigid and inappropriate for marketing decisions; (ii) too highly aggregated into summary results; and (iii) measuring in varying time frames (for instance, 'financial' as opposed to 'calendar' years).

Data are often variable in quality, for example, salesmen's call reports may be inaccurate or idiosyncratic, as may accounting records, for reasons of operational convenience or situational exigency, or simply because of the record-keepers. Non-business organizations – for example, those staffed by volunteers or non-professionals – are highly variable in their record-keeping.

This may lead to incompleteness, inaccessibility and indigestibility. Customer feedback – product returns, service records, customer correspondence, complaint letters – can form very useful data.

External data sources

External data sources – published by other researchers – fall into three main categories: (i) government statistics (census data, general household surveys, family expenditure data, business and investment data, farm income data, consumption data, international statistics); (ii) published data by trade associations, banks, stockbrokers, media owners (television and newspapers), local government and government-sponsored organizations; and (iii) data from syndicated market research organizations (retail audits, consumer panels). In the UK, retail audits are available through commercial agencies. These consist of an estimate of consumer sales based on the formula (opening stock + net deliveries) – (stock held at present). Consumer panels are compiled on the basis of interviews or diaries completed by households selected as a sample of the population, recording individual or household purchasing and consumption.

Other sources of secondary data

Other sources of secondary data might include: (i) consultation with experts (both within and outside of the client firm); (ii) observational studies (discussed later); and (iii) participation in an 'omnibus survey', in which a company undertakes to administer a very brief set of questions supplied by the client, as part of a general survey, often administered by telephone and processed immediately by computer.

The aim of secondary research is to: (i) formulate a problem in a way which can be tested using available marketing research techniques; (ii) locate the firm and those managers who face the decisions to be made more clearly in their business environment; and (iii) permit the development of preliminary hypotheses or a 'model' which can be refined in subsequent stages of the research. This may also serve to clarify information requirements.

Although, in principle, very large amounts of secondary data may be available to the researcher, in practice there are limits to how it can be used. The main problem is variability in quality, availability, cost, timeliness and relevance to needs of decision-makers.

Note, however, that distortions are involved with this kind of data and it should not be used without reference to supplementary material which will set it in context. Even when primary data are being used, it is desirable to consult secondary data to: (i) expand understanding of the problem; (ii) suggest hypotheses and research design alternatives to indicate preliminary a basis for choosing among alternative designs; (iii) help to plan the sample; and (iv) provide a basis for validating the obtained sample.

Finding secondary sources initially involves consulting *authorities*. It may be useful to consult existing experts, or refer to trade associations and specialized trade publications, which compile government data and collect additional information from their subscribers or members. For local information, a chamber of commerce or local authority provides a good starting point. A reference librarian is also vital.

General guides and indices may also be used. These might include guides of various kinds, from bibliographic sources, to general business sources and business periodical indices including database sources and international guides.

Compilations, which facilitate access to original sources, may also be very useful as a source of statistical information. *Marketing data guides* and *sales and marketing guides* might include such information as industrial incomes, sales at retail stores and market potential indices. Directories are useful for locating people or companies that could provide information (e.g. trade directories).

Computer retrievable databases may also be an important source here. These include:

1. Reference databases, which refer users to articles and news contained in other sources. They provide online indices and abstracts and up to date summaries of newspapers, business magazines and journals.
2. Source databases, which provide complete text or numeric information in contrast to the indices and summaries contained in the reference databases classified as: (i) full text information sources (complete stories and articles); (ii) economic and financial statistical databases (general economic information, specific industry analyses and forecasts); and (iii) online data and descriptive information on companies.

The main advantages of these sources are: (i) scope; (ii) speed of access and retrieval; and (iii) flexibility and efficiency in cross-referencing and searching. The main shortcomings include: (i) risky reliance on authority of anonymous authors; (ii) dependence on the selection policy of the database producer for articles, etc.; and (iii) variability of procedures used in such systems (e.g. search procedures) identifying relevant information in choosing between alternatives and in the costs involved, which may be significant.

The value of these sources for the agro-food researcher

For food research, the most important secondary data, for both market researchers and academics, are those produced by the government. In the UK, the Office of Population Census Statistics produces data on Household Consumption and Expenditure. The National Food Survey is a large panel study producing, for the last 50 years, data on food consumed within the home. Production consumption and attitude data are gathered by the Ministry of Agriculture, Fisheries and Food and by commodity organizations such as the Meat and Livestock Commission, the Sea Fish Industry

Authority, etc. Many market studies are combinations of official statistics with primary research into attitudes, behaviour and lifestyles, to produce profiles of consumers and estimates of market size and consumption. It is important, then, to distinguish between official statistics and the outcomes of analyses to which they are subjected.

Recently, however, Information Technology (IT), particularly the advent of optical codes and the means to store, handle and process the immense amounts of data which they generate, is revolutionizing marketing research and management information systems. In the area of food choice, the richness, accuracy and immediacy of the information which they offer about stock levels, ordering patterns, product distribution, customer profiles, etc. is readily apparent.

Database marketing

Database marketing, which uses individually addressable communications media (such as mail, telephone and sales force) to:

> (i) extend help to a company's target audience; (ii) stimulate their demand; and (iii) stay close to them by recording and keeping an electronic database memory of customer, prospect and all communications and commercial contacts, to help improve all future contacts (Fletcher *et al.*, 1990).

has grown from this. It aims

> ... to promote strategic improvements, identification of strategic advantage through better use of customer and market information leading to the development of new and unique products and services and development of long-term customer relationships.

These new developments are, to some degree, de-skilling market research and analysis, by making it possible for non-specialists to access and manipulate data very quickly and at very low cost. Strengths include the range, volume and detail in the data which are enormous and increasing, as new methods of analysis are developing fast. Data are also now becoming more accessible and widely used, as the cheapness and ubiquity of IT has encouraged the development of management information systems, generating a two-way flow of information.

Weaknesses arise from the fact that much of these data have been gathered by and for financial records systems. This limits their usefulness. Record systems (and consequently the quality and reliability of data) vary greatly. Data are often processed into reports rather than made available raw and this makes them more difficult to use.

QUALITATIVE RESEARCH

Qualitative research involves techniques intended to explore the meaning of consumer choices and patterns of usage, to gather information on opinions and attitudes and to develop a working knowledge of the words and phrases used by consumers in related consumer activities. There are two main typical methods: (i) focus groups; and (ii) individual (depth) interviews.

Focus groups involve between eight and twelve individuals and are used, in the case of food research, to test prototype food products, to provide a basis for attitude tests and to gain insight into food-using behaviour. Subjects are gathered together to discuss some aspect of the product. Schedules cover a clearly defined range of questions. Techniques, some borrowed from clinical psychology, for example, are used to spark discussion, encourage interaction and 'focus' the group around the issues, although discussions are free flowing. They are often written up into formal reports but, on many occasions, individual groups are used as the basis for presentations or 'de-briefs', in which only a bare outline of the discussions is included.

The data are similar to those of 'unstructured' field-work interviews and participant observation data, which are so important for social researchers and which have been criticized as 'soft' and subjective (Denzin, 1970).

Electronic recording has increased the usage of qualitative research. However, a permanent record of 'what happened' is held to provide 'harder' data, which can be analysed in new ways (using word counts, examining body language, doing micro-analysis of interaction, etc.). Also, changes in the commercial environment ('niche' marketing, faster turnover, increased competition) have made qualitative research more attractive because it provides insight into subtle aspects of 'consumer motivation' and it is fast and relatively cheap.

Another useful function of this kind of research is the gathering together of factors, statements and responses to products which can be further explored in quantitative stages of the research. For instance, in producing a quantitative comparison between the features of two products, or their image characteristics or any other areas on which they might be rated, the qualitative stage is used to draw up a list of consumer perceptions, behaviours or patterns of usage which can be incorporated into the questionnaires, diaries or experimental measures used in the quantitative stage. It can also be used to suggest ways of using or re-interpreting secondary data. Thus, for instance, we can make connections between the purchase patterns of seemingly unrelated product types by having insight into the minds of consumer groups. For example, EPOS data showed the USA retailer WALMART a parallel massive rise in the sales of beer and diapers in certain stores on Friday nights only; this is accounted for by one group – young fathers – getting in necessary provisions for the weekend. The sales figures alone would not tell us this, however.

The second major form, *individual (depth) interviews*, are usually carried out on specialists in a particular area, retailers and, sometimes, consumers, although these have to be carefully interpreted.

The main purposes of the quantitative research approach are quite distinctive. It aims to understand behaviour (by developing insight) and to evaluate reactions. This research is typically exploratory, concerning itself with issues of image, product usage, the associations of the product name, and possibly unexpected problems and usage of the product. Image and motivation may be explored in focus groups or individual interviews using a range of techniques derived from psychoanalysis. There are a number of different types. The approach also lends itself to experimental approaches – for instance, some techniques aim to explore subconscious or oblique associations of the product and how it is used.

Personification, for instance, requires subjects to provide imagined features to products which are fantasized as people. For example: if an apple were a person, what kind of suit would she or he wear? What job would he or she have? What car? This enables researchers to build up an image or a personality from the range of subject responses.

Story completion involves using story frameworks from which key details are omitted in order to build up ideas about consumer perceptions of the research topic. Key scenarios – such as product usage situations or related behavioural situations – can be used to generate insight and ideas about consumer usage behaviour.

Association tests ask subjects to react to stimuli related to the product or topic of the research (for example, the packaging used, the smell, the colour, the name, etc.). Some names may remind consumers of undesirable or unpleasant images, or suggest undesirable characteristics.

Projective techniques are intended to permit consumers to express ideas or aspects of their behaviour which may be sensitive or suppressed by putting them in the form of indirect expression – what a 'third person' might do, what 'most' or 'other' people think, how a certain person would react or behave. For instance, scenarios in which respondents are asked to supply the behaviour, thoughts, feelings, etc. of those taking part (typically, customers, salespeople, users of products, etc.).

Construction techniques, in which the story or scenario is produced by respondents, are also popular. This may involve, for example, cartoons in which speech bubbles are filled in; third person or picture accounts; or building up collages or pictures using ready made images from magazines.

Expressive techniques concentrate on how constructs are produced and involve play, drawing, role playing.

There are, according to Tull and Hawkins (1990), three main stages in the progress of these qualitative interviews or group discussions. First, the moderator seeks to *establish rapport* and lay down the rules which operate (for example, one person speaking at a time). Then, using various stimuli and techniques, he seeks to *provoke discussion* and, using minimal intervention, to *guide the discussion* along the appropriate lines, usually with the aid of a schedule drawn up carefully before the focus group. Finally, the moderator intervenes to determine levels of agreement by identifying issues and drawing out those who agree and those who do not. The moderator seeks to allow the group to generate the ideas and, as much as possible, to proceed under their own impetus, exercising minimal influence.

There are clearly problems which need to be addressed here. First, great skill is required of interviewers, making this technique tricky to use. Second, the representativeness of participants is always an issue, so findings must be rigorously and sensitively interpreted. Finally, the lack of measurement means that findings must always be conjectural and stand in need of empirical testing.

Qualitative research has traditionally had a limited role in marketing research, for example: (i) to gather ideas and concepts, for research hypotheses, attitude measure, scales, or questions about use or behaviour; (ii) to gain insight into the meanings of a product for consumers; and (iii) to gauge consumer reactions to changes in the marketing mix.

The main strengths of this approach are the production of rich and complex data, dealing with the meaning of behaviour, details of everyday behaviour, attitudes, values, motives, etc. It also focuses on behaviour, actions, choices and ideas which incorporate the world of commerce (different brands of products, retailers and their activities), advertising and so on – often bypassed by academic research. A wide range of methods are often used together in the same setting, borrowing from different academic traditions, e.g. psychology (projective techniques), sociology (attitude tests) and economics (trade-off tests).

Weaknesses are limited application of the data, which are often ready-interpreted. This limits the range and nature of application. At the same time, this may be of variable quality. In some cases there is a strong suspicion of 'psychobabble', with researchers using facile and inappropriate recipes for interpreting data. Compared with quantitative methods this is a cheap method but, in some cases, qualitative researchers, who are highly skilled, may charge high fees for what is, in many cases, a product with a restricted range of application. Quantification may be much more expensive, but arguably the data will have a much longer shelf life, more flexibility, and a wider range of application.

QUANTITATIVE RESEARCH METHODS

Experiments

Experiments are used in marketing research, for example, to examine consumer response to different food attributes. 'Sensory testing' or 'taste testing' may take place in a food laboratory with panels (who may be either 'naive' or trained); or in a 'Hall' test, in a supermarket or public place. Typically, such tests involve rating, comparing or evaluating a particular attribute – taste, smell, appearance or mouthfeel. Other experiments are carried out in the field by, for example, varying mix factors (such as promotion and advertising) between matched areas.

Taste panels for 'sensory testing' are particularly important for food marketing research and are the standard method used to test and evaluate new food products being brought onto the market. They are also used to evaluate the quality of produce and to test products which involve, for example, processing or blending.

Two different types of panel are employed:

1. *Trained panels* are composed of individuals who have been trained to make specialized or subtle discriminations or to test products which could not be tried out on a non-specialist experimental subject (for example, animal or children's food).

2. *Naive*, or *untrained*, *panels* are composed of individuals who are selected to represent 'ordinary' members of the population. They will be used for a relatively short period of time and are expected to provide 'naive' responses to products, and so to be indicative of consumer reactions in general. Different kinds of responses are expected from this type of panel and they are typically not required to provide subtle or specialized evaluations.

In food laboratories, panellists are placed in a tightly controlled environment (individual booths within which subjects are carefully shielded from smells and extraneous sights and sounds) and provided with carefully matched samples of the foods to be tested. This is intended to eradicate the influence of external factors which might distort or affect responses. The main objection to this kind of test is its artificiality. It is pointed out that foods are consumed in situations where many other factors are affecting response, apart from the organoleptic and olfactory aspects of the product, and these kinds of tests cannot take these into account.

Some tests may be carried out 'in the field' in order to get a truer 'feel'. These may take a number of forms:

1. *Hall tests* are conducted in malls or supermarkets, and consist basically of a table at which customers are given a sample. Then they are asked a few simple questions about the taste, mouthfeel, etc.

2. *Group tests* consist of focus group sessions at which new products are introduced to consumers, perhaps in stages, and then the moderator leads the group in discussion of the various dimensions of the product which may be relevant. These aspects include all the sensory dimensions of the product, but could also involve dimensions such as usage, acquisition, preparation, combination with other products or ingredients, characteristics of consumers or potential consumers, packaging, price, etc. Obviously this type of approach can produce wide ranging and detailed data and may provide the basis for important insights which can be crucial for effective product development work.

In general, it is best to approach the results obtained in both sensory tests and food panels with great caution. Scores about the taste of food are not a sufficient guide to the likely success of a product when it is on the market, since this will be affected by a wide range of other product features. Also, the level of involvement and risk is very different when foods are being sampled and responses are based on highly artificial and limited reactions. Food choice is likely to be much more complex in everyday life and to take into account a range of issues which cannot be accommodated into the test. For analysis of such factors in relation to fish see Gofton and Marshall (1992).

Observational studies

Observational studies may involve, for example, recording shopping behaviour according to specific criteria, details of eating behaviour and so on, or to explore or clarify the problem being investigated. Observation may be conducted under natural or contrived conditions – in a home setting or under the 'artificial conditions' of a focus group.

Studies of behaviour in food shops, and of food use, are regularly carried out to investigate the effects of store layout or promotional activity, or to gather information about the most appropriate product form or reformulation of the product concept.

Experimental and observational approaches using laboratory based or field tests with samples or panels are well established, and rigorously grounded methodologies, with good scientific and statistical procedures. In addition, there is commonality of statistical analysis procedures with many other types of academic research. However, the value of such data for the non-specialist tends to be limited. It is often tailored to provide guidance on specific aspects of product form (acceptability, comparison with alternatives, proto-product testing, etc.). It often involves reductionist assumptions (posing the primacy of physiological and psychological processes over social, cultural, even economic and commercial factors) in experimental work.

Interview survey methods

The same limitations and strengths apply to these methods as in academic social science. Market research surveys very often involve offering cash, gifts or other material inducements to obtain respondents' cooperation, which makes gathering respondents somewhat easier than for academic researchers, but raises new issues about the reliability of the data.

Vox Pops (street interviews) are also used to gather information quickly, using a 'convenience' sample and a very short and simple questionnaire. The outcome is often interesting, but highly unreliable.

Quantitative data indicate: (i) the total size of the market (volume, value); (ii) the segmentation of the market (different usage, users, different occasions or settings, etc.); (iii) market shares (by brands); and (iv) usage of a product (e.g. when eaten, in what combinations, how are they cooked, etc.).

In gathering primary data, the survey is the most well known and still the most popular method. It is important to consider the process of survey research in more detail, beginning with the selection of subjects to be studied.

Sampling

Decisions about markets usually relate to groups of individuals or households which may be large in number and, sometimes, dispersed across large geographical areas. Apart from a few special cases (for instance, some markets for industrial or scientific products or services, or in what is called a 'census study'), it is impractical and, in terms of cost, wholly uneconomic to attempt to reach and study every individual who consumes, or might consume, a product or service. In order to provide reliable information about such groups, researchers select a sample from amongst the population which is a fraction of the total number of elements (individuals, households, groups, towns, etc.) which concern decision-makers, with the objective of using them to make inferences about the behaviour or characteristics of the entire population. The population is the term used to describe those subjects who are considered relevant to the topic which is being studied and who may be asked questions in the course of the research.

Researchers may use either probability or non-probability samples:

1. Probability sampling is a method which aims at objectivity; the probability of selection is known in advance for each unit. The researcher plays no role in deciding which specific population units are chosen to be included in the sample. The researcher specifies some objective scheme for choosing these units, which is then independent of any selectivity or bias on behalf of the researcher.

2. Non-probability sampling is subjectively based and the probability of selection for each population unit is unknown beforehand. Selection is typically based on some kind of judgement by the researcher, with all the implications for possible bias. As a result, the objective probability of selection for each unit cannot be gauged.

Probability methods of sampling

Simple random sampling   Simple random sampling is the most basic form of probability sampling. The principle on which it is based derives from statistical theory; if we assign an equal probability of selection to every possible sample of a stated size within a population, then we can use such a sample to talk about the characteristics of the total population with a known probability. The actual selection can be carried out by randomly picking the desired number of units from the population, using, for example, random number tables to choose respondents from a census list. If it is the case that any member of the population has an equal chance of being selected, this is equivalent to identifying all possible samples and, then, choosing one at random.

Sampling error and sampling distribution   All sampling methods have a sampling distribution associated with them. This shows how the sample statistic varies between the random samples of a given size chosen from the same population. For a particular procedure, the nature of the distribution involved will indicate the possible range of sampling error which might arise in the sampling process. Sampling error, in simple terms, is the standard deviation of the sample statistic value obtained by gathering a number of different random samples. This figure is known as the 'standard error'.

Confidence intervals For a sample which has been randomly selected, the central limit theorem in statistics tells us that when the sample is large enough, the sampling distribution associated with a sampling procedure displays the properties of a normal probability distribution. We can, as a consequence, calculate confidence intervals for a population parameter (mean or proportion) from a sample statistic value.

Sample size selection The size of sample which is gathered must be derived from the desired precision level, in other words what margin of error can be tolerated, the desired confidence level or how robust the data need to be, the degree of variability in the population to be studied (which can only be estimated) and the resources available. Typical means of calculating sample size would be based on the formula:

$$n = \frac{Z_q^2 S^2}{H^2}$$

where n is the sample size; H is one-half of the confidence interval range; q is the confidence level; S is an estimate of the standard deviation; and Z is the population size.

Formulas such as these can be used to calculate the required sample size but, since it is difficult to estimate the standard deviation before the study has been carried out, results must be treated with caution, since they give an impression of precision and confidence which actually rests on a judgement which is highly subjective, in point of fact.

Note that resource constraints can be accommodated within this method by their effect on the confidence level – if the confidence level is lowered, then a smaller sample size will be required and this will involve less use of resources.

Problems of simple random sampling In order for simple random sampling to be used, it is necessary to have a complete list of the population involved, in order to draw up a complete sampling frame. This is often very difficult to achieve, given the necessary limitations of record-keeping procedures. In the case of population records, for instance, these are often based on census information and can quickly go out of date owing to changes of address, births, marriages and deaths. For more esoteric types of populations (for instance, the users of particular brands of product) it may be impossible to construct an exhaustive or accurate list and the identity of such group members will be hidden.

The size of such groups, and the geographical dispersion, may also make it impractical to gather a complete list. Even if one can be gathered, it would be very difficult to contact the members of the list because of the distances involved and the resources implied by gathering a very large sample to reach a reasonable level of precision.

Because of the basic requirement that every member of the population should have an equal chance of being selected by chance, the method also throws up extreme and unrepresentative samples from time to time, with the consequent danger that the data gathered about them will be misleading.

Because of these risks, the pure 'random sample' is rarely used in marketing research. Often, more subtle and complex methods are necessitated, such as *stratified random sampling*, in which a sample is forced to contain units from each of the segments or strata of the population. These may be *proportionate* or *disproportionate* to the characteristics of the population as a whole, depending on the nature of the population and the objectives of the researcher. Proportionate samples reflect the size of the different units within the population, while disproportionate sampling may be used to gather information on, for instance, the diversity of behaviour within a given sub-unit.

Non-probability methods of sampling

Convenience sampling   As the name suggests, this is done very much at the convenience of the researcher, with respondents being selected without regard for randomness or any objective methods being applied to try to avoid bias. This is the method typically used by naive or 'quick and dirty' researchers or by reporters from newspapers or mass media aiming to gauge public opinion. Using a convenient group of subjects, such research generalizes about an 'implied' population on the basis of data which may be very far from representative.

Judgement sampling   This is a slightly more refined technique, where subjects are selected in terms of some criteria which the researcher feels will make the sample more representative (for instance, consciously interviewing equal numbers of men and women, or picking out a 'spread' of people who appear to be from different social classes). This is also known as 'purposive sampling'.

Quota sampling   Quota controls are used in market research to 'spread' the samples and to make sure that we get a reasonable and proportionate representation from groups within the overall population.

The typical stages of obtaining a quota sample are: (i) divide the population into segments on the basis of control characteristics; (ii) determine a quota or number of people to be selected from each of these segments; and (iii) interviewers are typically instructed to fill the quotas assigned (e.g. 'interview at least five men and at least five women') but are also given some freedom in selecting the sample units.

On occasion, in marketing research we want to 'skew' our sample, so that we gather disproportionate numbers of a particular group, for instance, heavy users. We may want to talk about a specific sub-group more than others. Quota controls enable us to gather more of that group than if gathering a random or proportionate sample based on age, sex or normal socio-demographic variables (for instance, more heavy users, regular customers or more young women).

The practice is simple. Each field-worker is given a control sheet on which is printed the characteristics of the respondents which they must interview. If you must interview 30 respondents, for instance, the control sheet might specify:

Men   12
Women   12
Either   6

This means that, in the early stages, you can make sure you get 12 men and 12 women but, in the later stages, you have the freedom to accept either men or women. This is important because each variable is treated in the same way – the same control sheet would also say, perhaps:

Under 18 years   4
18–35 years   10
35–55 years   10
Over 55 years   2
Any age   4

If we want to pick out the behaviour of one specific sub-group (for instance, heavy users or light/infrequent users), then the sampling procedure must ensure that we gather enough of this group to be able to talk confidently of their characteristics. In market research, as opposed to other kinds of social scientific surveys, it is often impractical to stick to the strictest rules relating to the requirement for randomness in relation to representativeness, since this would result in large amounts of redundant information and make the gathering process unfeasibly lengthy.

Asking questions

In designing a questionnaire, a number of important decisions have to be taken and we can characterize the design process as seeking answers to these questions. They fall into a number of different categories, according to the stages of the research process (Table 11.2).

Questionnaire construction

Questionnaire construction involves developing an instrument which can test the hypotheses developed in the early stages of the research, in consultation with the client for the research information.

The stages of the construction process are as follows:

1. *Formulating research hypotheses* (preferably simple and actionable). Research hypotheses provide a focus for information gathering and check the information we gather.
2. *Decide levels of generality.* Groups and classifications must be relevant. The results of the qualitative research should alert us to the differences (and similarities) between groups and help to work out which groups should be identified by means of the classification, usage and other questions.

Table 11.2. Decisions to be taken when designing a questionnaire (source: Tull and Hawkins, 1990).

1. Preliminary decisions.
A. Exactly what information is required?
B. Exactly who are the target respondents?
C. What method of communication will be used to reach them?
2. Decisions about question content.
A. Is this question really needed?
B. Is this question sufficient to generate the required information?
C. Can the respondent answer the question correctly?
D. Will the respondent answer the question correctly?
E. Are there any external events which might bias the response to the questions?
3. Decisions concerning question phrasing.
A. Do the words used have but one meaning to all respondents?
B. Are any of the words or phrases loaded or leading in any way?
C. Are there any implied alternatives in the question?
D. Are there any unstated assumptions related to the question?
E. Will the respondents approach the question from the frame of reference desired by the researcher?
4. Decisions about the response format.
A. Can this question best be asked as an open ended, multiple choice or dichotomous (yes/no) question?
5. Decisions concerning the question sequence.
A. Are the questions organized in a logical manner that avoids introduced errors?
6. Decisions on the layout of the questionnaire.
A. Is the questionnaire designed in a manner to avoid confusion and minimize recording errors?
7. Decisions about the pretest.
A. Has the final questionnaire been subjected to a thorough pretest using respondents similar to whose who will be included in the final survey?

3. *Plan the tabulations.* Ways in which the data are likely to be analysed. The standard breakdown may be sufficient but we need to be aware of other possibilities. We must settle the tabulations before we write the final questionnaire.

4. It can be very effective in this process to *set up a matrix* with the topics in detail down one side, and standards and other groupings across the top. The cross points can be identified as questions which have to be asked (or question numbers in the draft). This ensures that nothing important is omitted.

5. *Ordering the topics.* Use a flow diagram to work out how the information is to be gathered.

It is good common practice to open with one or two general bland questions which the respondent will find easy to answer. If the objective of the research needs to be disguised, it might be useful to include 'dummy' questions at this stage. For example, if you are testing attitudes towards a particular product, it would be wise to include questions about competitors or alternatives. It is also useful to explore present behaviour in the market before looking at the past or future intentions (what is being done, used, bought and eaten now, before asking about earlier). We also need to record data about subjects' behaviour before asking questions about attitudes. This allows preliminary concentration process before scales, etc. can be tackled. It is also effective to take topics in a logical order to avoid confusion. For example, ask people if they ever eat a particular type of product, before asking questions about specific brands. Withhold topics which might be embarrassing, sensitive or personal until the interview is well under way. The more time is invested in an interview, the less likely is the subject to terminate because of sensitive or indeed difficult questions.

Be prepared, during the 'piloting' of the questionnaire, to try these sorts of questions in more than one place and in more than one form. Boring, repetitive sequences (e.g. runs of multi-choice questions or rating scales) are highly likely to produce bad data because subjects 'switch off' and give meaningless or inaccurate answers. The topic order should be carefully considered since early questions influence the answers to later questions.

Classification questions, which may be embarrassing or difficult, should be at the end of a questionnaire for probability samples, but at the beginning for quota samples, because it is necessary to identify where the respondent fits into the quota set. 'Show cards' may be used so that answers are not given 'out loud' (age, income).

Treatment of topics

A number of different question types may be employed. *Open* questions (for example, 'What do you think about brand X?') produce data which involve lengthy recording time, probably errors in recording, heavy coding and analysis time. *Closed* questions (for example, 'Do you think brand X is: (i) too expensive; (ii) too cheap; or (iii) the correct price?') are more lengthy and difficult to design, but are quick and easy to administer, more accurate and valid and easier to analyse. Unless carefully planned, they may also be more boring for the respondent but, from the point of view of the analyst, we should aim for as many closed and pre-coded questions as possible.

Pre-coding is very important. It involves: (i) anticipating possible answers (through the qualitative work and the piloting of the draft); (ii) supplying a code number or value range for each answer; and (iii) identifying with a reference number the location of the code on the code sheet or coding plan which is followed when data are entered, so that each answer is recorded at a unique and consistent place within the data file (e.g. male = 1, female = 2 recorded at, for example, column five of line number one on every code sheet).

Questionnaire layout

Each questionnaire needs identification of: (i) the (computer) job by means of a reference number; (ii) each individual questionnaire by means of a reference number; and (iii) the interviewer by means of a reference number.

Standardized introductory remarks are also required. For personal interviews these can be placed on a card, which can be shown to the respondent.

This may also include a photograph of the interviewer and, perhaps, an official identifier, such as a signed statement attesting the status of the field-work, assuring the respondent about the nature of the project and the confidentiality of findings.

Classification data about the respondents must also be gathered – the place and form depends on the nature of the sample and the form of enquiry. Sometimes, the name and address of respondents is taken in order to carry out checkbacks. (Checks on the honesty and reliability of field-workers are carried out routinely by commercial research firms.)

The layout must clearly distinguish questions from instructions. The usual way is to use capitals for instructions (often within square brackets), while the questions are in normal type (e.g. 'Do you eat meat?' Yes/No [IF NO; GO TO QUESTION 7]). These are sometimes called 'skips'.

Questions are best clearly separated from answers in the layout. The route through the questionnaire should be clear. The interviewer should be told whether or not to read out the pre-coded answers, and also when to show a card if one is to be used.

Layout generally follows a number of simple rules. The content is determined by the research objectives, the form is determined by the nature of the survey population and the method used to convey the questions, although these clearly interact and influence each other.

General guidelines to follow when deciding on questions involve considering, first of all, 'Can the respondent answer?'. Researchers should beware of assuming knowledge, behaviour or experience which the respondent may not have. Try to avoid embarrassment and fears about prestige, by asking questions which involve sensitive information or, indeed, simply asking the wrong person.

Second, 'Will the respondent understand what is wanted?'. Pitfalls include misunderstanding words and phrases (even simple ones such as the name given to foods or meals), asking about what people do 'generally', 'usually' or 'regularly'; these are very common sources of ambiguity or uncertainty. Rather ask what the respondent did yesterday, or this week. Also, unfamiliar words should be identified during the piloting process. Beware too of long explanations; these should be avoided whenever possible (although they can be important).

There are also dangers involved in the truthfulness of respondent answers. Problems may arise because of inability to verbalize answers. Pre-codes can be used to avoid this. Respondents may also suffer from defective memory. In order to avoid this problem, a checklist, diary records or mechanical recording devices could be used. Subjects may also show reluctance or unwillingness to answer the questions. In this case, indirect or subtle questions can be used (about other people, on hypothetical cases) to obtain true answers in survey research. Five main factors seem to be involved: (i) lack of awareness – respondents may be unaware of their own motives or attitudes; (ii) irrationality – respondents may feel inhibited by the value which society places on logic and good sense; (iii) inadmissibility and self-incrimination – these two refer to inhibition due to the perceived desirability or acceptability of behaviour; and (iv) politeness – consumers may not want to say bad things about others out of reserve or politeness.

Some researchers have used a variety of projective techniques or indirect methods of gathering such data to overcome this. Some of these approaches may be used in the design of questionnaires (for instance, words associated with a product or brand, or the image conveyed by a particular marketing strategy).

Asking questions about attitudes

Attitudes are learned pre-dispositions to respond in a consistent manner, favourable or unfavourable, towards a given object. Opinions are the expressions of these attitudes, although the two words tend to be used interchangeably. These may be general (about eating) or specific (about brand of bread), about physical or functional properties (colour, speed, taste) or subjective or emotional dimensions (style, gender, mood). They are produced by experience or background and function as a speedy and effective way to process information and form a response or orientation towards it. In order to develop ways of measuring attitudes, we need to go through a number of stages in the marketing research process.

Establishing the 'universe of content' or forming research hypotheses refers to the body of ideas held by the relevant population about the 'attitude object' (e.g. cheeses, wines, burgers, grocery shopping in general). Depth interviews/group discussions and general qualitative research in the exploratory phase will have established a variety of statements about products or service in the market in question and the context in which they are used.

The universe of content is formulated in a list of statements or hypotheses drawn from this qualitative research. These include: (i) ideas held about the attitude object by the survey population; and (ii) expressions used by the population when talking about these ideas. Note the variety of ways in which the same idea may be expressed. This is sometimes referred to as the 'attitude battery'. Decisions are then taken about the order of topics/ideas and the number and variety of statements associated with each topic. Note that some are likely to be more important than others. A scale could thus record 'strong agreement' with both: (i) 'convenience foods are a necessity for modern housewives'; and (ii) 'convenience foods make it possible to give more time to the family'. But the latter is clearly more important as a basis for actions. It is important also to bear in mind that agreement with (i) may express a belief ('I accept this as a statement') or an evaluation ('I identify with this point of view').

Types of attitude scale

Scales can be very simple, as follows:

	Agree	Neither agree nor disagree	Disagree	Don't know
(code)	1	2	3	4

This is a *nominal scale*. Analysis consists simply of a counting of the responses for each category and then a comparison across categories and with the total response. Going further, we might ask the respondents to rank the order of importance for a series of statements. For example:

'Soft cheese is very healthy'
'Soft cheese is very good value for money'
'Soft cheese is a very versatile food'
'Soft cheese is very good for children', etc.
[RANK THESE STATEMENTS IN ORDER OF IMPORTANCE]

This uses an *ordinal scale*. To summarize the responses, we allocate a number to each rank depending on the number asked for (for 4, top scores 4 and bottom scores 1). This is useful but the scale is not sensitive – there is no indication of the intensity with which attributes are viewed, nor of 'distance' between attributes in terms of this intensity. One may be very important and the rest relatively trivial.

To indicate 'strength/weakness' of feeling, we use rating scales. Some of these scales are listed below.

Likert rating scales   A statement is put to the respondent and he or she is then asked 'Please indicate your level of agreement or disagreement'. Commonly a five-point scale is used:

	Strongly agree	Agree	Neither agree nor disagree	Disagree	Strongly disagree
(points)	5	4	3	2	1

A seven-point scale may be constructed by adding 'very' to the phrase at each pole. Note that these weights are reversed for negative statements and that the wording can vary.

The verbal rating scale recommended by the Market Research Society of Great Britain is as follows: 'I am going to read out some of the things which people have said about ... (topic) ... Please tell me how much you agree or disagree with each one (show card). Pick your answer from this card':

Agree strongly
Agree slightly
Neither agree nor disagree
Disagree slightly
Disagree strongly

A diagrammatic rating scale looks like this:

	Strongly agree	Slightly agree	Neither agree nor disagree	Slightly disagree	Strongly disagree
	0	0	0	0	0

The statement is read out by the interviewer or written on the question-naire in the postal survey. The respondent is invited to point at the position that expresses his or her 'feeling' in response to the statement or tick in a mail survey.

For the analysis, weights are allocated to the scale positions. Numbers should be consistent, if both positive and negative statements are mixed (e.g. 'hamburgers are very tasty food' – 'hamburgers are very unhealthy food'). Care should be taken to make 'strongly agree = 5' for the first statement but 'strongly agree = 1' for the second statement, and to switch between high scores for negatives and high scores for positives in the course of the questionnaire. This prevents responses becoming automatic and encourages respondents to listen to the question that is being asked of them.

These scales can be used to compare the total response of the sample to individuals' statements, with response to the attitude battery as a whole. Also, we can compare the summed score of individual statements and see how they correlate (are there consistencies which enable us to identify psychographic segments for instance?).

Osgood's semantic differential scales   Likert scales are commonly used for general subjects or to rate specific attributes of brands or products. Semantic differential scales are attractive because they are faster and easier to administer, also because they have proved more meaningful to the respondents when it comes to rating responses to statements about the specific attributes of named products and services.

A product or service is usually designed to have certain specific desirable attributes. We want to find out whether or not, and how strongly, these desirable attributes are associated with our product, compared perhaps with competitive products. The following scales might be used to compare and rate refrigerator displays in competitive grocery stores:

Good product range	○ ○ ○ ○ ○ ○ ○ ○	Poor product range	
Attractive layout	○ ○ ○ ○ ○ ○ ○ ○	Unattractive layout	
Good quality brands	○ ○ ○ ○ ○ ○ ○ ○	Poor quality brands	
Easy to get at goods	○ ○ ○ ○ ○ ○ ○ ○	Difficult to get at goods	

(commonly seven- or nine-point scale)

The respondent might be asked to rate a number of different stores which have refrigerator displays, using these scales. If a list of stores were used, the order should be rotated so that the same store does not always come first.

The attributes can be either 'mono-polar' (sweet – not sweet) or 'bi-polar' (sweet – sour). Bi-polar scales appear more natural but, sometimes, there are no generally agreed opposite adjectives which the general population would easily recognize.

Constant sum scales   These may be used when we want to force respondents to choose an order of importance for the attributes. Thus, we might ask the respondent to allocate, in order of importance, 13 points only between the following food attributes:

Safety and hygiene	with	4 points for an essential attribute
Value for money		3 points for a very important attribute
Ease in preparation		2 points for an important attribute
Good tasting		1 point for a not very important aspect
		0 points for not important at all

These can be very useful since they prevent respondents from making facile and unthinking evaluations. They are also more difficult to administer and require more work from the respondent and greater concentration. They can also cause some resentment, since people may feel that they have been forced to make a choice.

Often, combination scales can be used, with respondents being asked to look at a list of statements or attributes, rank them in order of importance and, then, evaluate the strength of agreement or rating of each statement in relation to a particular product. Other approaches involve, for instance, asking the respondent to rank in order of importance and, then, record level of agreement/disagreement (code 7 = strongly agree; code 1 = strongly disagree).

'Soft cheese is very healthy'
'Soft cheese is very versatile'
'Soft cheese is very tasty'
'Soft cheese is good for children', etc.

This approach is more economical but also more complex to administer. Note that too many scale positions confuse subjects, while too few mean that we get a clustering around the centre as subjects avoid extremes of opinion. An even number of points forces a choice since there is no neutral point. While researchers like this, some subjects resent it and the final differences may be artificial.

If a brand is generally liked, it may score high on all attributes (the halo-effect). Scoring positions can become patterned or habitual; it is useful to mix positive and negative scales (but be sure to reverse the ratings, for example).

The data generated are used for sophisticated statistical analyses, but scale positions and weight attached to them are arbitrarily fixed. With the scale in common use in market research, distance between positions appears equal; this is not necessarily (or usually) the case for the respondent. Thus, it is dangerous and misleading to treat scores here as having some of the attributes of real numbers (e.g. measuring intervals or intensity).

Pre-testing or piloting

This is carried out in order to identify with the form, meaning, order, structure, etc. of a survey questionnaire. It may be used to: (i) develop lists of possible answers, so that open questions can be closed and pre-coded; (ii) examine the form of data which a question will provide and the likely usefulness of questions; (iii) probe for differences between consumer groups and, perhaps, further refine the classification schemes and data which may be used to segment the sample; and (iv) identify problems with the way in which the data can be analysed by computer, possibly entering the data, and trying out some analytical techniques.

The main purpose of the pilot, however, is to fix the form of the questions. Very often, simple words or phrases may be misinterpreted or may be open to a range of different interpretations by different respondents. We try to identify these problems in the pilot stage by administering the draft instrument (and we must, at some stage, use respondents with the same characteristics as those of the sample, although friends and colleagues are useful 'guinea pigs' in the early stages). Then, we must go through again carefully, asking of each question 'What do you understand by this?', 'Is this clear?', 'Is there any other possible meaning for this question, word, or phrase?'. Responses are recorded alongside the questions and, then, the reactions of all pilot subjects are checked against the intentions and objectives of the project. Questions are then modified, taking problems into account.

A further use of the pilot is to check on the form and conditions of administration – How long does this take? Are people likely to be prepared to answer these questions? Where do we need to show cards? Can we carry these interviews out in the street? Will we need special items such as photographs, samples, display material, or perhaps pencil and boards for the subjects to use?

If we want to develop a list of possible responses, it may be useful to do a larger pilot sample and to take more care about the representativeness of the sample chosen. This may also be the case if we want to test the quality or the usefulness of the questions that we ask by entering dummy data (or test data from the pilot). In special cases, such data can be aggregated with the final survey material and so provide a larger final sample, but note that this must be carefully done and can only apply to some questions.

Coding

Questionnaire coding involves assigning a means of identification (usually a number) to each variable about which questions are asked or observations made (classifying men or women, or identifying a class, an income group or a residential neighbourhood may not involve questions, for instance) which is then counted during the process of analysis. In former times, this might have been done by hand; more recently early computers used punched cards which were mechanically sorted and counted according to whether or not they had a hole in a specific place or not. Modern electronic computers actually use a very similar principle, sorting coded numbers in data files in the form of binary numbers. A wide range of software is now available for this task.

CONCLUDING REMARKS

The best way to learn about marketing research, however, is to do it. It is a logical process, the seemingly tedious checks and classifications are all necessary to make it efficient – they all save work and produce better results in the long-run. Market research data are, as we have seen, very special, and we must be aware of what they can and cannot do. The next chapter discusses some techniques for analysing and interpreting survey data.

FURTHER READING

Chisnall, P.M. (1992) *Marketing Research*. McGraw-Hill, London.

Churchill, G. (1987) *Marketing Research: Methodological Foundations*. Dryden, New York.

Day, A. (1991) *Marketing Research*. Prentice-Hall, New York.

Gill, J. and Johnson, P. (1991) *Research Methods for Managers*. Chapman Publishing, London.

Hawkins, T. (1990) *Marketing Research: Measurement and Method*. Macmillan, New York.

Malhotra, M. (1993) *Marketing Research: An Applied Orientation*. Prentice-Hall, New York.

Zikmund, W.G. (1991) *Exploring Marketing Research*. Dryden, New York.

REFERENCES

Aaker, G. and Day, A. (1991) *Marketing Research*. Prentice-Hall, New York.

Beardsworth, A. and Keil, T. (1990) Putting the menu on the agenda. *Sociology* 24(1), 139–151.

Denzin, N.K. (1970) *Sociological Methods: A Sourcebook*. Aldine, New York.

Drucker, P. (1985) *Innovation and Entrepreneurship*. Heinemann, London.

Drucker, P. (1990) *The New Realities*. Heinemann Professional Publishing, London.

Fletcher, C., Wheeler, C. and Wright, J. (1990) The use of database marketing. *Quarterly Review of Marketing* 4, 14–23.

Gofton, L. and Marshall, D.W. (1992) *Fish: A Marketing Problem*. Horton Publishing, Horton.

Kinsey, J. (1993) Changing societal demands: consumerism. In: Padberg, D.J. (ed.), *Food and Agricultural Marketing Issues for the 21st Century*. The Food and Agricultural Marketing Consortium, College Station, TX.

Malhotra, M. (1993) *Marketing Research: An Applied Orientation*. Prentice-Hall, New York.

Perloff, J.M. and Caswell, J.A. (1993) Implications of new industrial organization and demand models for marketing research. In: Padberg, D.J. (ed.), *Food and Agricultural Marketing Issues for the 21st Century*. The Food and Agricultural Marketing Consortium, College Station, TX.

Peters, T. (1987) *Thriving on Chaos*. Warner, New York.

Sanderson, M. and Winkler, J. (1983) Chewing over a healthy diet. *The Health Services* 21, 10–11.

Tull, G. and Hawkins, T. (1990) *Marketing Research: Measurement and Method*. Macmillan, New York.

Zikmund, W.G. (1991) *Exploring Marketing Research*. Dryden, New York.

12 Multivariate Analysis in Marketing Research

Mitchell Ness

Department of Agricultural Economics and Food Marketing, Faculty of Agriculture and Biological Sciences, University of Newcastle upon Tyne, Newcastle upon Tyne NE1 7RU, UK

INTRODUCTION

The subject matter of this chapter concerns the role of multivariate analysis in marketing research and, subsequently, its contribution to marketing decision-making. Marketing is a management process concerned with the anticipation and satisfaction of consumer needs, for products, services or ideas, in order to meet the objectives of an organization. The adoption of a marketing orientation requires that the consumer is placed at the centre of attention; and that the integrated effort of an organization addresses the satisfaction of consumer needs through the marketing mix elements of product, price, promotion and place (Lancaster and Massingham, 1988).

The link between the market-oriented organization and its consumers lies in the existence of a marketing information system (MIS). The MIS is a framework for the management of data, from both internal and external sources, and its transformation into information, provided in the right place, at the right time and to the right person to inform marketing decision-making (see Chapter 9). The MIS is usually defined to comprise the functions of information analysis, internal records, intelligence and research (Kotler, 1984).

Within the research component it is possible to identify three specific areas of interest. The first area is concerned with the organization as a whole in terms of its image or identity. A second area deals in business or market research, which considers characteristics of markets using a market structure–conduct–performance framework. Finally, a third area considers consumer research in terms of the marketing mix elements concerned with existing or potential products. This area is the one which is given most emphasis here.

The application of multivariate analysis in academic and commercial marketing has grown substantially over the last three decades due to advances in research design, statistical methodology and the availability of increasingly friendly computer software. Sheth (1971), writing about the evolution of marketing research, described the 1960s as a period in which a

'Multivariate Revolution in Marketing Research' had taken place. The significance of the period was in the rapid diffusion of the techniques of multivariate analysis in marketing research, as quantitative methodologies from other disciplines were adopted by marketing academics and managers (Sheth, 1971). Sheth regards this as an indication of the transition of marketing from a 'speculative philosophy' to an 'empirical science'.

The structure of the chapter is as follows. First, it deals with the nature of marketing research data. This is of interest because marketing research data consist of several different types of measurement properties and it is important to understand what these are before proceeding to analysis. Following this, a classification of some multivariate techniques is introduced to indicate the link between the properties of the data and the broad objectives of the techniques. From this perspective several of the multivariate techniques will be discussed within the limitations of space and, finally, some concluding remarks are presented.

MARKETING RESEARCH DATA

Marketing research typically gathers data on demographic, behavioural and attitudinal profiles of consumers. This results in a variety of data types with respect to the measurement properties of the numerical systems in the data. It is usual to classify data as one of four types. These are respectively: (i) nominal or categorical; (ii) ordinal or rank; (iii) interval; and (iv) ratio (see for example Tull and Hawkins, 1990). Data may be further classified as non-metric (nominal and ordinal) and metric (interval and ratio). It is important that the analyst is able to identify, or make reasonable assumptions about, the measurement properties of various subsets of the data because these properties govern the range of analytical techniques which can be employed, initially in the form of statistical inference using summary statistics and subsequently in more complex multivariate analyses. The measurement properties of each of these data types will be explained briefly below.

Nominal or categorical data are used to categorize people or objects into various kinds of groups, for example, on the basis of gender, age group and product usage. This form of data is the weakest form of measurement in terms of the information which can be derived from the numerical system. If people or objects are allocated the same number, they belong to the same group, if they are allocated a different number they belong to a different group. This is illustrated below with respect to gender:

Consumer	Gender	Measurement (1 = male; 2 = female)
1	male	1
2	male	1
3	female	2
4	male	1
5	female	2

Ordinal or rank measurement allocates numbers to people or objects in a way which expresses preference or status. In marketing data such measurement systems frequently occur in association with the measurement of preferences, when, as in the example shown below, people are required to rank products in order of their preference:

Consumer	Preference (1 = most preferred; 3 = least preferred)		
	A	B	C
1	1	3	2
2	2	1	3
3	1	2	3
4	1	2	3
5	3	1	2

The information contained in this type of data, as well as containing the special properties of an ordinal system, provides nominal information. Thus if products are allocated the same number they are in the same preference category (most preferred category, least preferred category, etc.). In addition, the organization of the number system indicates that lower numbers represent higher levels of preferences.

Interval measurement allocates numbers to things in a way which conveys information about the distances between them. This is illustrated below with an example using calendar time associated with the year consumers first experienced a product:

Year	Interval
1980	–
1982	2
1986	4
1990	4

In addition to the special property of interval measurement, this numerical system contains nominal and ordinal properties. First, if events are allocated the same number they belong to the same annual category, otherwise they belong to different ones. Second, there is an implied ordinal system because the bigger the number, the more recent is the event in question. Finally, the special property of interval data is that the intervals between the numbers also convey information which is not contained in either nominal or ordinal data. Thus the interval from 1982 to 1986 is twice the interval from 1980 to 1982.

Ratio data are the most powerful form of data because they contain the properties of the previous three data types as well as the special property of ratio numbers. The example below deals with beer consumption:

Consumer	Beer consumption (litres per week)
1	5
2	4
3	2
4	1
5	0

The data are nominal because they categorize people on the basis of beer consumption. They are of an ordinal nature because the bigger the number, the higher the consumption rank. The differences between the numbers also reveal information which is of value. But the property which really distinguishes ratio measurement is that the ratios between numbers are also meaningful. Thus one can conclude that consumer 1 consumes five times as much beer as consumer 4 and so on. This kind of information is only contained in ratio data because the number zero indicates that the property being measured is absent, in this example a non-beer drinker.

A further aspect of marketing research data which merits some attention concerns attitude scaling. Attitudes are mental states that structure the way individuals perceive and respond to their environment. They are typically depicted as having cognitive, affective and intention components. These components reflect knowledge, overall feelings and future intentions respectively. Since attitudes cannot be measured directly, they have to be inferred from some sort of measurement device in the form of a numerical scale (Aaker and Day, 1980).

For example, it is often desirable to measure consumer awareness, of products or advertising, consumer evaluations of existing or new products, or consumer preferences for actual or ideal products. In product research, it is common practice to define products in terms of their attributes. Thus soft drinks could be defined as a series of attributes concerned with physical properties (colour, flavour, sweetness, fizziness, thirst quenching), usage (with meals, for relaxation) and image (youthful, health conscious).

For example, if we take the attribute of sweetness then a rating scale could be devised to evaluate an existing product in the following manner:

Very sweet	Sweet	Neutral	Bitter	Very bitter
1	2	3	4	5

Alternatively, the scale could be devised so that the respondent rates the importance of sweetness in choosing the product. 'In my choice of soft drink sweetness is':

Not very important	Not important	Neither important nor unimportant	Important	Very important
1	2	3	4	5

Another use of scales is in the specification of attribute levels for ideal products. For example, 'My ideal soft drink would be':

Very sweet	Sweet	Neutral	Bitter	Very bitter
1	2	3	4	5

In this way attitudes can be measured on three main aspects: (i) in terms of the evaluation of existing products; (ii) in terms of the importance of attributes; and (iii) in terms of preferences for ideal products.

The problem faced by the marketing researcher is to decide what measurement properties are embodied in scales. From the example given above, it is evident that scales do not have ratio properties because they are not organized on an absolute scale but on a rather arbitrary basis. However, it is also evident that under appropriate assumptions, scales may be associated with nominal, ordinal or interval properties.

THE CLASSIFICATION OF THE TECHNIQUES OF MULTIVARIATE ANALYSIS

Multivariate analysis concerns the simultaneous analysis of data involving two or more variables. Marketing research, like many other social sciences, involves the influence of many variables in complex relationships. In the early stages of marketing, researchers were restricted by the limitations of research design, statistical methodology and computer technology to the analysis of very simple relationships based on univariate or bivariate analyses.

A useful way of understanding the purpose of multivariate techniques employs a method of classification based on the characteristics of the data: specifically the nature of dependency amongst variables and the measurement properties of the data (Hair *et al.*, 1992). Thus, the suitability of any particular technique can be assessed by reference to these dimensions. First, it is necessary to consider dependency, whether there is an a priori basis for a model based upon one or more variables being dependent upon other variables or, whether the variables are interdependent. Second, it is necessary to consider the measurement properties of the data. A secondary scheme of classification on the basis of data measurement properties defines nominal and ordinal data as non-metric data and interval and ratio data as metric data. This scheme is summarized in Table 12.1.

In the sections which follow, the most important of the techniques of multivariate analysis are described. For each of the techniques it is intended to discuss the objective, the underlying model or principle and, finally, to provide some examples of the application to marketing decision-making.

Table 12.1. The classification of multivariate techniques (source: Green and Tull, 1978).

Data	Type of relationship	
	Dependent	Interdependent
Metric	Regression analysis Canonical analysis Analysis of variance	Factor analysis Cluster analysis Metric multidimensional scaling
Non-metric	Discriminant analysis Conjoint analysis	Non-metric multi- dimensional scaling

FACTOR ANALYSIS

Factor analysis is an exploratory technique which is applied to a collection of inter-correlated metric variables with the objective of data reduction and interpretation.

The factor model assumes that each of the original observed variables $(x_j, j = 1,2, ...p)$ are determined by a linear combination of non-measurable common factors $(f_i, i = 1,2, ...k)$ and a unique factor (e_j):

$$x_j = \lambda_{j1} f_1 + \lambda_{j2} f_2 + ... + \lambda_{jk} f_k + e_j$$

The aim of the analysis is to estimate the factor loadings $(\lambda_{ji}, j = 1,2, ...p; i = 1,2, ...k)$, coefficients which indicate the strength of association between each common factor and the dependent variable.

The technique assumes that the common factors are like composite variables and condense the information contained in the original variables. Thus data reduction is achieved by summarizing the information contained in the original variables into a smaller number of factors. Interpretation relies on the factor loadings which provide an understanding of what the factors represent. Thus a successful application would reveal that each factor is strongly associated with one or more of the original variables.

Factors are derived sequentially in a descending hierarchical scheme according to the contribution each factor makes in explaining the total variance of the original variables. Thus the first factor contributes the major share and the last factor contributes least. The maximum number of factors which can be derived is equal to the number of original variables. If the maximum number are derived, the set of factors will totally explain the total variance of the original variables.

However, the usual strategy is to derive fewer factors in order to achieve data reduction. Thus the researcher faces a basic trade-off, as the number of derived factors increase, the larger is the share of variance explained but a lower degree of data reduction is achieved. The ideal is to achieve the maximum amount of data reduction whilst at the same time preserving as much variance information as possible.

The estimation of the factor model involves two stages. In the initial stage the researcher determines the appropriate number of factors and, on estimation of the model, evaluates the model in terms of goodness of fit. The second rotates the initial solution to facilitate interpretation of the model.

The validation of a factor model concentrates upon two summary measures of performance, communality and cumulative variance. Communality indicates the proportion of variance of each original variable explained by the complete set of derived factors. Cumulative variance reflects the proportion of total variance of all original variables explained by the complete set of factors.

The rotated solution establishes the significance of the coefficients and facilitates the interpretation of the factors. The test for the significance of the coefficients is based upon a null hypothesis that the true coefficient is zero. It is usually implemented using a rule of thumb, for example where the null hypothesis is rejected if the absolute value of the coefficient is greater than

0.3. Since the estimated coefficients are correlations between factors and original variables, an alternative approach is to employ a *t* test for the significance of a correlation coefficient. The interpretation of each factor is deduced by deriving a collective name for those variables most strongly correlated with the factor.

A common application in the area of consumer marketing is to identify the main dimensions of consumers' perceptions of products or services. Marketing research often results in the use of many variables to describe a single phenomenon. For example, products and services may be described as a series of attributes concerned with physical characteristics, service characteristics, usage characteristics and user image. The appropriate attributes are usually identified from desk research but the researcher may not appreciate whether some attributes may be more important than others, for example whether price/quality characteristics are more important than physical characteristics. In addition, some attributes may measure, to varying degrees, the same phenomenon, although this may not be apparent to the researcher prior to the analysis. With respect to consumers' perceptions of beer, for example, colour and strength are synonymous attributes (darker = stronger).

A study by Baron and Eagle (1980) identified a series of attributes on which UK consumers perceived meats. They found that it was possible to understand these perceptions using five dimensions: nutritional quality, preparation, taste, value for money and versatility. Meidan (1976) conducted a study which addressed the importance of attributes concerning pub choice and Gofton *et al.* (1983) established the dimensions on which beer drinkers perceive beer. Carsky *et al.* (1994) employed a comparative study of shopping involvement between Californian and Parisian shoppers.

In addition, the factor model may provide an input into subsequent analyses such as cluster analysis or multidimensional scaling where the scores on original variables are replaced with scores on each factor to achieve data reduction. Askegaard and Madsen (1995) employed this approach in a cluster analysis of European food consumers to establish consumer segments based upon food behaviour, attitudes and preferences. In the study 138 original variables were reduced to 41 factors and the cluster analysis was conducted on factor scores. This study is described in more detail in the section on cluster analysis. Van Dam and Vollebregt (1995) examined Dutch supermarket buyers' perceptions of six types of pork ranging from intensively produced to organic. They established 13 attributes on which buyers evaluate pork and employed factor analysis to reduce these to three factors: (i) reliability; (ii) sustainability versus commercial attractiveness; and (iii) sensory quality, respectively. Using the first two factors as axes, average scores on the factors for each type of pork were then plotted on the axes to obtain a perceptual map of the products.

Research conducted by Ness and Walker (1993) established a set of 13 physical and service attributes associated with compound dairy feeds. In a subsequent survey, dairy farmers were asked to score their regular compound feed firm on a series of semantic scales. Factor analysis, conducted using the procedure factor in SPSS (SPSS, 1990) reveals four main dimensions underlying these attributes (Table 12.2).

Table 12.2. Dairy farmers' perceptions of compound feeds (source: Ness and Walker, 1993).

Attribute	1	2	3	4	h^2*
		Factor number			
Nutrition	0.724	−0.247	0.359	0.127	0.730
Hardness	0.862	0.153	0.056	0.022	0.769
Dustiness	0.729	0.170	−0.354	−0.055	0.689
Palatability	0.540	0.446	0.176	−0.207	0.564
Range of technical advice	0.115	0.115	0.853	0.092	0.762
Helpfulness of technical advice	−0.387	0.459	0.533	0.109	0.657
Speed of delivery	0.139	0.868	−0.271	0.180	0.879
Reliability of delivery	−0.023	0.867	0.115	−0.087	0.772
Cooperation of driver	0.163	0.786	0.279	0.049	0.697
Competitive price	0.046	−0.017	−0.884	−0.024	0.784
Value for money	0.818	0.027	−0.110	0.181	0.716
Range of other feeds	0.428	0.270	−0.062	0.707	0.759
Range of other products	−0.091	−0.089	0.184	0.826	0.732
Variance (%)	27.6	20.7	15.5	9.4	
Cumulative variance (%)	27.6	48.3	63.8	73.2	

* h^2 refers to communality.

From Table 12.2 the cumulative variance measure indicates that the four factors explain 73% of the variation in the original variables and the communalities indicate that the four factors explain substantial proportions of the variances of most of the original attributes.

In order to interpret the model it is necessary to identify the factors. To accomplish this, it is necessary to identify which of the original attributes are most strongly associated with each factor. For example factor 1 is most strongly associated with 'hardness', 'value for money', 'dustiness' and 'nutrition'. Thus it could be identified as a factor concerned with the value for money of the physical qualities of the feed. By the same process factor 2 is a delivery service factor, factor 3 represents a price premium associated with technical advisory service and factor 4, the range of other products.

The results of the analysis reveal that there are four main dimensions which underlie the original 13 attributes associated with compound animal feeds. Thus the analysis has clarified and simplified our understanding of farmers' perceptions of compound dairy feeds. This would be very useful in understanding market segments and establishing marketing mix to target some or all segments.

NON-METRIC MULTIDIMENSIONAL SCALING

The aim of multidimensional scaling (MDS), also known as perceptual mapping, is to represent data which measures judgements of similarity as a map on two or more dimensions. The MDS solution presents a configuration of objects, for example, brands or products, in such a way that distances between objects reflect the degree of similarity. Objects which are perceived to be most similar are located closer together than objects which are perceived to be least similar. Thus, the map presents a spatial representation of the objects which indicates the extent to which objects are perceived to be similar or dissimilar. In addition, the positioning of the objects can be explained relative to the interpretation of the dimensions of the map.

MDS may be conducted through a wide range of specific methods which differ according to the method of data collection and assumptions about the measurement properties of the data. Data collection may be conducted using non-attribute- or attribute-based methods. A non-attribute method requires consumers to evaluate the similarity of objects through a direct comparison method. For example, respondents may be presented with a list of all possible pairs of brands and asked to rank each according to their perception of similarity. Attribute-based methods require consumers to evaluate each brand on a series of attributes so that the similarity measures are derived subsequently. The measurement properties of the data may be assumed to be non-metric or metric.

The basic non-metric model typically uses non-metric non-attribute data. The original data are represented by a dissimilarity matrix in which pairs of objects are ranked according to perceived similarity. The model incorporates a monotonic transformation which assumes the distances between objects in the solution are a function of the non-metric ranks. For a specified number of dimensions, the model solution is obtained through an iterative procedure which minimizes a goodness of fit measure called stress.

The applications of MDS to marketing emphasize product and brand positioning and new product development. Lautman and Kordish (1978) discussed the derivation of promotional strategies from the use of MDS, whilst Johnson (1971) included a study of consumer perceptions and segmentation of the Chicago beer market. Sengupta (1990) provided excellent examples of such applications to various Indian consumer markets. Doyle and McGee (1973) applied the principles of MDS to alternative convenience foods, whilst Wierenga (1980) investigated the analysis of consumers' perceptions and preferences with respect to agricultural and food products. A recent study by Marshall *et al.* (1993) used an attribute-based approach to research Spanish attitudes to meat and fish.

For the purposes of illustration we present an application adapted from data from Green *et al.* (1989) which addressed consumers' perceptions of American soft drinks. Consumers were asked to rank all possible pairs of ten soft drink brands according to their perceived similarity. The data, averaged over all respondents, are presented as a lower triangular dissimilarity matrix in which Pepsi and Coke are perceived to be most similar and 7-Up and Tab least similar (Table 12.3).

The application of the Alscal MDS procedure within SPSS produced a solution in two dimensions shown in Fig. 12.1. The goodness of fit measure is represented by a stress value of 0.038 which is evaluated as good according

Table 12.3. Dissimilarity data for ten brands of soft drinks* (source: Green *et al.*, 1989).

Brand code†	Brand code†									
	1	2	3	4	5	6	7	8	9	10
1	–									
2	1	–								
3	3	2	–							
4	7	9	8	–						
5	27	28	32	22	–					
6	41	42	43	29	13	–				
7	18	17	19	20	30	40	–			
8	24	25	26	31	5	16	23	–		
9	35	36	38	44	15	6	37	14	–	
10	12	11	10	4	33	34	21	39	45	–

* 1 = most similar brand pair, 45 = least similar brand pair.
† Key to brand codes: 1, Pepsi; 2, Coke; 3, Coke Classic; 4, Diet Pepsi; 5, Diet Slice; 6, Diet 7-Up; 7, Dr Pepper; 8, Slice; 9, 7-Up; 10, Tab.

to the rule of thumb criterion suggested by Kruskal (1964) in which a stress value of zero is 'perfect' and a value of 0.20 is 'poor'.

Figure 12.1 illustrates that the aim of MDS is to reflect the interrelationships between brands as closely as possible, although less than perfectly. Thus, Coke and Pepsi are positioned so close together as to be effectively in the same location, whilst 7-Up and Tab are positioned some distance apart.

The interpretation of the dimensions in the basic non-metric model has to be achieved through the use of extraneous information, for example, using subjective judgement. In this case interpretation is possible by examining the

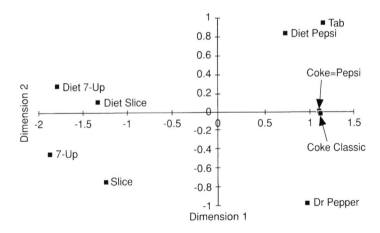

Fig. 12.1. Soft drinks: non-metric MDS solution (source: solution produced by author from data in Green *et al.*, 1989).

characteristics of brands located at extremes of each dimension. In this analysis dimension 1 was interpreted as a 'Non-Cola-Cola' dimension, whilst dimension 2 was interpreted as a 'Non-Diet-Diet' dimension.

In this context it is evident that the brand positions indicate four segments: (i) a Diet-Cola segment (Diet Pepsi and Tab); (ii) a Non-Diet-Cola segment (Coke, Pepsi and Dr Pepper); (iii) a Diet-Non-Cola segment (Diet 7-Up and Diet Slice); and (iv) a Non-Cola-Non-Diet segment (7-Up and Slice).

The solution provides marketing decision-makers with a representation of consumers' perceptions of brand similarities. This example represents a picture of a perceived market. Second, it could aid the repositioning of existing products to provide unique appeal. Third, it may form an initial basis for the generation of new product ideas to confer product benefits not currently available.

CLUSTER ANALYSIS

The objective of cluster analysis is to classify objects on the basis of measured characteristics into a series of groups such that there is within-group homogeneity and between-group heterogeneity.

Cluster analysis may be applied to non-metric or metric data. Generally, the data is required to be of one type or the other although some software allows mixed data types. From the original data, measures of inter-object similarities or distances are computed. The basic principle of cluster analysis relies on the rationale that objects which are closer together should be allocated to the same group, whilst objects which are far apart should be allocated to different groups.

Cluster analyses are many according to the way in which inter-object similarity is measured and the particular clustering algorithm. The main types of algorithm are hierarchical or non-hierarchical.

The hierarchical method begins with every object allocated to its own cluster. The objects are then merged into a successively smaller number of groups until, on completion of the procedure, there is a single cluster of all objects. The procedure employs the information on inter-object distances and, at any stage, merger takes place between the objects which are most similar (nearest). The researcher faces the task of deciding the appropriate number of clusters from summary information.

With non-hierarchical procedures objects are grouped into a pre-specified number of clusters relative to an objective. The procedure begins with an initial grouping of data and the subsequent application of a clustering criterion to reach a final solution. The clustering criterion may be based upon the minimization of within-cluster variance or the maximization of between-cluster variance. During the clustering procedure objects are typically relocated amongst clusters.

The main areas of application to marketing are in test market selection and in the identification of consumer or product segments. Test market selection attempts to identify groups of cities and towns with similar demographic profiles, such as population size, ethnic composition, income, education and lifestyle characteristics. Towns could then be selected from each cluster type and alternative marketing strategies could be tested and compared (see for example Green *et al.*, 1967).

Consumer segments may be identified on the basis of their perceptions of products, the benefits they seek from products or their lifestyles (Saunders, 1980).

An application of cluster analysis presented here is based upon a Danish study by Askegaard and Madsen (1995). The aim of the study was to identify homogeneous consumer segments with respect to food behaviour and preferences throughout 79 regions within the 12 EU countries and four EFTA countries (Norway, Sweden, Austria and Switzerland).

The data were obtained from a lifestyle survey of 20,000 consumers conducted in 1989 by the 'Centre de Communication Avancé', a market research agency based in Paris. The data were collected on 138 variables concerned with general food behaviour (food style, trends and preferences), product-related food behaviour (nibbling habits and drinking habits) and health-related food behaviour and attitudes (diet willingness and diet behaviour).

Given the size of the data, it will be appreciated that the data management task was formidable. Therefore, prior to the cluster analysis the data were reduced to 41 dimensions with the use of factor analysis. Subsequently, the observations on the 138 variables were replaced by 41 factor scores for each respondent.

The cluster analysis was conducted on the factor scores and employed a hierarchical method which produced 12 clusters. In terms of regional composition, the clusters represented seven nation states (Denmark, Norway, Sweden, Portugal, Spain, Italy and Greece), four transnational regions (British Isles, Netherlands and Flanders, France and the French-speaking part of Switzerland, Germany, Austria and the rest of Switzerland) and, finally, a cluster comprising the regions of Brussels, Wallonia and Luxembourg.

On examination of the factor scores, the authors were able to profile the behaviour and attitudes of each segment, to reveal some interesting insights (Table 12.4). The analysis reveals that the cultures most concerned with food are to be found in the Latin countries of Europe, whilst the greatest health consciousness is found in the Germanic countries. Although the authors present some qualifications, the study establishes a link between language, culture and food which is of interest to international marketers.

DISCRIMINANT ANALYSIS

Whereas cluster analysis attempts to identify natural groupings in data, discriminant analysis aims to explain and predict the group membership of things on the basis of measurements on explanatory variables. Analysis concerns estimation of the coefficients (a_i, $i = 1,2, ...k$) in the discriminant function(s) for an appropriate set of variables (x_i, $i = 1,2, ...k$) which best discriminate between groups:

Table 12.4. Cluster membership and characteristics (source: compiled from Askergaard and Madsen, 1995).

Cluster	Characteristics	Cluster	Characteristics
Germany Austria Rest of Switzerland	Health conscious Fast food not popular Enjoy cooking and spicy food Healthy nibbling habits Heavy use of healthy drinks Below average use of spirits except schnapps	Netherlands Flanders	Traditional habits, attitudes Preference for heavy meals Light use of alcoholic drinks Heavy use of port, tea and schnapps
Brussels Wallonia Luxembourg	Sophisticated tastes High interest in food Emphasis on sensory enjoyment Heavy use of chocolate Heavy use of beer, red wine, mineral water Light use of tea, white wine Increasing use of fast food and take-away food	France French-speaking Switzerland	Emphasize sensory enjoyment Heavy use of convenience meals in the week Enjoy week-end cooking Heavy use of pastis, red wine and mineral water Light use of beer, white wine, colas and tea
Spain	Preference for natural products Naturally healthy diet Less inclined to reduce intake of fat, fried food, salt, sugar Preference for salty snacks Heavy use of red wine, aperitifs and spirits Light use of tea, sparkling mineral water	Portugal	Meals are social events Prefer small, light dishes Light use of convenience food Prefer soft-textured food Heavy use of wines and spirits, fruit-flavoured drinks and instant coffee Heavy use of pies, biscuits, cheese, popcorn
Italy	Emphasize sensory enjoyment Limited health consciousness Light use of convenience food Heavy use of wines, aperitifs Light use of instant coffee, fruit juice and syrups	Greece	Prefer several light meals Like to eat out Light use of convenience food Heavy use of retsina and ouzo Heavy use of beverages Light use of fruit-flavoured drinks, sparkling mineral water
British Isles	Weak food culture Prefer pastries and sweets Prefer sour tastes Heavy use of tea, instant coffee Light use of mineral water, red wine, aperitifs	Denmark	Heavier use of fast food Low enjoyment of cooking Prefer heavy meals Heavy use of beer, schnapps, filter coffee Light use of snacks apart from fruit
Sweden	Weak preference for natural products Prefer heavy meals Emphasize sensory enjoyment Heavy use of syrups, filter coffee, sodas Light use of alcoholic drinks	Norway	Strong food culture High interest in food Health conscious Like convenience foods Dislike fast food, take-away food Heavy use of fruit syrups, sparkling mineral water Low use of other beverages

$$D = a_1 x_1 + a_2 x_2 + \ldots + a_k x_k$$

With g groups a maximum of $g - 1$ discriminant functions are necessary although successful discrimination may be achieved using fewer than the maximum number of functions.

The estimation procedure employs canonical correlations to derive the appropriate number of functions in descending hierarchical order, such that the first function discriminates most and the last function discriminates least. It is usual for most software to produce both standardized and unstandardized versions of the coefficients.

The choice of the appropriate number of functions is based upon Wilks' lambda test. The test is applied sequentially to each function in turn for the significance of the information which the function can explain. The null hypothesis is that the information is not significant. Thus, the derivation of a particular function is justified if the null hypothesis is rejected at the chosen level of significance.

The estimated model provides for identification of the relative importance and direction of influence of the explanatory variables on the basis of magnitude and sign. Having established the explanation of group membership the following stage involves an evaluation of the classification performance of the function.

For this purpose, group and individual scores are calculated from the unstandardized function. The group scores are obtained from group average values on the explanatory variables whilst individual scores are obtained from the observations on each individual's explanatory variables. The classification procedure compares the individual scores with the group scores and classifies the individual as a member of the nearest group. Morrison (1969) suggested that evaluation of the classification performance of the discriminant function should be undertaken using data which has not been used for estimation to avoid an upward bias in performance (Frank *et al.*, 1965).

Classification provides a predicted group membership which can be compared to actual membership. A cross-classification of actual and predicted groupings is used to derive a summary measure of performance (C) which calculates the percentage of cases correctly classified. This measure may be compared to the performance of an alternative random classification method.

Discriminant analysis has been applied to areas of product/service usage, store site selection and prediction of company failure.

The application to product usage classifies consumers by their degree of product use (user vs. non-user; heavy, medium, light user) or the time lapse which evolves before they try a product (early adopter, late adopter, non-adopter). Psychographic and demographic variables are then used as discriminating variables. Within the literature there are examples of applications to product innovators (Robertson and Kennedy, 1968), buyers of a new supermarket product (Montgomery, 1975), generic brand grocery products (McEnally and Hawes, 1984) or buyers of a new detergent

Table 12.5. Discriminant function coefficients (source: Cunningham and Cunningham, 1973).

Variable	Coefficient	F ratio*
Family life cycle	0.34	2.85
Occupation of head of household	0.63	10.13**
Total family income	0.40	4.09**
Education level of head of household	0.40	4.15**
Social class	0.69	12.34**
Trust in people	0.02	0.01
Cosmopolitanism	0.51	6.50**
Attitude towards credit	0.47	5.64**
Attitude towards impulse buying	0.12	0.36
Adventuresomeness	0.23	1.28
Conservatism	−0.44	4.96**

* The *F* ratios are associated with a test for significant differences between the population group scores on each variable, based on a null hypothesis that population group scores are equal; those values which carry a double asterisk indicate variables which are associated with a rejection of the null hypothesis at a significance level of 5% or lower.

(Pessemier *et al.*, 1967). In store site selection, stores are classified on the basis of performance, and demographic characteristics of the population are used to discriminate between good and bad sites (Sands and Moore, 1981). The objective of the analysis is to formulate a screening policy for new store sites. The application to the prediction of company failure classifies firms on the basis of their current performance and uses financial ratios for previous years to discriminate between good and bad performers. The objective is to provide a decision framework to anticipate company decline and institute policies to prevent failure. See for example Taffler (1982) and Steele *et al.* (1985).

A study by Cunningham and Cunningham (1973) focused on an increasing feature of food buyer behaviour which involves shopping from home. This aspect of buyer behaviour takes the form of mail-order catalogue shopping, direct response to advertising in print media, direct sales door to door and, lately, shopping through specialist TV channels.

The study was conducted in the context of research which attempts to profile consumer segments on the basis of the degree of usage of this method of marketing.

The data were collected from a sample of 570 shoppers in Lansing and East Lansing, Michigan, USA. For the purpose of the study shoppers were defined as 'active in-home shoppers' or 'inactive in-home shoppers'. The discriminating variables comprised socio-demographic, motivational and attitudinal variables.

With two groups a maximum of one discriminant function is required. An important issue is thus to test whether the derivation of the function is

Table 12.6. Classification matrix (source: Cunningham and Cunningham, 1973).

	Predicted group		
Actual group	Inactive in-home shopper	Active in-home shopper	Row total
Inactive in-home shopper	152	127	279
Active in-home shopper	101	190	291
Column totals	253	317	570

justified. The basis for this test is a Wilks' lambda test based on the null hypothesis that there is no significant variation between groups for the function to explain. The application of this test leads to a rejection of the null hypothesis at a significance level of 0.8%.

The discriminant function coefficients are presented in Table 12.5. The magnitudes and signs on the discriminant function coefficients indicate the degree of influence and the direction of influence of each variable. Larger coefficients indicate greater importance. Thus it is evident that social class is the most important variable which discriminates between groups, the occupation of head of household is second most important, and so on.

The discriminant scores for each group indicate that 'active in-home shoppers' have a higher group average score compared with 'inactive in-home shoppers'. Interpretation of this result with respect to the discriminant function indicates that, compared with 'inactive in-home shoppers', 'active in-home shoppers' tend to have higher status, as indicated by higher social class and higher status occupations. They also tend to be more cosmopolitan in their attitudes, have more positive attitudes towards the use of credit and are less conservative.

A further test on the performance of the function employs a summary measure of classification performance. A summary of this performance is presented in Table 12.6 as a confusion matrix, a cross tabulation of actual and predicted group membership. The summary measure of performance is based upon the percentage of all cases correctly classified:

$$C = 100(152 + 190)/570 = 60\%$$

To determine whether this performance is acceptable, C is then compared to the percentage classification of all cases using an alternative random method such as a lottery or gambling game (C_{pro}).

Morrison (1969) shows that such a measure can be obtained from the sum of the squared proportions of actual group membership. Thus, the information in Table 12.6 provides a C_{pro} value of 50%.

Since the performance of the function is better than the performance using the alternative random device, the application of the analysis seems worthwhile.

The implications of this analysis reveals some interesting differences between 'active' and 'inactive' in-home shoppers which could inform decisions regarding promotion or targeting of direct marketing activities.

CONJOINT ANALYSIS

Conjoint analysis is a technique which models the nature of consumer trade-offs amongst multi-attribute products or services. The model assumes that alternative product concepts can be defined as a series of specific levels of a common set of attributes. It also assumes that the total utility the consumer derives from a product is determined by the utilities (part-worths) contributed by each attribute level.

The aims of conjoint analysis are: (i) to identify attribute combinations which confer the highest utility to the consumer; and (ii) to establish the relative importance of attributes in terms of their contribution to total utility. Subsequent analyses provide a means of identifying consumer segments with similar preferences and the simulation of choice amongst alternative product concepts using choice simulation models.

Data collection requires consumers to evaluate alternative product concepts described in terms of a set of attribute levels. A variety of specific data collection methods are possible. For example the researcher may use a verbal description or, where visual presentation is important, for example in packaging or merchandising, physical, graphic or photographic presentations may be used. The most common methods of data collection are the trade-off method and the full profile method (Hair *et al.*, 1992).

The trade-off method requires respondents to compare attributes two at a time, ranking all combinations of levels. However, this method has become less popular, for example, because it restricts the researcher to nonmetric analysis and because it sacrifices realism in the process of evaluating alternative products for the sake of simplicity. The criticism is that, in real choice situations, consumers are confronted with all attributes and their levels simultaneously and not in pairs. The full concept method presents respondents with a series of full descriptions of the product concepts and requires them to rank (using card sorting techniques) or score each concept according to their preference or willingness to buy.

The number of product concepts is equal to the product of the number of levels associated with each attribute. Consequently, even in fairly modest experiments, the number of concepts can become too great to expect respondents to make meaningful evaluations. Fortunately, researchers can take advantage of orthogonal factorial experimental design which enables the main effects to be estimated with fewer cards than the full number of product concepts (Addelman, 1962). A recent development is Adaptive Conjoint Analysis (ACA) in which data are collected directly from the respondent using interactive software (Sawtooth Software, 1992).

There are two aspects of the conjoint model which require specification prior to estimation. These concern the composition rule and the part-worth relationship. The composition rule specifies how attribute levels are combined to obtain an overall preference rating for alternative products, whilst the part-worth relationship specifies the relationship between attribute levels and their respective utilities.

The composition rule may be additive or interactive. The additive model assumes that the respondent adds the part-worths of attribute level combinations to obtain an overall preference or utility rating. In the interactive model the respondent is also assumed to add the part-worths to obtain an overall valuation of the product but the influence of attribute combinations may make the overall rating smaller or larger than the sum of the individual part-worths. This would occur, for example, if a particular attribute level had a low part-worth unless it was combined with a particular level of another attribute.

There are three types of part-worth relationship. These are the linear or vector model, the quadratic or ideal model and the part-worth model. In the linear model it is assumed that the attribute levels are linearly related such that increasing (decreasing) the attribute level changes utility proportionally. For example, it may be that the utility associated with preparation time of a food product may increase as preparation time decreases.

In the ideal point model it is assumed that utility and attribute level are determined by a concave or convex curve such that the utility is related to the squared distance of an attribute level from a subject's ideal point. This would be the case, for example, if an individual attached a low part-worth to small and large package sizes but high utility to a medium size.

The part-worth model assumes that the relationship between utility and attribute levels is entirely flexible and is not assumed to follow any set functional relationship. Thus the part-worth model is the most general case and the linear model is the most restrictive.

The additive model is the most commonly employed model. Hair *et al.* (1992), for example, reported that this model is often preferred because the interactive model often results in low predictive power because the reduction in statistical efficiency, arising from the need to estimate the individual part-worths and the interactive effects, is not offset by the increase in predictive power gained from the interactions. In addition, the use of the interactive model is deemed to be more suitable if the attributes are intangible, for example, in the case of aesthetic or emotional attributes. The additive model also requires fewer evaluations from the respondent and is easier to estimate.

The additive part-worth model assumes the part-worth of each attribute level is independent and that total utility is the sum of attribute level part-worths. For example, assuming four attributes (A to D inclusive), a consumer's preference for a particular product combining attribute i from A, attribute j from B, attribute k from C and attribute l from D is:

$$\text{Pref}_{ijkl} = a_i + b_j + c_k + d_l$$

where: Pref_{ijkl} = consumer's total utility or preference rating for a product combining attribute levels $ijkl$ from attributes A, B, C, and D respectively; a_i

= the utility or part-worth of attribute level *i* from attribute *A*; b_j = the utility part-worth of attribute level *j* from attribute *B*; c_k = the utility or part-worth of attribute level *k* from attribute *C*; and d_l = the utility or part-worth of attribute level *l* from attribute *D*.

The aim of estimation is to estimate the part-worths for every attribute level. There are five commonly used estimation methods: (i) Monanova (Kruskal, 1965); (ii) Linmap (Srinivisan and Shocker, 1973); (iii) dummy variable regression (Johnston, 1972); (iv) Logit analysis (McFadden, 1976); and (v) Probit analysis (Goldberger, 1964).

For a given sample, the researcher faces the choice of estimating models for each respondent, or an aggregate model for the whole sample. Unless one is satisfied that preferences are homogeneous, it is usual to estimate individual models first. This enables one to evaluate the fit of the model to individual respondents and perhaps to discard cases which produce poor results. Most software provides estimates for each individual and an aggregate model for the whole sample. In addition, researchers may establish aggregate models for consumer segments based upon a priori criteria (age, sex, social class, etc.) or *ex post* criteria based upon the results of the analysis (Marr, 1987).

Applications of conjoint analysis have been varied but have tended to concentrate on the analysis of consumer preferences for various products. In the USA, Tantiwong and Wilton (1985) analysed food store preferences of the elderly, June and Smith (1987) considered service and situation aspects of restaurant dining, and Cattin and Wittink (1982) provided a survey of the extensive commercial use of conjoint analysis. European studies include an application to ham quality evaluation (Steenkamp, 1987) and a study by Loader (1990) which applies conjoint analysis to the purchase of fruit and vegetables using photographic representations of product concepts. An excellent example applied to the Indian soft drink market is to be found in Sengupta (1990).

The application presented here concerns consumers' preferences for attributes which indicate quality and freshness of eggs (Ness and Gerhardy, 1994). Qualitative research revealed several possible indicators of quality and freshness which buyers evaluate. In the event it was decided to specify four attributes, defined in terms of production method, origin, freshness information and price, each with three levels.

This attribute specification gives rise to 81 possible product concept definitions. Thus, even with a research design of modest dimensions, the data collection task is quite formidable. However, assuming that there are no interaction effects, the use of the fractional orthogonal design procedure Orthoplan in SPSS generated nine product concepts from which it is possible to estimate the model.

The questionnaire was designed to solicit information on egg buying behaviour and preferences, lifestyle and demographic indicators and consumer preferences for the nine alternative egg concepts. The sample targeted egg buyers in the 18–65 age group with quotas imposed for sagacity lifestyle groups (NTC Publications, 1991). The questionnaire was administered to 171

Table 12.7. Representative results from sample.

Attribute and level	Part-worth estimates for consumer		
	1	2	3
Production method			
Battery	−0.4444	−0.8889	−0.1111
Barn or perchery	0.2222	0.7778	−0.1111
Free-range	0.2222	0.1111	0.2222
Range	0.6666	1.6667	0.3333
Relative importance	28.57	45.45	14.29
Origin			
Local	0.2222	0.1111	0.2222
British	−0.1111	0.1111	0.2222
Imported	−0.1111	−0.2222	−0.4444
Range	0.3333	0.3333	0.6666
Relative importance	14.29	9.09	28.57
Freshness information			
Date laid	−0.1111	−0.5556	0.2222
Date packed	−0.4444	0.1111	−0.1111
Sell-by-date	0.5556	0.4444	−0.1111
Range	1.0000	1.0000	0.3333
Relative importance	42.86	27.27	14.29
Price			
(pence per half dozen)			
52	0.2222	0.4444	0.5556
72	−0.1111	−0.2222	−0.1111
84	−0.1111	−0.2222	−0.4444
Range	0.3333	0.6666	1.0000
Relative importance	14.29	18.18	42.86
Constant	2.1111	1.8889	2.1111

respondents. The preference data for the egg concepts was based upon a nine-point preference scale (1 = very low preference, 9 = very high preference).

Conjoint models were estimated for individual respondents using the conjoint procedure in SPSS which is based on a dummy variable approach. For obvious reasons the complete set of individual results cannot be presented here. Consequently, the explanation and discussion which follows focuses upon three representative results (Table 12.7) and deals in particular with the interpretation and use of the estimated conjoint relationships.

First, one can use the results to identify a consumer's ideal product. A consumer's ideal or most preferred product is identified from the magnitudes of the part-worths for each attribute. The product is ideal because it confers the highest utility, determined by summing the part-worths and constant term. For example for consumer 1 the utility of a local free-range egg with a sell-by-date at 52 pence per half dozen is equal to 3.3333 units whilst a local battery egg with the same freshness information and price provides 2.6667 units. Thus a respondent's ideal product is defined in terms of the attribute levels with the highest part-worths.

From the representative results in Table 12.7 the ideal product for consumer 1 is a barn or free-range egg produced locally with a sell-by-date at 52 pence per half dozen. The ideal for consumer 2 is a barn egg, which is British, with a sell-by-date at a price of 52 pence per half dozen. Finally, for consumer 3 the ideal product is a free-range egg, which is British, with a date laid at a price of 52 pence per half dozen. It will be appreciated that ideal products may not necessarily be commercially feasible. For example, barn eggs and free-range eggs carry a price premium so that it would not be feasible to offer these at a price which is associated with battery eggs. Therefore, in the simulation results which are presented later, consumer preferences for actual products are considered.

Further analysis reveals that the relative importance of attributes may vary between consumers. Relative importance of the attributes is calculated from the range of part-worth values for each attribute. The range indicates the impact of changing the attribute level within a particular attribute and enables one to identify the relative importance of attributes. The bigger the range is, the greater is the impact on total utility of changing the attribute level. For example, for consumer 1 (Table 12.7) the attribute with the largest range is associated with freshness information. This attribute has the greatest impact on overall utility because changing from 'date packed' to 'sell-by-date' increases total utility by one unit. The relative importance measure is calculated as the percentage of each attribute range of the sum of all ranges.

The range or relative importance measures enable one to rank the attributes in order of importance to identify each respondent's priorities in terms of product attributes. The marketing implications of these results are in the way in which the eggs should be promoted or positioned in the market. For consumer 1 the priority is freshness information followed by production method and origin and price. Consumer 2 regards production method as the most important attribute followed by freshness information, price and origin, whilst consumer 3 is most sensitive to price followed by origin, production method and freshness information.

The estimated conjoint relationships reveal an understanding of the relative importance of attributes and the impact of particular attribute levels for individual respondents. Subsequently, it is possible to employ these relationships in further analyses.

Estimated conjoint models are typically employed in further analyses for the purposes of segmentation, profitability analysis and market share simulation. In simulation analysis, choice simulation models are used to estimate the market shares of specified products. Particular forms of analysis could analyse the impact of product introduction to a market, product deletion from a market, or the impact of multiple brand or product strategies, including cannibalism (Hair *et al.*, 1992).

Table 12.8. Simulated market shares of product concepts: maximum utility model (source: Ness and Gerhardy, 1994).

Product concept*	Percentage market shares for social class segment		
	AB	C1C2	DE
1	47.62	59.52	61.11
2	40.48	14.29	23.61
3	11.90	26.19	15.28
Total	100.00	100.00	100.00

* Product concepts are defined in Table 12.9 in terms of production method, origin, freshness information and price.

Market share simulation for alternative, competing products employs the estimated conjoint models, which capture information on part-worths and their relative importance, for each individual. In addition it requires: (i) the specification of alternative products in terms of attribute level combinations; (ii) individuals' preference ratings (utilities) for each alternative product calculated from the sum of appropriate part-worths and the constant term; and (iii) a simulation model which determines how individuals choose amongst alternative products and how individual choices are aggregated to simulate market share.

The process of simulation typically employs either a maximum utility model and/or a probability model (Green and Krieger, 1988). Each model assumes different styles of purchase decision behaviour. The maximum utility model assumes that the individual consumer chooses the product which confers the highest preference rating. Bearing in mind that individual preferences will not be homogeneous, but that there may be groups of individuals with similar preferences, individual first choices will vary. The simulated market share is obtained from a relative frequency distribution over all individuals based upon the proportion of times that each product is nominated as the most preferred product.

The example of post-estimation analysis presented here employs the maximum utility model to simulate market shares within three social class groups (AB, C1C2 and DE respectively) of three egg concepts. The simulation defines three alternative egg concepts which are commonly found in UK supermarkets, defined in terms of the attribute levels as: (i) British battery

Table 12.9. Definition of product concepts presented in Table 12.8.

Concept	Production method	Origin	Freshness information	Price (pence per six eggs)
1	Battery	British	Sell-by-date	52
2	Barn or perchery	Local	Sell-by-date	72
3	Free-range	Local	Sell-by-date	84

eggs carrying a sell-by-date at 52 pence; (ii) barn or perchery eggs produced locally carrying a sell-by-date at 72 pence; and (iii) free-range eggs produced locally carrying a sell-by-date at 84 pence.

The results of the simulations are presented in Tables 12.8 and 12.9. With respect to the base simulation it is evident that the appeal of the three concepts varies somewhat between the segments. The segment comprising consumers in social category AB has a greater preference for eggs which are produced less intensively than battery eggs (concept 1) than social categories C1C2 or DE. Social category DE has a much greater preference for battery eggs than the others.

This analysis is intended to give an insight into the use of conjoint analysis in estimating relative market share. More complex simulations could be conducted by varying several attribute levels simultaneously. In addition, it is possible to analyse the impact of new products to an existing set of products, for example, to the three concepts which form the basis for the base simulation.

CONCLUDING REMARKS

The aim of this chapter has been to provide a brief introduction to the use of multivariate analysis in marketing research. As with the application of all quantitative methods, it is important for the practitioner to understand the conditions under which particular techniques should be applied. The chapter attempts to address this point by identifying the intended purposes of each technique and the type of data which is required. However, it should be emphasized that the attention given to each technique has been brief and that, since the subject area is continually developing, there are several hybrid versions of techniques which extend applicability to a wider range of data types.

A further point is that within each technique there are many variations of particular methods which require the practitioner to make critical decisions. In general, these decisions concern the form of data collection, whether the data are to be transformed prior to the analysis, and which particular algorithm to use. In this respect, for the practitioner, in contrast to the theorist, the application of multivariate analysis extends from the pure science of statistics into the art of application. The art of multivariate analysis of course is acquired partly through experience. But it is equally important that the practitioner understands the data and, more generally, the research problem under investigation. In the application of software, a practical approach to methodological decisions is to be guided by the default options which are available. Such options are usually 'safe' in that they do not imply extreme assumptions about the data or the model.

The other point which should be emphasized is to understand the role of multivariate analysis within marketing decision-making in a wider context. Marketing decision-making is a very complex strategic decision process which relies on many forms of decision input. Thus it is important to appreciate that multivariate analysis makes an important contribution to marketing decisions but that this contribution should not replace other vitally important decision inputs.

FURTHER READING

Aaker, D. and Day, G.S. (1980) *Marketing Research.* Wiley, New York.
Bagozzi, R.P. (ed.) (1994) *Advanced Methods of Marketing Research.* Blackwell, Oxford.
Bagozzi, R.P. (ed.) (1994) *Principles of Marketing Research.* Blackwell, Oxford.
Hair, J.F., Anderson, R.E., Tatham, R.L. and Black, W.C. (1992) *Multivariate Data Analysis with Readings,* 3rd edn. Macmillan, New York.
Johnson, R.A. and Wichern, D.W. (1988) *Applied Multivariate Statistical Analysis.* Prentice-Hall International, Englewood Cliffs, NJ.

REFERENCES

Aaker, D. and Day, G.S. (1980) *Marketing Research.* Wiley, New York.
Addelman, S. (1962) Orthogonal main effect plans for asymmetrical factorial experiments. *Technometrics* 1, 21–46.
Askegaard, S. and Madsen, T.K. (1995) Homogeneity and heterogeneousness in European food cultures: an exploratory analysis. In: *European Marketing Academy Conference Proceedings,* Cergy-Pontiose, France, 16–19 May, pp. 25–47.
Baron, P.J. and Eagle, R. (1980) *Consumer Attitudes to Different Cuts and Types of Meat.* Report No. 27. Department of Agricultural Economics and Food Marketing, University of Newcastle upon Tyne, Newcastle upon Tyne.
Carsky, M., Smith, M.F. and Dickinson, R.A. (1994) Measuring the involvement construct: A cross-cultural examination of food shopping behaviour. *Journal of International Agribusiness Marketing* 64(4), 71–102.
Cattin, P. and Wittink, D.R. (1982) Commercial use of conjoint analysis. *Journal of Marketing* 45(3), 44–53.
Cunningham, I.C.M. and Cunningham, W.H. (1973) The urban in-home shopper: Socio-economic and attitudinal characteristics. *Journal of Retailing* 49, 42–50.
van Dam, Y.K. and Vollebregt, Y.C.J. (1995) Marketability of green pork: retail purchasers' attitudes toward different types of pork in The Netherlands. In: *European Marketing Academy Conference Proceedings,* Cergy-Pontiose, France, May 16–19, pp. 1195–1204.
Doyle, P. and McGee, J. (1973) Perceptions of and preferences for alternative convenience foods. *Journal of the Market Research Society* 15(1), 24–33.
Frank, R.E., Massy, W.F. and Morrison, D.G. (1965) Bias in multiple discriminant analysis. *Journal of Marketing Research* 2, 250–258.
Gofton, L.R., Lesser, D., Ritson, C.R. and Ness, M.R. (1983) *The North East Beer Drinker: A Study of Attitudes to Beer in Tyne and Wear.* Report to Newcastle Breweries Ltd, Vols 1–5. Department of Agricultural Economics and Food Marketing, University of Newcastle upon Tyne, Newcastle upon Tyne.
Goldberger, A.S. (1964) *Econometric Theory.* J. Wiley and Sons, Inc., New York.
Green, P.E. and Krieger, A.M. (1988) Choice rules and sensitivity analysis in conjoint simulators. *Journal of the Academy of Marketing Science* 16, 114–127.
Green, P.E. and Tull, D.S. (1978) *Research for Marketing Decisions,* 4th edn. Prentice-Hall, Englewood Cliffs, NJ.

Green, P.E., Frank, R.E. and Robinson, P.J. (1967) Cluster analysis in test market selection. *Management Science* 13, 387–400.

Green, P.E., Carmone, F.J. and Smith, S.M. (1989) *Multidimensional Scaling: Concepts and Applications*. Allyn and Bacon, London.

Hair, J.F., Anderson, R.E., Tatham, R.L. and Black, W.C. (1992) *Multivariate Data Analysis with Readings*, 3rd edn. Macmillan, New York.

Johnson, R.M. (1971) Market segmentation – A strategic management tool. *Journal of Marketing Research* 9, 13–18.

Johnston, J. (1972) *Econometric Methods*. McGraw-Hill, London.

June, L.P. and Smith, S.L.J. (1987) Service attributes and situational effects on consumer preferences for restaurant dining. *Journal of Travel Research* 26(2), 20–27.

Kotler, P. (1984) *Marketing Management: Analysis, Planning and Control*. Prentice-Hall International, Englewood Cliffs, NJ.

Kruskal, J.B. (1964) Multidimensional scaling by optimising goodness of fit to a non-metric hypothesis. *Psychometrika* 29(1), 1–27.

Kruskal, J.B. (1965) Analysis of factorial experiments by estimating monotone transformations of the data. *Journal of the Royal Statistical Society, Series B* 27, 251–263.

Lancaster, G. and Massingham, L. (1988) *Essentials of Marketing*. McGraw-Hill, London.

Lautman, M.R. and Kordish, G.R. (1978) Campaigns from multidimensional scaling. *Journal of Advertising Research* 18(3), 35–40.

Loader, R. (1990) The use of conjoint analysis in the purchase of fruit and vegetables. Unpublished Monograph, University of Reading, Department of Agricultural Economics and Management.

Marr, N.E. (1987) A method for the aggregation of the results in a conjoint measurement study. *European Research* 15(4), 257–263.

Marshall, D.W., Currall, J. and Rodriguez, D. (1993) Food provisioning and Spanish attitudes. *Food Quality and Preference* 4(4), 177–185.

McEnally, M.R. and Hawes, J.M. (1984) The market for generic brand grocery products. *Journal of Marketing* 47(4), 75–83.

McFadden, D. (1976) Quantal choice analysis: A survey. *Annals of Economic and Social Measurement* 5, 363–390.

Meidan A. (1976) Pub selection criteria. In: *Marketing Education Group Annual Conference Proceedings*, University of Strathclyde, Glasgow, July 1976, pp. 41–59.

Montgomery, D.B. (1975) New product distribution: An analysis of supermarket buyer decisions. *Journal of Marketing Research* 12(3), 255–264.

Morrison, D.G. (1969) On the interpretation of discriminant analysis. *Journal of Marketing Research* 1, 156–163.

Ness, M.R. and Gerhardy, H. (1994) Consumer preferences for quality and freshness attributes of eggs. *British Food Journal* 96(6), 26–34.

Ness, M.R. and Walker, C.S. (1993) Dairy farmers' perceptions of compound feeds. Unpublished Report.

NTC Publications (1991) *Marketing Pocketbook*. NTC Publications, Henley on Thames.

Pessemier, E.A., Burger, P.C. and Tigert, D.J. (1967) Can new product buyers be identified? *Journal of Marketing Research* 4, 349–354.

Robertson, T.S. and Kennedy, J.N. (1968) Prediction of consumer innovators: Application of multiple discriminant analysis. *Journal of Marketing Research* 5(1), 64–69.

Sands, S. and Moore, P. (1981) Store site selection by discriminant analysis. *Journal of the Market Research Society* 23(1), 40–51.

Saunders, J.A. (1980) Cluster analysis for market segmentation. *European Journal of Marketing* 14(7), 422–435.

Sawtooth Software, Inc. (1992) *ACA System Version 3.1*. Sawtooth Software, Inc. Evanston, IL.

Sengupta, S. (1990) *Brand Positioning*. Tata McGraw-Hill Publishing Company, New Delhi.

Sheth, J. (1971) The multivariate revolution in marketing research. *Journal of Marketing* 35, 13–19.

SPSS, Inc. (1990) *Statistical Package for the Social Sciences*. SPSS, Inc., Chicago, IL.

Srinivisan, V. and Shocker, A.D. (1973) Linear programming techniques for multidimensional analysis of preferences. *Psychometrika* 38, 337–369.

Steele, P., Storey, D. and Wynarczyk, P. (1985) *The Prediction of Small Company Failure Using Financial Statement Analysis*. Discussion Paper No. 19. Centre for Urban and Regional Development Studies, University of Newcastle upon Tyne, Newcastle upon Tyne.

Steenkamp, J. (1987) Conjoint measurement in ham quality evaluation. *Journal of Agricultural Economics* 38, 473–480.

Taffler, R.J. (1982) Forecasting company failure in the UK using discriminant analysis and financial ratio data. *Journal of the Royal Statistical Society, Series A* 45(3), 342–358.

Tantiwong, D. and Wilton, P.C. (1985) Understanding food store preferences among the elderly using hybrid conjoint measurement. *Journal of Retailing* 61(4), 35–64.

Tull, D.S. and Hawkins, D.I. (1990) *Marketing Research*, 5th edn. Macmillan, New York.

Wierenga, B. (1980) Multidimensional models for the analysis of consumers' perceptions and preferences with respect to agricultural and food products. *Journal of Agricultural Economics* 31(1), 83–98.

13 Product Policy

Marianne Altmann
67, rue de la Gare, 3377 Leudelange, Luxembourg

INTRODUCTION

The product is the pivot point of all decisions made in marketing policy. Without a really good product all the other marketing measures will fail in the long-run. If a company has the wrong or a bad product, the most marvellous advertising campaign and even the best selling system will not help it stay in the market.

The key to company success is to enter the market with the right product. This does not only include the physical nature of a product but also the use and advantage the product can give. This use can be defined as the potential or *capability of a product to solve problems*. As most products are sold indirectly to the consumer, the product has to solve at the same time the problems of the middlemen (e.g. wholesaler, retailer) as well as those of the producer in view of the increasing capital intensity of production.

The product moves around in this triangle: consumer, middleman, producer. Ideally, the requirements of all three market areas go hand in hand. But in reality ways of compromise have to be found whereby, following the principles of marketing philosophy, the requirements of the consumer get the highest priority followed by the trade and lastly by the producer.

To summarize, the product is the bundle of satisfaction that is transferred in the exchange process and it includes every activity that provides satisfaction to consumers. A successful product can be considered as a product with a large capability of solving the problems of each market section.

With regards to product policy and offering the right product, there is another important aspect to be pointed out: product policy is always a *dynamic process*. The necessity to ensure a company's success with concrete marketing measures is based on the principle that every launched product, after reaching an individual peak of success, loses its attractiveness to the consumer in the course of time. The product switches from the profit to the loss side of the company.

Every product in its 'lifetime' has to pass through several stages of growth, maturity, saturation and decline. Old products have to be continuously replaced by new ones. This special process is called the *product life cycle*. Every company constantly has to bring in new products to remain successful in the market which is the reason why *new product development* is so important.

If product policy is the key to a company's success it is obvious that really good market research has to be performed to find out what are the needs of the consumer, what kind of capability has the product for solving problems, at which 'stage of life' is the product, and in what direction the new product development has to be carried out.

As regards product policy and other marketing instruments, the producers of agricultural products and their sales organizations in Europe have not been very active in this field so far. This is due to the atomistic production structure of small family enterprises and the large capital sums required for professional marketing. Nevertheless, in recent years in Europe there were observed tendencies towards the development of marketing strategies for regions or for individual businesses in the agricultural sector. In comparison, in the food industry sector the activities in the field of product policy attract the same professionalism as in all the other consumer good markets. The food industry, with its many multinational orientations, invests a lot in product development and promotion. In 1990, in the USA, of the twelve companies spending most on promotion, nine belonged to the food industry.

As more than 80% of food products reach the consumer via food retailers, structural changes, i.e. a growing concentration and the worldwide interconnection of big companies, have a high influence on the product policy of the producer. Economies of scale and information technology have pushed the development of food retailers enormously, so that a changing purchasing behaviour of the mostly chain-type organized companies must inevitably lead to a changed production pattern (Statistical Office of the European Communities, 1993).

Food retailers are increasingly involving themselves further back in the food production business. They are investing a lot in processing, food logistics and are even looking for direct contact at the farm level to reduce costs and minimize risks in food quality (Jahn, 1991).

OBJECTIVES OF PRODUCT POLICY

The basis of product policy is determined by the main management decision of a company on the question of whether to pursue existing or new strategies, in existing or new markets. It is to decide whether products are to be eliminated, modified or newly designed.

Apart from the decisions on product design, the main decisions within product policy are those concerning the product range: should the current

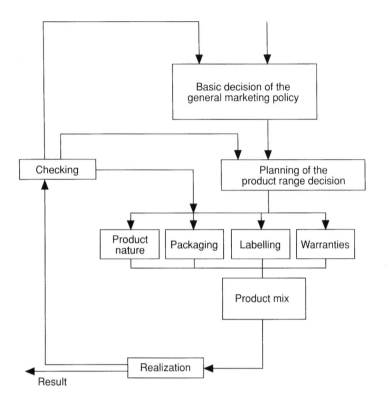

Fig. 13.1. Process of product policy (source: Strecker *et al.*, 1990).

production and selling programmes be expanded or reduced to match the requirements of the market? The careful and exact decision on the product range represents a prerequisite for the design of the products. Figure 13.1 illustrates the product decision process.

Based upon the basic marketing planning the general decisions on objectives and strategies concerning the product range are made; from this point specific aims and measures of the product design are developed which are then harmonized with each other, realized, checked and planned once more. Possible targets and objectives of a company's product policy may be (Strecker *et al.*, 1990): (i) to use the capacities of production, manpower and capital in an optimal way; (ii) to utilize fully the selling system; (iii) to offer a product range matching the trade needs; (iv) to offer a product range matching the consumer needs; (v) to increase the product capacity of solving consumer or trade problems; (vi) to improve product quality; (vii) to improve the product image; and (viii) to lengthen or shorten the lifetime of a product. This list is not exhaustive, there are many other objectives one can think of. But it shows that aims have to be specified very precisely if the measures of product policy are to contribute to the company's goal.

Table 13.1. Characteristics of the traditional product life cycle.

	Stage in life cycle			
Characteristics	Introduction	Growth	Maturity and saturation	Decline
Sales	Increasing	Rapidly increasing	Stable	Decreasing
Competition	None or small	Some	Much	Little
Profits	Negative	Increasing	Decreasing	Decreasing
Profit margins	Low	High	Decreasing	Decreasing
Customers	Innovators	Adopters	Majority	Laggards

DYNAMICS OF PRODUCT POLICY

As mentioned before, product policy is a dynamic process. As the 'lifetime' of a product becomes shorter, product development becomes more and more important.

Product life cycle

The product life cycle is defined in terms of two dimensions, *sales volume* and *time*. The traditional life cycle follows an S-shaped sales curve which is made up of the following stages (see Table 13.1 and Fig. 13.2): (i) introduction, a period of slow growth and almost non-existent profit; (ii) growth, a period of rapid market acceptance and substantial profit improvement; (iii)

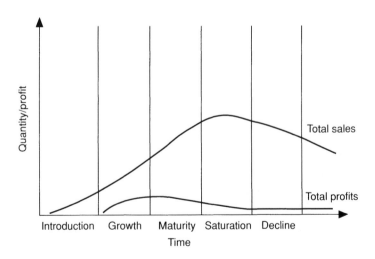

Fig. 13.2. Product life cycle (source: Strecker *et al.*, 1990).

maturity and saturation, a period of a slight slow-down in sales growth and declining profit; and (iv) decline, a period of decreasing sales and profit.

The product life cycle patterns vary a lot in length of time and in shape. The reason for the different curves of different products is the variation in behaviour of the consumer, the middlemen and the competitors. Due to today's consumer requirements (the consumer wants more changes in product supply) the length of a product's lifetime is becoming shorter.

The product life cycle is based on the assumption that all groups consist of innovators, early adopters, early and late majority and laggards. Here is an example. In the 1970s New Zealand launched a product to the Northern European market which was new to the European consumer – the kiwi fruit. After the introduction stage, the country was able to make high profits and increased the export quantities very quickly. Soon after, other countries (Italy, Southern France, Spain, South America) started kiwi fruit production as well to participate in this boom. The supply of kiwi fruit increased and the prices declined in competition. As a consequence the profit of New Zealand decreased heavily. Kiwi fruit became a mass product. New Zealand had missed the chance to differentiate from other origins by product policy (e.g. branding and name protection). Today kiwi fruit of different origins is interchangeable in the eyes of consumers.

The product life cycle model is an interesting and useful concept for marketers. But although it provides a framework for product planning, its practical use can be overestimated. The model is not workable for forecasting, as the stages of the life cycle, the time span and the shape vary by product and because of the unpredictability of external influences such as economic growth, inflation and life style, which disturb the life cycle. But in spite of the above arguments, the model is not completely useless. It shows very clearly the development of sales and profit in the course of time and emphasizes the necessity of market research and new marketing strategies (Altmann, 1994).

New product development

There are three kinds of new products: real innovations, adaptive replacements and me-too products.

Real innovations are unique products for which there are true needs but no existing satisfactory substitutes. In the case of agricultural products real innovations are very rare. New agricultural products are often *adaptive replacements*, which introduce significant changes that can replace existing products, like new varieties, fruit and vegetables of bio-organic production or cereals of controlled production.

Real innovations are however a feature of the food industry (e.g. dried milk, instant cappuccino). New product development can be a company strategy. Nestlé, for example, launches new products very frequently, such as instant cappuccino, instant espresso, chocolate bars, etc. Following this strategy Nestlé always remains new and interesting for the consumer.

Table 13.2. Failure rates of new products (source: Lebensmittelzeitung, various editions).

Product category	Number of new products	Number of new products 5 years later	Failure rate (%)
Preserved meat and sausage	44	1	97
Preserved fruit and vegetable	43	1	97
Bread and baker's ware	41	2	95
Spread (sweet)	15	1	93
Instant meal	51	5	90
Diet food	59	6	89
Sweets and cakes	56	7	87
Dairy products	94	18	80
Spices and sauce	101	20	80
Meat and sausage	49	10	80
Frozen food	42	10	76
Cheese	26	9	65
Other	447	68	84
Total	1134	169	85

The *me-too products*, which are imitative products, are new to the producer but not to the market. The imitative products are the least risky type of new products. The more innovative a product, the greater the risk.

Despite improved marketing technology the failure rate of new products remains high. The failure rate (or so-called 'flop rate') is estimated at up to 90%. Table 13.2 shows some failure rates of food products. The main reasons why products fail are: (i) poor planning or bad organization; (ii) poor timing; (iii) a lack of a differential advantage; (iv) an excessive enthusiasm; (v) increasing competition; and (vi) ever-increasing consumer demands in an affluent society.

New product development is a process that involves seven steps (Fig. 13.3): (i) idea generation; (ii) product screening; (iii) concept testing; (iv) business analysis; (v) product development; (vi) test marketing; and (vii) commercialization. During the process, the company generates several potential opportunities, evaluates them, weeds out the least attractive ones, obtains consumer perceptions, develops the product, tests it and introduces it to the market. The loss and termination of an idea can occur at any step. Costs increase from step to step.

In the agro-food sector it is mainly the food manufacturer who invests in product development in order to become attractive for the consumer through the development of brands. Real innovations or adaptive replacements are created in the research and development departments of companies.

In recent years the food trade has used a new strategy to attract the consumer by presenting so-called own label products (known as private label in America). These are mainly me-too products, i.e. initiatives modelled on a market that has already been opened at substantial expense by innovative

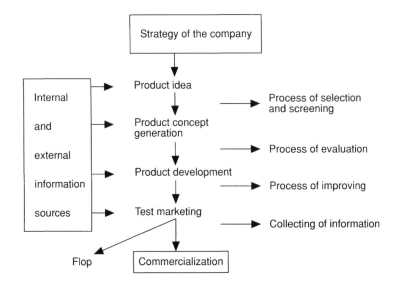

Fig. 13.3. Stages of new product development.

producers. As the food retailer does not have to incur high expenses on product development as well as on the creation and maintenance of the brand, he is able to offer such products at a considerably lower price (Jafri *et al.*, 1993; Padberg *et al.*, 1993).

PRODUCT RANGE DECISIONS

A product range is determined by its width (the number of different product lines), its depth (the number of different products, varieties in each line) and its consistency (the relationship among the product lines). The decision on the best width and depth of a product range of a company is based on the following conditions. In order to meet consumer demand for a large range of products, it is necessary to offer a wide and deep range. On the other hand, to reduce production and selling costs, it is useful to produce large quantities of one single product (economies of scale). This leads to specialism which implies a narrowing and flattening of the product range.

There are two different ways to solve the conflict between economy and marketing requirements:

1. Indirect selling companies pass the task of fulfilling the product range to the downstream level (e.g. marketing organization, wholesalers, retailers) and can concentrate on the production of a number of products which are optimal from a point of view of economy as well as of marketing. They are able to produce with lower costs but have to accept lower prices due to the margins the other marketers need.

2. Direct selling companies are forced to offer a broad and deep product range. They produce with high production costs. Often these companies split their range into articles of their own production and articles bought from other producers. Companies selling on wholesale markets or on weekly markets normally do not need such a broad range, because their offer is complemented by that of their colleagues (Altmann, 1994).

In the agricultural sector the decision on the right product range is connected with the decision on the distribution system. Concerning the direct selling company cost analyses are very important in order to establish whether a product should be self-produced or bought. As to the indirect selling company, it should not only take into account cost analyses, but also the capacity of marketing organizations, the selling potential and the marketing strategies of the middlemen.

The most important objective of every product range decision should be to create a close link to both the trade and the consumer in order to be successful in competition. In comparison to the agricultural sector, there is a much wider scope of freedom of action in the food industry as far as product range decisions are concerned. The food manufacturer replies to the consumer preferences by developing strong brands implying high investments in the promotion of the product image. This behaviour promises at the same time good sales figures to the trade.

In some cases the relation between the client and the brand is so close that the trade is practically unable to refuse the insertion of a certain product into its product range. A typical example is Coca Cola which is expected to be found in every supermarket, except for extreme discounters where the client accepts a reduced product range.

Product range policy of the food trade has changed in the last few years. The mere competition of prices has lost importance in favour of a competition of image. What is considered essential today is a diversification of the product range and quality. Since shops look all more or less the same throughout the world, exclusive products and product ranges can help to make the difference[1].

PRODUCT DECISIONS

Quality

The quality of a product is a subjective, individual evaluation criterion for the consumer. Quality summarizes all the other product characteristics, which are: (i) the nature of the product; (ii) the packaging; (iii) the labelling and branding; and (iv) the warranties and legal protection.

The evaluation of product quality is the main part of the image the product has (Strecker *et al.*, 1990). Figure 13.4 shows the interrelationship.

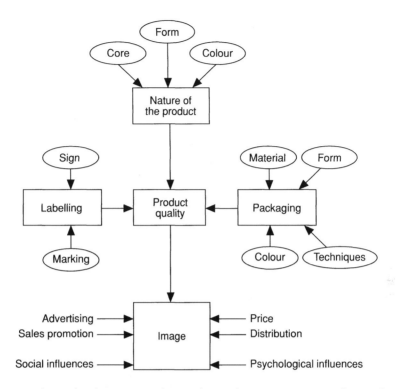

Fig. 13.4. Relationship between product quality and image (source: Strecker *et al.*, 1990).

We all know that image plays the major role within the purchase decision process. If asked 'what is the quality of a product?' people would enumerate a range of criteria with different meanings and importance for each person.

When evaluating food quality, we have to distinguish between two different views: (i) objective quality which can be measured by chemical analysis (factors investigated are the ingredients like vitamins, minerals and so on); and (ii) subjective quality, which includes the taste, the enjoyment and satisfaction a consumer experiences when using the product. Other factors in evaluating quality of food are the freshness and the absence of toxic agents which can be interpreted in an objective and subjective way. The range of consumers' tolerance concerning freshness and toxics is broad (*cf.* Fig. 13.5).

Looking at the consumption patterns of the European market, where people on average are accustomed to affluent society, there are three main trends which are becoming more and more important:

1. *Ecology and conservation.* Increased education, affluence and environmental pollution have led people to be concerned about environmental problems and diminishing natural resources. Demand for recycled products, returnable bottles or containers and for products which contribute less to pollution (less fertilizer, less agrochemicals) is increasing.

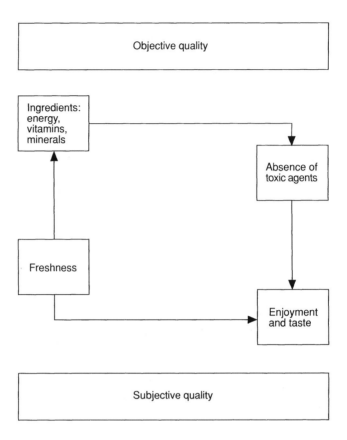

Fig. 13.5. Evaluation of quality.

2. *Health.* Affluence, more leisure time and the focus on youthfulness all contribute to concern for feeling and looking healthy. Demand for health food, diet food and biodynamic food is becoming important. This nutrition trend is supported by increasing illness due to malnutrition or allergies.

3. *Luxury needs and pleasure.* Increased income and a lifestyle orientated towards self-satisfaction, creative expression and self-fulfilment have led people to want products with which they can express themselves and show their ways of life. Demands for luxury products are increasing. Brand loyalty is becoming more important. Only high quality is demanded. Shopping and eating have to be great fun and enjoyable.

These trends suggest that a successful new food product would need to be associated with health, and it must taste well and give enjoyment, and at least be neutral to ecology and natural resources.

To succeed in today's markets some producers believe they must develop and supply a product of the highest quality. High quality involves: (i) ingredients of a high standard in accordance with current chemical analysis; (ii) the least possible number/amount of toxic ingredients; (iii) as much freshness as possible; and (iv) a taste and enjoyment matching the needs and wishes of the target markets. Today consumers take high quality for granted. Therefore we no longer have so much a competition of quality, but rather a competition of promotion, communication and information.

Following the worldwide trend towards more stringent consumer expectations concerning quality, the EU is harmonizing the series of International Standards in order to rationalize the many and various national approaches. The ISO 9000 is a guideline referred to in Chapter 9 for quality management and quality assurance standards. It gives more transparency to the consumer about the way the product is produced, so that a comparison between several competitors is possible. However, it does not define the level of quality. ISO norms and Quality Management are now heavily discussed and will become a prerequisite for staying in and selling to the EU market.

The individual farmer who sells fresh, unprocessed products has far more difficulties to accept the ISO norms than the food industry which has become accustomed to a process control. But in the long-run farmers will have to accept the application of norms for their production too, as the trade in Europe will ask for it more and more in the future.

Packaging

Packaging as a part of product policy has become of increasing importance in recent years. The reasons for this are: (i) the growth of supermarkets and discounters has led to more self-service and more pre-packaging; (ii) wholesalers and retailers follow efficient and easy handling systems which require standardized packaging; (iii) increasing consumer income has developed a high demand for convenience, appearance and the attractiveness of better packaging; (iv) companies have recorded a better return on investment by developing new packaging of a product; and (v) increasing competition increases the necessity of attractive packaging to gain consumer attention.

Packaging has two main functions:

1. The technical function. The package enables the product to be shipped, stored and handled in a safe and reliable manner (transport capability, storage capability, protection capability). The package allows the product to be easily used. Sometimes a package is reusable after the product is depleted. Reutilization is becoming more important.
2. The communication function. The package communicates a company image through its design, label, colour, brand and display. The package is a major promotion tool. Communication functions include information, identification, advertising, quantity of contents and attractiveness.

In the case of agricultural products, the packaging is the only element which allows customers to distinguish between different production origins or marketers. Looking only at the core of the product, neither consumer nor middlemen can recognize and identify the origin to build up preferences and brand loyalty (Strecker *et al.*, 1990).

Packaging has been criticized in recent years because of its negative impact on the environment and scarce resources. Criticism refers mainly to the costs involved and raises questions of its necessity. Consumers are now starting to refuse affluent packaging. Companies can gain advantages by offering their products in reusable, ecologically beneficial packaging. In the case of vegetables and fruit, consumers may prefer non-packaged products.

According to a new packaging law in Germany, producers are responsible for taking back all the packaging material and for its recycling. Other European countries will probably follow.

Branding

Besides its core and packaging, each product is marked in a certain way to be easily identified. The main purpose of branding food products is: (i) to give a product a name to differentiate it from those of competitors; (ii) to allow a product to be easily identified; (iii) to enlarge the attractiveness by creative labels and brand names or marks; and (iv) to assure the consumer of a certain level of product quality.

Usually the simple fact of giving a name to a product does not automatically include a legal quality control, nor does it provide any legal guarantee or warranty for the consumer. A 'real' trademark or brand does. If the producer or trade has found a really good brand name or mark he tries to give it legal protection. In this case a registered product mark is followed by ® copyright. The product mark is protected for ten years; the protection time can be prolonged.

There are three different kinds of trademarks or brands (Altmann, 1994): (i) mark of origin; (ii) mark of quality; and (iii) 'real' trademark or brand.

The *mark of origin* reveals the national, regional or local origin of a product. The utilization of these marks is based on the assumption that the consumer has special regional preferences. The mark of origin is a certain kind of mark of quality.

The *mark of quality* cannot be given by a single company, but by an association. The members (several companies) agree on special production conditions and controlling systems and are allowed to use a special sign. The product checks are made by independent institutes. This can be particularly useful for agricultural products.

The *'real' trademark* or *'real' brand* policy plays a major role in marketing performance. It promises the consumer a certain quality standard. The most important features of a trademark product or a brand product are: (i) unmistakable name and sign; (ii) promises as regards quality; (iii) reputation in the market for being a high quality product; (iv) well known in the market, a feature which requires an intensive information policy; and (v) supra-regional sales, i.e. high production potential.

No matter whether a product carries a mark of origin, quality or trademark, the main concern in each case is to break through the anonymous mass production and to create a connection between producer and consumer.

Table 13.3. Market share of own labels, 1992 (source: Morgan, 1992).

Country	Market share (%)
United States	14.0
Japan	20.0
United Kingdom	37.0
Switzerland	30.0
Germany	29.8
Sweden	20.0
Belgium	19.8
Netherlands	17.1
France	16.5
Spain	7.7
Italy	6.8
Greece	3.5
Portugal	1.7

As far as brands are concerned a further distinction has to be made between *producer brands* and *dealer brands* (own label, known as private label in the USA, as mentioned before). In recent years own labels have become more and more important and competed heavily with producer brands. This is due to the changed marketing strategy of the food retailer groups which are facing increasing competition. Retailers are more interested in a constant relationship between the consumer and the outlet rather than in brand loyalty of consumers to the producer. On the other hand, the producers want to distinguish themselves through their brands and images in order to reduce their dependence on single dealers or to convince the trade to offer their well-known products because of the high consumer demand.

As margins drop and competition intensifies, food retailers are becoming more and more aware that selling products under their own brand name has important advantages. Own labelling provides a higher influence in product innovation, quality and in larger margins. The cost of goods makes up 80% of the retailer's total costs, which means that buying cheaper own label products can lead to higher profits.

Within the EU, the own label strategy is most developed in the United Kingdom where it makes up 37% (1992) of the market share. Marks & Spencer, for example, sells *only* own label products, and in Sainsbury, Britain's largest supermarket chain, these products account for 60% of the product range offered. Own labels have helped British food retailers to achieve profit margins of 6–8%, which is very high in comparison with other countries (France and USA 1–2%). In Germany own labels accounted for 30% in 1992, with an estimated growth of 35% and more in the following two or three years (Table 13.3; Rabobank, 1994).

Following this trend, the so-called no-name products have lost in importance in favour of high quality own label products which promise higher profits. No-name products are cheap, quickly sold without expensive packaging, with little added value and little consumer identification.

Today, own labels have become an essential marketing tool for the large retail chains. Retailers offer a full range of grocery products under own labels and quality has increased to the level of the manufacturers' brands, with higher cost of investment and commitment (Jafri *et al.*, 1993).

Even in America, the development is the same as in Europe. In America, private label was long regarded as a cheap generic substitute for the real thing. Private label goods sold 20–30% cheaper. The share of private label products increased to 14% in 1992 and the growth came from premium private label goods that compete in quality with manufacturers' top brands. The share of cheap generics is tiny and declining[1].

Brand policy is becoming more and more important because the consumer's awareness and appreciation of brands is still increasing, due to the growing requirements as regards quality, insecurity concerning food, and high safety requirements. In addition, brands simplify purchase decisions in the face of a multitude of information.

In comparison to the food industry, brand policy is much more complicated for the agricultural sector, especially in Europe. This is mainly caused by: (i) the relatively small structured companies, which cannot supply the necessary quantities to cover an entire region/market segment; (ii) the high costs for a single company to carry out the necessary promotion campaign; (iii) the seasonal production aspect of agricultural goods, which causes problems concerning constant delivery and constant quality and product homogeneity throughout the year; and (iv) the complicated handling of fresh products.

In spite of these problems it is possible to sell fresh products under trademarks. Good examples are bananas (e.g. Chiquita, Dôle) and oranges, with intensive, strong export marketing activities, the marketing of prince de Bretagne in France, or bio-products sold under labels such as Demeter, Bioland and Naturland in Germany.

The existing atomistic producer structure requires cooperation to solve the above problems. Important prerequisites for brand building are: (i) the cooperation of growers with similar objectives (production potential and know-how) and/or a cooperation with a marketing organization; (ii) a clear definition of quality guidelines and production pattern which easily differentiate the product from others; (iii) a strict control system throughout the whole production process to guarantee the quality statement; (iv) clear sanctions in the case of ignoring the rules; and (v) good cooperation with middlemen/traders.

Experience shows that both the trade and consumers welcome the efforts of the production side. The previously homogeneous market is splitting up and provides new opportunities for product diversification and better profits for agricultural products.

CONCLUDING REMARKS

This chapter has shown that product policy, as the pivot point of all decisions made in marketing policy, is more and more influenced by structural changes in food retailing and consumer preferences.

The following developments seem to be important for the future:

1. The growing concentration and the worldwide interconnection of big food retail companies lead to a changing purchasing behaviour and to a changed production pattern of the producers.

2. Food retailers are increasingly involving themselves further back in the food production business. (Direct contact at the farm level reduces costs and minimizes risks in food quality).

3. Product life cycles are becoming shorter. Flexibility and creativity gain importance.

4. New product development and brand policy gain importance in order to attract consumer awareness for both the manufacturer and the retailer (with own or private labelling).

5. Quality competition is replaced by competition of promotion and communication.

6. Quality is perceived by the consumer as associated with aspects of ecology, health and pleasure. High quality is taken for granted.

7. Quality is seen by the retailer as a way of gaining advantages in competition.

In future, food product policy will increasingly be driven by the attitudes and requirements of the middlemen and the consumers.

NOTES

1. The Economist (1995) Survey retailing. *The Economist*, March 4, pp. 3–18.

FURTHER READING

Booz, Allen & Hamilton, Inc. (1982) *New Product Management for the 1980's*. Booz, Allen & Hamilton, Inc., New York.

Buchholz, H.E. and Wendt, H. (1991) *Food Marketing and Food Industries in the Single European Market*. FAL, Braunschweig.

Jones, G. (1994) *Adding Value: Brands and Marketing in Food and Drink*. Routledge, London.

Kuczmarski, T.D. (1992) *Managing New Products: The Power of Innovation*, 2nd edn. Prentice-Hall, Inc., New York.

Rabobank Nederland, Agribusiness Research (1994) *The Retail Food Market*. Rabobank Nederland, Agribusiness Research, Frankfurt.

Steenkamp, J.–B.E.M. (1990) *Product Quality: An Investigation into the Concept and how it is Perceived by Consumers*. Books International, Herdon, VA.

Wheelwright, S.C. and Kim, B.C. (1992) *Revolutionizing Product Development*. The Free Press, New York.

REFERENCES

Altmann, M. (1994) Marketing. In: Storck, H. (ed.), *Taschenbuch des Gartenbaues*. Ulmer Verlag, Stuttgart, pp. 51–90.

Jahn, H.–H. (1991) Strategic adjustments of agro-food-firms. In: Buchholz, H.E. and Wendt, H. (eds), *Food Marketing and Food Industries in the Single European Market.* FAL, Braunschweig, pp. 51–66.

Jafri, S.H., Rogers, T. and Padberg, D.I. (1993) Prices competition between national brand and private label products. *Journal of Food Distribution Research* September, 49–55.

Lebensmittelzeitung (various editions) Deutschen Fachverlag, Frankfurt.

Morgan, J.P. (1992) *Nielsen, Points de Vente.* Rabobank Report. Rabobank Nederland.

Padberg, D.I., Knutson, R. and Jafri, S.H. (1993) Retail food pricing: horizontal and vertical determinants. *Journal of Food Distribution Research* 24(1), 48–59.

Rabobank Nederland, Agribusiness Research (1994) *The Retail Food Market.* Rabobank Nederland, Agribusiness Research, Frankfurt.

Statistical Office of the European Communities (1993) *Der Einzelhandel im EG-Binnenmarkt 1993.* Eurostat, Luxembourg.

Strecker, O., Reichert, J. and Pottebaum, P. (1990) *Marketing für Lebensmittel.* DLG Verlag, Frankfurt.

14 Pricing Policy

George G. Panigyrakis

Department of Business Administration, The Athens University of Economics and Business, 76 Patission Street, Athens 104 34, Greece

INTRODUCTION

This chapter covers the second major element of the food marketing mix, pricing. The material is organized around four main issues.

First, we give an overview of food price planning. Each of the factors affecting price decisions is considered, grouped into company internal factors and external environmental factors. The second section explains how a pricing strategy is developed. It distinguishes among sales, profit and *status quo* objectives. The use of a broad price policy and the utility of a multistage approach to pricing are presented. The three basic types of pricing strategy are outlined: cost, demand and competition. The various methods for adjusting food prices are noted. The third section shows how cost, demand and competition techniques of pricing may be applied. The attributes and variations of each technique are evaluated. Finally, the chapter ends with a section on the managerial pricing issues related to food products for industrial markets.

NATURE AND IMPORTANCE OF PRICE TO FOOD MARKETERS

Pricing is part of the marketing mix and pricing decisions must therefore be integrated with other aspects of the marketing mix. Price is one of the most visible variables to the food buyer and, besides being controllable, it is usually one of the more flexible and independent variables.

From a marketing point of view, price is the exchange of something of value between the parties involved in a transaction. Something of value is exchanged for satisfaction or utility. Purchasing power depends on a buyer's income, credit and wealth. Money is most often used as the common denominator to establish value. Price, however, is not always paid in money. In fact, in the trading of food products, barter is the oldest form of exchange. Thus, money may or may not be involved in the process.

Food buyers' concern for the interest of price is related to their expectations about the satisfaction associated with the product. In effect, buyers must decide whether the utility gained in an exchange is worth the buying power sacrificed.

From a consumer's point of view, price is often used to indicate value (Monroe, 1990). Consumer's value assessments are often comparative. Here value involves the judgement by a consumer of the worth and desirability of a product relative to substitutes that satisfy the same need. In this instance a 'reference value' emerges, which involves comparing the costs and benefits of substitute items. For example, although 'Equal', a sugar substitute containing NutraSweet, might be more expensive than sugar, some consumers value it more highly than sugar because 'Equal' contains no calories.

Price can also add symbolic value to the marketing mix. Witness the 'snob effect' created by the price of prestige, speciality food products referred to in Chapter 10. By raising a price, food marketers can emphasize the quality of a product and try to increase the status associated with its ownership. By lowering a price, they can emphasize a bargain and attract customers who go out of their way to save a small amount.

Price is a key element in the marketing mix because it relates directly to the generation of total revenue and thus has a direct effect on a firm's profits. This is apparent from a firm's profit equation:

$$\text{Profits} = \text{Total revenue (Prices} \times \text{Quantities sold)}$$
$$- \text{Total costs (Fixed costs} + \text{Variable costs)}$$

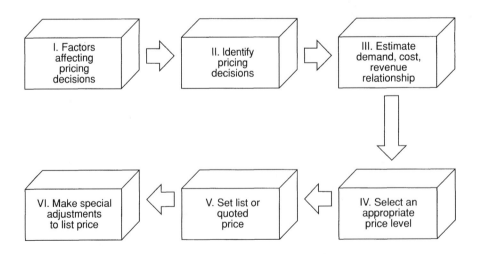

Fig. 14.1. A six-step pricing model.

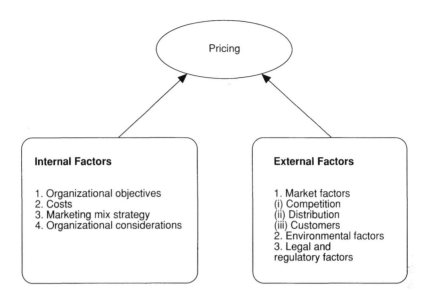

Fig. 14.2. Factors affecting pricing decisions.

The price variable affects the profit equation in several ways. It directly influences the equation because it is a major component. It has an indirect impact because it can be a major determinant of the quantities sold. Even more indirectly, price influences total costs through its impact on quantities sold. An analysis of the roles of price and product performance on the success or failure of 100 new grocery brands indicated that successful brands tended to offer better value for the money than did those of their competition; unsuccessful brands did not (Davidson, 1976).

The importance of pricing in the marketing mix necessitates an understanding of a multistage approach (six major steps, Fig. 14.1) involved in the process organizations go through in setting prices. These six steps are covered in this chapter. However, no one procedure, generally acceptable by all companies, has yet been developed for determining base prices. One critical reason for the lack of exact pricing models or formulas is that sufficient detailed information on costs at various volumes, and on demands at various prices, is not available.

STEP I: IDENTIFY PRICING CONSTRAINTS

To define a pricing problem it is important to consider both the objectives and constraints that narrow the range of alternatives available to solve it. Pricing constraints are those factors that limit the latitude of price a food manufacturer may set. They can be grouped into two major categories: internal company factors and external environmental factors (Fig. 14.2).

Internal company factors

Most companies begin pricing deliberations based on their own internal considerations. The most important internal company factors are considered below.

Organizational and marketing objectives

Food marketers should set prices consistent with the organization's objective and mission. Thus, before setting price, the company must decide on its product strategy. If the company has selected its target market and positioning carefully, then its marketing mix strategy, including price, will be fairly straightforward. For example, a supermarket chain trying to impose itself as value oriented may wish to set prices that are quite reasonable relative to product quality. The clearer a firm is about its objectives, the easier it is to set prices. A food company's marketing objectives are likely to be as follows: (i) financial (cash generation, profit, return on investment); (ii) marketing (maintain–improve market share, skim/penetrate depending on stage in product life cycle); (iii) competitive (prevent new entry, follow competition, market stabilization); (iv) product differentiation (high price aids perception of product differences).

A company may have different objectives in different markets and thus need to adopt different pricing policies. Examples of the most common objectives are:

1. Survival. In some cases, objectives such as profits, sales and market share are less important objectives of the firm than mere survival. Companies set survival as their major objective when they face serious difficulties with competition or with changing consumer wants.
2. Social responsibility. A firm may also forgo higher profit on sales and follow a pricing objective that recognizes its obligations to customers and society in general. Gerber, for instance, supplies a specially formulated product free of charge to children who cannot tolerate foods based on cow's milk[1].
3. Profit. Many firms want to set a price that will maximize current profits. The objective of price maximization is rarely, however, operational because its achievement is difficult to measure.
4. Market share. This is a firm's sales in relation to total industry sales and it is a very meaningful benchmark of success. Companies often pursue a market share objective when the industry sales are flat or declining.
5. Product quality leadership. A company might have the objective of product quality leadership in the market. This decision normally calls for charging a high price to cover the high product quality.

Costs

Costs set the lower limit for the price that the company can ask for its products. Thus, a firm may temporarily sell products below cost to match competition, to generate cash flow, or even to increase market share, but in the long-run it cannot survive by selling its products at such a price. Companies with lower costs can set lower prices resulting in greater sales and profits. Understanding the role and behaviour of costs is critical for pricing decisions as it acts as a floor limit of price. Four concepts of costs are important in pricing decisions (these were introduced in Chapter 3):

1. Total cost is the total expense incurred by a firm in production and marketing the product. Total cost is the sum of fixed cost and variable cost.
2. Fixed cost is the sum of expenses of the firm that are stable and do not change with the quantity of product that is produced and sold.
3. Variable cost is the sum of the expenses of the firm that vary directly with the quantity of product that is produced and sold.
4. Marginal cost is the change in total cost that results from producing and marketing one additional unit. The message of marginal analysis is to operate up to the quantity and price level where marginal revenue equals marginal cost.

Marketing managers often make use of break-even analysis, which is a simple but efficient technique for looking at cost volume and profit relationships. This method is a technique that analyses the relationship between total revenue and total cost to determine profitability at various levels of output. The break-even point (BEP) is the quantity at which total revenue and total costs are equal and beyond which profit occurs.

Marketing mix strategy

Price decisions must be coordinated with all the other elements of the marketing mix to form a consistent and effective marketing programme. Thus, the food marketer must consider the total marketing mix when setting prices. If the product is positioned on non-price factors, then decisions about quality, promotion and distribution will strongly affect price. If price is a key positioning factor, then price will strongly affect decisions on the other marketing mix elements. A food manufacturer's pricing discretion varies with the length and complexity of the chain of distribution, and so does the discretion of any organization within the trade. The longer the chain, the less flexible the price structure. The effect of distribution channels on pricing discretion has yet another aspect: the more complex the channel, the less difference changes at the manufacturer's level may make at the retail level. The complex role of the distribution channel in the marketing mix is discussed in Chapter 17.

Organizational considerations

These considerations include: firm and product positioning, available resources (financial, material, human, experience in business, corporate reputation and image), product range, substitutes, product differentiation and unique selling propositions. Finally, management must decide who within the organization should set prices. In small companies, prices are often set by top management rather than by the marketing or sales department. In large companies pricing is handled by divisional (strategic business unit (SBU)) or product line managers. In industrial markets, sales people may be allowed to negotiate with customers within certain price ranges.

Company external factors

Market factors affecting pricing

Companies cannot establish pricing policies in a vacuum. Although cost information is essential, prices are affected by competition, distribution and income.

Competition   When considering food product pricing, the author finds it helpful to delineate three types of competitive markets. These are: (i) open competition, i.e. markets with a large number of small firms; (ii) loose oligopoly, with at least five strong competitors in the market; and (iii) tight oligopoly, characterized either by the presence of one leading firm or by a small dominant group of companies and a very restricted number of competitors. Examples of food products where these kinds of competition apply in at least some advanced countries are: (i) open competition (table wines, canned vegetables, bacon); (ii) loose oligopoly (coffee, pasta, cigarettes, chocolate and confectionery, soft drinks, biscuits, mineral waters, sugar); (iii) tight oligopoly (whiskies, baby foods, pet foods, breakfast cereals, cola beverages, instant coffee and margarine).

The type of competition dramatically influences the latitude of price competition and, in turn, the nature of product differentiation and extent of advertising. A firm acting as the sole supplier of a product in a given market enjoys greater pricing flexibility. The opposite is true if that same company has to compete against several other local or international firms. A firm must know what specific price its present competitors charge and what they and potential competitors will charge in the future. When the NutraSweet Company planned the market introduction of the 'Simlesse' all natural fat substitute, it had to consider the price of fat replacements already available, as well as potential competitors such as Procter and Gamble's Olestra and a fat substitute product being developed by Kraft General Foods, Inc.[2].

Competition analysis   Competition helps set the price within the limit of cost and demand. Depending on the marketer's objectives and competitive position, it may choose to compete directly on price or elect for non-price measures. If a company's position is being eroded by competitors who focus on price, the marketer may have no choice but to respond. In most cases, marketers are in a better position to establish prices when they know the prices charged for competing brands. Learning competitors' prices may be a regular function of marketing research, as discussed in Chapter 11. Some grocery chain stores, for example, have full-time comparative shoppers who systematically collect data on prices. Food companies may also purchase price lists, sometimes weekly, from syndicated marketing research services. Present and anticipated competition is an important influence on price determination.

The threat of potential competition is greatest when the field is easy to enter (low barriers to entry) and the profit prospects are encouraging.

Competition can come from three existing sources: (i) directly similar products (Coca-Cola Light versus Diet Pepsi); (ii) available substitutes (dairy proteins versus vegetable proteins); (iii) unrelated products (biscuits versus drinking yogurt).

Distribution Prices have to be set keeping in mind, not only the ultimate consumers, but also the intermediaries involved. Each channel member seeks to play a significant role in setting prices to generate sales volume, obtain adequate profit margins, derive a suitable image, ensure repeat purchases, and meet specific goals. The amount of profit expected depends on what the intermediary could make if it were handling a competing product instead. Also, the amount of time and the resources required to carry the product influences intermediaries' expectations. A producer must also consider costs of discounts expected by distributors and the cost of several support activities, because failure to price the product so that the producer can provide some of these support activities may cause resellers to view the product less favourably.

It is important to note that today there is an enormous pressure on food manufacturers' margins from the side of distributors (also known as supermarkets, multiples or food chains). Several chapters in this book have referred to their growing importance in agro-food marketing in terms of both size and global presence. These intermediaries such as the French Carrefour demand low-cost, direct-supply contracts, which many manufacturers may not be willing or able to furnish (Treadgold, 1990). The recent trend in most advanced countries is towards the increase of concentration in the distribution sector. The growth of the large distributors (like the French Auchan, Promodes, Carrefour, Leclerc, Intermarché; the Belgian GB-INNO-BM; the Dutch Ahold group; the British Sainsbury, Tesco, Argyll (Safeway), ASDA; the German Edeka, Aldi, Tengelman-Kaiser, REWE; the Italian Rinacente and Citta, Standa, Iperstanda, Euromercato; and the Swiss Metro, Glencore, Micros) and increasing competition among them, has limited the margin for manoeuvring of the food manufacturer.

Customers Customer characteristics, behaviour and attitudes are the ultimate determinants of what actually occurs in the market-place. Consumers rely on reference prices as the basis of comparison against which an observed price is compared. There are two kinds of reference prices: internal and external. External observed prices are typically posted at point of purchase as the 'regular retail price'. Internal reference prices are mental prices used to assess an observed price. A large number of internal reference prices have been proposed (Winner, 1988), including: 'fair' price (or what the product 'ought to cost' the customer); the price frequently charged; the last price paid; the upper amount someone would pay (or reservation price); the lower threshold (or lowest amount a customer would pay); the price of the brand usually bought; the average price charged for similar products; and the expected future price.

Environmental factors

A number of environmental factors (external/macroeconomic variables) influence pricing. These external variables, uncontrolled by any individual company, include macroeconomic and fiscal trends such as: changes in the price of basic raw materials; changes in labour productivity and labour costs; changes in the rate of inflation; changes in the government policy towards Value Added Tax; use of price controls as a tool of government policy; and exchange rate movements. These factors restrict company decision-making authority and become dominant concerns for food marketing managers.

Legal and regulatory factors

Pricing decisions are further complicated by legal and regulatory restrictions. They must avoid four major pricing approaches that have received the most public scrutiny.

Price fixing Price fixing is a conspiracy among firms to set prices for a product (Nagle, 1987). When two or more competitors explicitly or implicitly set prices this practice is called horizontal price fixing. Vertical price fixing involves controlling agreements between independent buyers and sellers whereby seller's are required not to sell products below a minimum retail price. Price fixing is illegal *per se*. The only exception is where price agreements are carried out under the supervision of a government or a certain legal body. In the international market, occasionally, price levels are manipulated by cartels or other agreements among local competitors. Cartels are forbidden by European Union legislation, but many Non-EU governments allow cartels provided they do not injure the consumer. The cartel formed by Swiss chocolate manufacturers provides an excellent example of the type of arrangement that companies can legally enter (Kennedy, 1977).

Predatory pricing Predatory pricing is the pricing of goods at a level that will drive competitors out of business. Its objective is the destruction of competition.

Deceptive pricing Deceptive pricing refers to price deals which mislead consumers. The most common deceptive methods are: bait and switch[3], bargains conditional on other purchases, comparable value comparisons, comparisons with suggested prices and former price comparisons.

Dumping Inexpensive imports often trigger accusations of dumping, the practice of selling a product at a price below actual costs. Dumping ranges from predatory dumping to unintentional dumping. Predatory dumping refers to a tactic whereby a firm intentionally sells at a loss in another country to increase its market share at the expense of domestic producers, which amounts to an international price war. Unintentional dumping is the result of time lags between the dates of sales transaction, shipment and arrival. Most governments have adopted regulations that prevent dumping because of potential injury to domestic manufacturers.

STEP II: IDENTIFY PRICING OBJECTIVES

Types of pricing objectives

Pricing objectives are overall goals that describe what the firm wants to achieve through its pricing efforts. Pricing objectives normally derive from overall marketing strategy and are subject to change as marketing strategy changes in response to market conditions. Numerous objectives have been cited for pricing. For the food industry, however, the dominant pricing objectives may be grouped into three major categories from which a firm may select: (i) sales oriented, to increase sales volume or to maintain or increase market share; (ii) profit oriented, to achieve a target return or to maximize profit; (iii) *status quo* oriented, to stabilize prices and to meet competition.

Knowing its objective, a company then can move to the actual determination of the base price of a product. Base price (list price) is the price of a unit of the product at its point of production or resale. This is the price before provision is made for discounts, freight charges, or any other modification, such as those discussed in the next section.

Sales (market share)-based objectives

With sales-based objectives the firm is interested in sales growth or maximizing market share. One factor that makes market share a workable goal is that a company can usually estimate what share of the market it achieves. A firm would focus on sales-based objectives for the following reasons: (i) it is interested in market saturation or sales growth as a major step leading to market control and sustained profits; (ii) it seeks to maximize unit sales and is willing to trade low per-unit profits for larger total profits; (iii) it assumes that higher sales will enable the firm to have lower per-unit costs.

Large food manufacturers use market share as a major objective in establishing prices. The assumption, of course, is that in the long-run an expanding or substantial market share will serve as a source of high profit.

Profit-oriented objectives

They are used when a firm designates high monetary profit as an objective.

Return On Investment (ROI) Some food manufacturers in their effort to establish pricing objectives that they hope will maximize long-range profits use a profit target in pricing a product or a product line. The target is usually established by specifying a rate of return consistent with the company's product investment policy, and then pricing to achieve that rate of return. This type of pricing objective is commonly referred to as Return On Investment (ROI) pricing, which will be discussed in greater detail later in this chapter. With ROI objectives the firm states that profits must relate to investment costs. The implication is that although another price might produce an even larger return over the long-run, the firm is satisfied with a return that is conventional for the given level of investment and risk.

Profit-maximization objectives   Although profit is a major consideration in almost every business situation, the idea of short-run profit maximization is in most instances an oversimplification. Few firms are interested only in short-run success, and few know enough about market demand at various price levels to price strictly on a profit-maximization basis. Their concern, rather, is to establish pricing objectives that will maximize long-range profits. The trouble with this objective is that the term 'profit maximization' has an ugly connotation. It is connected in the public mind with profiteering, high prices and monopoly. In the market-place for food products, however, it is difficult to find many situations where a monopolistic situation has existed over an extended time. Substitute products are available, or competition increases, and prices are thus kept at a reasonable level.

Status quo *pricing objectives*

The objective of stabilizing prices is often sought by a firm interested in stability or in continuing a favourable climate for its operations. It is often found in industries that have a price leader. Companies seeking stability in their prices are very eager to avert price wars, even when demand is declining. This type of pricing objective can be defined as a low-risk objective. It seeks to avoid 'rocking the boat' and bringing on retaliatory price warfare, and tends to promote industry-wide price stability. Price leaders, in particular, tend to take a long-run point of view in achieving stability. Their objective, in a sense, is to 'live and let-live'. However, it should not be inferred that *status quo* objectives require no effort on the part of the firm. To retain customers, a wholesaler may have to match the price cuts of its competitors. To maintain channel cooperation, a food manufacturer may have to lower its mark-up in the face of rising costs. A food retailer may have to charge low prices to discourage competitors from stocking certain product lines.

STEP III: ESTIMATE THE DEMAND FOR THE PRODUCT

Basic to setting a product's price is the extent of consumer demand for it. Understanding demand requires a look at how both economists and business people view it. The people who make up a market must have the ability to buy a product. Buyers must need a product, be willing to use their buying power, and have the authority to buy. Regarding the estimation of the demand, economists stress three key factors: consumer tastes, price and availability of the product, and consumer income. The first of these three factors influences what consumers want to buy, and the third affects what they can buy. Along with price, these are often called demand factors that determine consumers' willingness and ability to pay for goods and services. While economists talk about 'demand curves' (as in Chapter 3), marketing executives are more likely to speak about 'revenues generated'. For them, demand curves would ideally lead directly to three related revenue concepts critical to pricing decisions: total revenue, average revenue and marginal revenue.

Determining the demand for a product is the responsibility of the marketing managers with the help of marketing researchers. Marketing research techniques of the kind discussed in Chapter 11 yield estimates of sales potential or the quantity of a product that could be sold during a specific period. These estimates are helpful in estimating the relationship between a product's price and the quantity demanded.

The two primary steps in demand estimation are:

1. To determine the 'expected price'. The expected price for a product is the price at which customers consciously or unconsciously value it. It is what they think the product is worth. Expected price usually is expressed as a range of prices, rather than a specific amount.

2. To estimate sales at various prices. It is important in price determination to estimate what the sales volume will be at several different prices. These estimates in effect involve a consideration of the demand elasticity of the product. A product with an elastic demand should usually be priced lower than an item with an inelastic demand. By estimating the demand for its product at different prices, management is implicitly determining the demand curve for the item. These volumes' estimates at different prices are important in relation: (i) to determining break-even points, the point at which cost of producing a product equals revenue made from selling the product, a topic that we discuss later in this chapter; and (ii) to determining the best price for profit maximization by analysing both demand and costs (marginal analysis).

Food marketing executives also recognize that the price elasticity of the demand is not always the same for product classes (such as tobacco, cereals, dairy, chocolate, cola) and brands within the product class (such as Marlboro, Kellogg's, Yoplait, Nestlé). Finally, marketers cannot base prices solely on elasticity considerations. They must also examine the costs associated with different volumes and see what happens to profits. These considerations are included in step IV which follows.

STEPS IV, V AND VI: SETTING A SPECIFIC PRICE

In the pricing procedure outlined, the final steps were to narrow the range of expected prices to the point where a specific selling price is established. The price the company charges will be somewhere between one that is too low to produce a profit and one that is too high to produce significant sales. Over the years many different pricing methods have been used by individual companies to accomplish this task. Most of these approaches to price setting, however, are based on one of the following major methods: cost-based, demand-based, or competition-based (Fig. 14.3).

All three approaches should be considered when establishing a price strategy. They do not operate independently of one another: pricing methods, like pricing objectives, are not necessarily mutually exclusive.

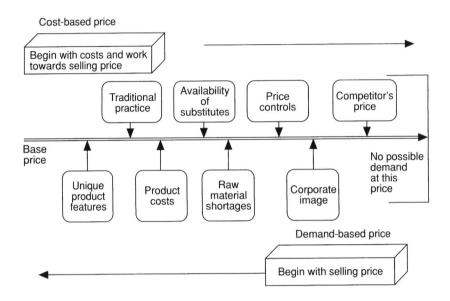

Fig. 14.3. Cost-, demand- and competition-based pricing methods and the determinants of price.

Cost-based pricing

Cost-based pricing is probably the most widely used pricing rule and it is based on all costs. It is simple, based on relative certainty, and tied to a reasonable profit. Cost-oriented pricing methods do not consider market conditions, plant capacity and competition[4]. Four common cost-oriented pricing methods are considered below.

Cost-plus pricing

Cost-plus pricing is the easiest form of pricing, based on units produced, total costs and profits. The formula for cost-plus pricing is:

$$\text{Price} = \frac{\text{Total fixed costs} + \text{Total variable costs} + \text{Projected profit}}{\text{Units produced}}$$

Although the method is easy to compute, it has the following shortcomings: (i) profit is not expressed as a per cent of sales but as a per cent of cost, and price is not tied to consumer demand; (ii) adjustments for rising costs are poorly conceived, and there are no plans for using excess capacity; (iii) there is little incentive to improve efficiency to hold down costs, and marginal costs are rarely analysed.

Cost-plus pricing is most effective when price fluctuations have little influence on sales and when a firm is able to control price. Agricultural marketing boards and cooperatives sometimes adopt this method as it can be said to protect sellers.

Mark-up pricing

Mark-up pricing is the practice of setting a price by adding a percentage to the cost of the product. This percentage is called a mark-up, and various alternative ways of calculating mark-ups are discussed in Chapter 8. The conditions required for applying the method are costs and demand elasticity which are fairly constant over time.

Mark-up is quite prevalent, particularly among wholesalers, retailers and the giant grocery chains who can determine costs rather precisely from their invoices. For managers of wholesale stores, supermarkets or grocery stores have such a large number of products that estimating the demand for each product as a means of setting price is impossible. The percentage mark-up that they charge depends on the type of retail store (grocery, supermarket, delicatessen) and on the product involved. High-volume products usually have smaller mark-ups than do low-volume products. Supermarkets have different mark-ups for staple items and discretionary items. The mark-up on staple items like sugar, eggs and dairy products can vary from 5% to 25%, where mark-ups on discretionary items such as snack food and candy may range from 25% to 50%. Some common mark-ups in supermarkets are 10% on baby foods, 20% on bakery products, 30% on dried foods and vegetables.

The method has the important virtues of simplicity of operation, ease of understanding, competitive harmony and social fairness to both the buyer and the seller. However, many food manufacturers consider mark-up pricing as an inefficient pricing method due to the following reasons: (i) it ignores both cost variations and the influence of demand factors on the particular product; (ii) it stresses the cost of the product or acquisition cost which is irrelevant to the buyer; (iii) it leads to incorrect profit projections, price has an effect on demand, in turn this reacts upon sales; (iv) it ignores demand elasticity that changes seasonally or cyclically over the product life cycle, in setting prices; (v) it does not take into consideration consumer or competitive reaction; (vi) this method cannot be adopted for a new product to recover the R&D costs.

The break-even concept is the basic tool for the analysis of cost-based pricing decisions. Figure 14.4a shows the basic relationship of the break-even concept. Figure 14.4b illustrates the effect of mark-up percentages on revenue, the higher the mark-up, the steeper the slope of the revenue line, and therefore the lower the break-even quantity. For example R_1 has the highest percentage mark-up and correspondingly the lowest break-even point Q_1, while R_2 and R_3, representing lesser mark-ups, have high break-even points. Of course, the revenue lines R_1, R_2 and R_3 are hypothetical as it may not be possible to sell beyond a certain quantity at any particular mark-up.

Experience curve pricing

Experience curve pricing is the result of new research and computerized improvements in information assembly and analysis which have focused on deeper understanding of costs than simply categorizing them as fixed or variable. Experience analysis has two aspects: scale and learning. The scale

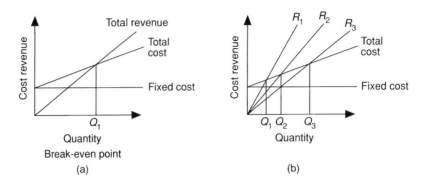

Fig. 14.4. (a) Break-even concept; (b) effect of mark-ups on revenue.

effect is the traditional concept of economies of scale – large operations process products at lower costs per unit, because fixed costs can be spread over a greater number of units. Nevertheless, size alone does not guarantee economies of scale; it simply provides the opportunity for them. Achievement requires proper managerial decisions and action – experience. This second effect is reflected by the costs which fall with cumulative production in a systematic way.

Figure 14.5 shows a typical experience curve with cumulative units of production on the horizontal axis and cost per unit on the vertical axis. The graph shows that as production doubles, the cost falls to 85% of the original level (Alberts, 1989). Experience effects explain the observed fact that larger organizations and/or those which are the leaders in new product innovation tend to have lower product costs that offer to those manufacturers a price-competitive advantage.

Return On Investment (ROI) pricing – target pricing

According to this method, the firm tries to determine the price that would give it a specified target rate of return on its total costs at an estimated standard volume. ROI is, by definition, a ratio – the ratio of profit to equity. To see the role played by price in ROI, the basic ROI ratio can be decomposed as follows:

$$\text{ROI} = \underbrace{\frac{\text{Revenues}}{\text{Assets}}}_{\text{Turnover}} \times \underbrace{\frac{\text{Profits}}{\text{Revenues}}}_{\text{Earnings}} \times \underbrace{\frac{\text{Assets}}{\text{Equity}}}_{\text{Leverage}}$$

Thus, the three components of ROI are themselves ratios: turnover, earnings and financial leverage. Pricing decisions are the integral component of two of these ratios, since revenues are obtained by multiplying price per unit times the number of units sold. As a result, ROI can be improved through pricing decisions that increase turnover while maintaining earnings. Grocery

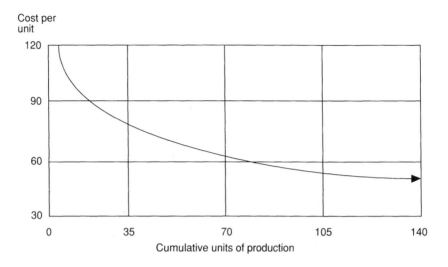

Fig. 14.5. The learning curve concept: an 85% learning curve.

stores, for example, operate on a thin earnings ratio; they do it by relatively rapid turnover. Conversely, if turnover can be maintained, ROI can be improved by increasing the earnings ratio; and of course any marketing manager would be pleased with increases in both ratios.

This method of pricing presents advantages and disadvantages. The advantages are: (i) it yields stability in prices over a period of time by accepting lower and higher returns in poor and good years respectively; (ii) it is a good method when the existence of uncertainty about the marginal relationships of the demand and cost functions makes it too risky to move away from full cost pricing in practice; (iii) the firm does not want to maximize profits. The disadvantages are: (i) like cost plus, it ignores demand and other market factors; (ii) an estimate of sales volume to derive the price is used, but price is a factor that influences sales volume; (iii) it is not useful for food manufacturers with low capital investment because it will understate selling price; (iv) production problems may hamper output and the standard volume may not be attained; (v) price reductions to handle overstocked inventories are not planned under this approach.

Despite these conceptual flaws, target-return pricing has been relatively common practice, especially in food manufacturing organizations.

Demand-oriented pricing methods

Demand-oriented approaches look at the intensity of the demand. This method results in a high price when demand for the product is strong and a low price when demand is weak. With demand-based pricing, a firm determines the prices consumers and channel members will pay for products, and then determines the maximum it can spend to produce its offering. In this way, prices and costs are linked to consumer preferences and channel needs, and a specific product image is sought. Demand-based methods require consumer research regarding the quantities that will be bought at various prices, the elasticity of demand and the existence of market segments and consumer willingness to pay. With demand pricing, highly competitive situations result in small mark-ups and low prices because consumers will buy substitutes. In these cases it is necessary for costs to be held down or else prices will be too high, which might result under cost-based pricing. In non-competitive situations, food firms can achieve large mark-ups and high prices because demand is relatively inelastic.

A common form of demand-oriented pricing is price discrimination, in which a particular commodity is sold at two or more prices. Price discrimination can be based on such considerations as type of customer, type of distribution channel used, or the time of the purchase.

Price skimming

In this case the firm seeks to take advantage of the fact that some buyers are always prepared to pay a much higher price than others because the product has high perceived value to them. The objective of skimming pricing is to gain a premium from these buyers and only gradually reduce the price to draw in the more price-elastic segments of the market. It is a form of price discrimination over time rather than over space. Skimming pricing gets its name from skimming successive layers of 'cream', or customer segments, as prices are lowered in a series of steps. Price skimming is more applicable to consumer durables than to food products and makes sense when any of the following conditions is present: (i) there are enough buyers whose demand is relatively inelastic; (ii) the unit production, promotion and distribution costs of producing a smaller volume are not so much higher that they cancel the advantage of charging what some of the traffic will bear; (iii) there is little danger that the high price will stimulate the emergence of rival firms.

Two advantages of this approach are that it is usually easier to reduce price than to increase it, if a mistake is made, and that a high price usually creates an impression of a high-quality product. This strategy may be consistent with branding although it is a dangerous strategy when concerned with a new brand aiming at long-run survival. A high initial price raises quality expectations amongst consumers. Once disappointed in a brand's quality and performance, it may be very difficult to persuade consumers to try again.

Penetration pricing

In order to achieve high sales, food producers frequently use a penetration pricing strategy, which is the opposite of skimming pricing. A penetration price is a low price intended to stimulate the growth of the market and to capture a large share of it. Any of several conditions might favour setting a low price: (i) the market appears to be highly price-sensitive; that is, many additional buyers would come into the market if the product were priced low; (ii) the unit costs of production, distribution and promotion fall with increased output; and (iii) a low price would discourage actual and potential competition.

Penetration pricing recognizes that a high price may leave a product vulnerable to competition. The policy is frequently used in supermarkets to introduce new food products. Once the consumer is a repeat buyer (of a new yogurt or snack, for example) the price is raised. To use this strategy, a firm would need to be convinced of the long-run viability of the market, of the high level of price sensitivity of the market and that low prices would discourage competition. In addition, if unit costs fall with increased output due to economies of scale this would be a bonus to the supplier. This objective is not really consistent with a policy of branding. Price, rather than non-price competition, is made the focus of marketing strategy and the implication is that all brands in the market are similar. The success of a brand creation policy leads to the reduction of the price sensitivity in a market through brand differentiation and higher perceived values by consumers. Low prices may encourage competition if they are perceived as illustrating financial weaknesses. The concept of value for money may lead consumers to take a broader view of price that includes quality connotations. Other things being equal, they may equate a low price with low quality. Thus, this pricing strategy holds potential pitfalls, in particular where successful branding is possible.

Step pricing

In some situations penetration pricing may follow skim pricing. The basic idea in this case is to set a relatively high initial price for the product to attract price-insensitive consumers and then gradually reduce the price over time to appeal to a broader segment of the population and increase market share. Demand is assumed to be inelastic at low sales, but elastic at a large volume, which allows penetration pricing to follow a price skimming policy. For example, a food manufacturer might sell a new recipe meat at a high price at first and it would be bought by high-income consumers. The novelty to them would then wear off and the food manufacturer would reduce the price to penetrate the large ready meals market.

Psychological pricing

Price has a psychological aspect if it influences customer reaction to the product. The seller in this case attempts to create an illusion about prices in the minds of buyers. It encourages purchases based on emotional rather than rational responses. Here, one must be able to discriminate between consumers' ability to pay and how willing they are. It is used most often at retail level. The procedure relies on the assumption that people prefer to make comparative judgements rather than absolute ones, as it is the problem of price differences that count rather than price levels.

Odd–even pricing Odd pricing has developed from the belief that consumers are more likely to buy a product at an odd price (99 FF) than at an even price (100 FF). Odd prices serve to create the illusion of low prices (Schindler and Wiman, 1989). In food retail advertisements the use of odd prices is quite widespread. Whether odd prices have the assumed psychological effect on consumers has never been adequately studied. The small reduction, however, in price may involve much money if sales volume is large. Even prices are used to give a product an exclusive or upscale image. An even price supposedly will influence a customer to view the product as being a high-quality premium brand.

Price lining   Often a firm that is selling not just a single product but a line of products may price them at a number of different specific pricing points, which is called price lining. For example, in city centres there is often a strong demand from office workers for ready-made sandwiches at lunchtime, and a particular outlet might offer 20 or 30 varieties. A mark-up pricing policy would produce a confusing range of prices because of differences in the cost of the ingredients used. Instead, prices are likely to be 'lined' into – say – three categories.

Prestige pricing   Prestige pricing is where price is used as a proxy for value. Thus, a high price is used to add prestige. Consumers, for example, may associate quality in beer with high price. The beer industry caters to this perception, as evidenced by its jargon: popular (to mean inexpensive), premium (to mean higher priced) and super premium (to mean expensive)[5]. Typical food and drink product categories in which selected products are prestige priced include: exotic food, liquor, whiskey, champagne. Here, part of the demand curve may be 'backward' sloping – that is, a reduction in price would lead to a reduction in volume (because of the reduction in 'prestige').

   This policy is applicable when: (i) large quality differences exist between available goods; (ii) the typical consumer cannot judge these quality differences; (iii) a high level of risk would be associated with making the wrong decision and obtaining a poor quality product; (iv) the product in question is visible to people who have an influence on the buyer.

Customary pricing   In customary pricing, the producer or the retailer must adjust the product to conform with tradition. Product, not price, is the basis of competition. Nestlé, for example, has changed the amount of chocolate in its candy bars depending on the price of cocoa, rather than its customary retail price, so it can continue selling through vending machines.

Product line promotion

Price, being an element of the marketing mix, can be coordinated with promotion. The two variables sometimes are so interrelated that the pricing policy is promotion oriented. Product line promotion occurs when the price setting of a brand is undertaken to attempt to maximize profits from the entire product range. Thus, the brand-pricing strategy will be more flexible than if the profit from each brand separately is to be maximized. The cross-elasticities need to be considered. These involve the degree of complementarity or substitutability within the product range. Again this strategy may have quality implications for individual brands, which should be considered in the decision process. The price level for a brand also needs to reflect the type of the distribution outlet used. One would expect to pay a higher price for an item in a delicatessen shop than in a certain supermarket or according to the degree of exclusivity of distribution outlets (e.g. vegetarian food boutiques).

Basket pricing

One approach to basket pricing is to take a set of products, offer them to customers in a package, and then price the package lower than the sum of the individual components. An alternative approach takes the opposite view; the basket can be priced higher than the sum of the components because it is attractive or convenient.

Competition-oriented pricing

So far, we have stressed the importance of cost factors and product demand in setting prices. In some situations a third factor – competition – must also be included. In using competition-oriented pricing, an organization considers costs and revenue secondary to competitors' prices. The importance of this method increases if competing products are almost homogeneous. For, when products are relatively homogeneous, the market may range from almost pure competition to oligopoly. In the first case, if a market tends towards pure competition, no individual seller can influence the price of the product. In an oligopoly situation, on the other hand, each organization knows the price set by its competitors.

Competition-based pricing is popular for several reasons. It is simple, with no calculations of demand curves, price elasticities, or cost per unit. Pricing at the market level is assumed to be fair for both consumers and companies. Pricing at the market level does not disrupt competition and, therefore, does not lead to retaliations. However, it may lead to complacency. Two aspects of competition-based pricing are discussed in the following sub-sections: follow-the-leader and bidding.

Follow-the-leader

In this situation a price leader exists in the market and most competitors set their prices in relation to this, either above, below or at the same level. The firm maintains the price if competitors do so even if cost or demand has changed. It can be used in markets characterized by little product differentiation (margarine, vegetable oils, sugar). In these cases, price is used in the battle for survival, dominance and market share. Follow-the-leader pricing happens when one or two firms are so big and powerful that their prices cannot be ignored.

Competitive bidding

Competitive bidding is a specialized competition-oriented pricing method used in industrial buying. The objective of the seller is to set prices low enough to get contracts and yet still generate an adequate profit. Various mathematical models have been applied to competitive bidding. All of them utilize the expected profit concept, which states that as the bid price increases the profit to a firm increases but the probability of its winning the contract decreases. Thus, prices are set on the basis of what you expect competitors to bid rather than on what the consumer will pay or cost considerations (Morse, 1980). Competition-oriented pricing methods may be combined with cost approaches to arrive at price levels necessary for a profit.

Price adjustments

Discount prices and allowances

Most companies will adjust their basic price to reward customers for certain activity that is favourable to the seller. Three special adjustments to the list of quoted price are: discounts, allowances and geographical adjustments.

Discounts A single base price is also used to quote different prices to various channels of the trade (that is, trade discounts to jobbers, distributors, wholesalers, retailers), to vary price by the quantity of purchase (quantity discounts), to vary price by season (seasonal discounts) and to specify cash discounts.

Discounts may also be used as temporary price cuts to retaliate against the inroads of a superior new product. The types of differentials and discounts vary by industry. They are numerous and constantly being added to. Four kinds of discounts are especially important in food marketing: quantity, seasonal, trade (functional) and cash discounts.

1. Quantity discounts. This is a price reduction to buyers who buy large volumes (for example, order-size discounts and package-size differentials). Quantity discounts are of two general kinds: non-cumulative and cumulative (Wilcox *et al.*, 1987). Non-cumulative quantity discounts are based on the size of individual orders; they encourage large individual orders, not a series of orders. Cumulative quantity discounts apply to the total purchases of a product over a given time of period, usually one year; they encourage repeat buying by a single customer to a far greater degree than do non-cumulative quantity discounts. A recent decision by Burger King to replace Pepsi-Cola with Coca-Cola in its outlets was partly based on the cumulative discounts offer by Coca-Cola on its syrup[6].
2. Seasonal discounts. This is a price reduction to buyers who buy merchandise or services out of season or advance orders. Seasonal discounts allow the seller to keep production steady during the year.
3. Trade (functional) discounts. These are the traditional discounts provided to wholesalers and retailers to compensate them for the distribution they perform. Traditional trade discounts have been established in various food product lines.

4. Cash discounts. This is a price reduction to buyers who pay their bills quickly, for example, a US$10 case, with 2.5% off for payment within 30 days. Retailers have recently started to provide cash discounts to consumers as well to eliminate the cost of credit granted to consumers.

Allowances Another type of reduction from the price list is an allowance, which is a concession in price to achieve a desired objective.

Trade allowances are price reductions given for turning in a used item when buying a new one. Promotional allowances are payments or price reductions to reward dealers for participating in advertising and sales-support programmes. Various types of allowances include an actual cash payment or an extra amount of 'free goods', as with a free case of Coca-Cola to a retailer for every 22 dozen cases bought. Frequently a portion of these savings is passed to the customer.

Geographical pricing

Geographical pricing involves reductions for transportation costs or other costs associated with the physical distance between the buyer and the seller. The underlying analysis will be found in Chapter 4. The five general methods for quoting prices related to transportation costs are: Free-On-Board (FOB) origin pricing, uniform delivery pricing, zone pricing, basing-point pricing and freight absorption pricing.

In FOB origin pricing, the buyer selects the transportation form and pays all freight charges. The delivery price to the buyer depends on freight charges. Supporters of FOB pricing feel that is the fairest way to assess freight charges because each customer picks up its own cost. However, buyers farthest from the seller face the big disadvantage of paying the higher transportation costs.

In contrast, uniform delivery pricing is the opposite and results in a fixed average cost of transportation. The major advantage is that it is fairly easy to administer and lets the firm advertise its price nationally.

Zone pricing falls between FOB origin pricing and uniform delivered pricing. It provides for a uniform delivered price to all buyers within a geographic zone. In a multiple-zone system, delivered prices may vary by zone. The delivered price to all buyers within any zone is the same, but prices across zones vary depending on the transportation cost of the zone and the level of competition and demand within the zone.

Basing-point pricing involves selecting one or more geographical areas (basing point) from which the list price for products plus freight expenses is charged to the buyer regardless of the city from which the goods are shipped. Using a basing-point location other than the factory raises the total price to customers near the factory and lowers the total price to customers far from the factory. The sugar and starches industry has used basing-point pricing for years. Some food companies set up multiple basing-points to create more flexibility.

Finally, freight absorption pricing is where the seller agrees to absorb all or part of the actual transportation cost to get the business. This strategy is

Table 14.1. Price–quality policies.

	High price	Medium price	Low price
High quality	Premium strategy	Penetration strategy	Super value for money
Medium quality	Over-pricing strategy	Average price –quality strategy	Value for money strategy
Low quality	Hit and run strategy	Shoddy goods strategy	Cheap goods strategy

used to improve market penetration and to retain a hold on to increasingly competitive markets.

Price–quality interaction

The interaction between price and quality is implicitly considered in the sections above on pricing research and on demand-oriented pricing approaches. Consumers and buyers will normally expect that there will be a consistent relationship between the quality of a product or service, and its price (Panigyrakis, 1989). The extent to which suppliers can satisfy this expectation will, however, be a function of the price–quality policy that they follow. Some examples of price–quality strategies are shown in Table 14.1.

The choice of price–quality strategy will depend upon such factors as: the market segment upon which the product is targeted, the stage of the product life cycle, the likelihood of repeat buying behaviour and the competitive circumstances.

CONCLUDING REMARKS

In spite of the increasing role of non-price factors in the modern food marketing process, price remains an important element in the marketing mix. Factors which food manufacturers must consider when deciding on a price strategy are: the customer, competition from other manufacturers and suppliers, government regulation, and institutional differences. Not all customers buy with price as their first priority; perceived value is often the criterion for purchase decisions. Current competition in the major markets has forced food manufacturers to intensify further their market-share battles. The pricing strategy is largely determined by the company's target market and positioning objectives. Costs set the floor for the company's price. A price strategy may be a cost-, demand- and competition-based approach. Implementing a price strategy involves a variety of separate but interlocking specific decisions. List or quoted price is often modified through discounts, allowances and geographical adjustments to account for differences in consumer segments and situations.

NOTES

1. Times (1990) Made just for him. *Times*, April, p. 49.
2. USA Today (1990) Monsanto's Simlesse weights. *USA Today*, February 23, p. B1.
3. This pricing practice exists when a firm offers a very low price on a product (the bait) to attract customers to a store. Once in the store, the customer is persuaded to purchase a higher-priced item (the switch) using a variety of tricks, including (i) downgrading the promoted item and (ii) not having the item in stock or refusing to take an order for the item.
4. As cost-oriented pricing is based essentially on considerations within the organization, cost-based pricing decisions are said to be 'administered'.
5. Hall, T. (1986) Miller seeks to regain niche as envy of beer industry. *Wall Street Journal*, December 3, p. 3.
6. The Wall Street Journal (1990) Burger King, in big blow to Pepsi, is switching to Coke. *The Wall Street Journal*, May 2, pp. B1–B6.

FURTHER READING

Deshayes, G. (1990) *Logique de la Co-operation et Gestion des Cooperatives Agricoles.* Skippers, Paris.

Leuthold, R., Junkns, J. and Cordier, E. (1989) *The Theory and Practice of Future Markets.* Lexington Books, New York.

Meulenberg, M. (ed.) (1994) *Food and Agri-Business Marketing in Europe.* International Business Press, London.

Monroe, K. (1990) *Pricing Making Profitable Decisions*, 2nd edn. McGraw-Hill, New York.

Nagle, T. (1987) *The Strategy and Tactics of Pricing.* Prentice-Hall, Englewood Cliffs, NJ.

Oxenfeldt, A. (1960) Multistage approach to pricing decisions. *Harvard Business Review* 38, 125–133.

Stobart, P. (1994) *Brand Power.* Macmillan, London.

Ward, K. (1989) *Financial Aspects of Marketing.* Heinemann Professional Publishing, The Marketing Series, London.

REFERENCES

Alberts, W. (1989) The experience curve doctrine reconsidered. *Journal of Marketing* July, 36–49.

Davidson, H. (1976) Why most new consumer brands fail. *Harvard Business Review* March–April, 119.

Kennedy, J. (1977) *Chocolate Division, Nestlé.* IMEDE, Management Development Institute, Lausanne.

Kohls, R. and Downey, D. (1972) *Marketing of Agricultural Products*, 4th edn. MacMillan, New York.

Monroe, K. (1990) *Pricing Making Profitable Decisions,* 2nd edn. McGraw-Hill, New York.

Morse, W. (1980) Probabilistic biding models: A synthesis. *Business Horizons* 18, 67–74.

Nagle, T. (1987) *The Strategy and Tactics of Pricing,* Prentice-Hall, Englewood Cliffs, NJ.

Panigyrakis, G. (1989) Modèle global du comportement d'achat du consommateur et la perception de la qualité. *Economie et Gestion Agro-alimentaire* 14, 23–31.

Schindler, R. and Wiman, A. (1989) Effects of odd pricing on price recall. *Journal of Business Research* November, 177–183.

Treadgold, A. (1990) The developing internationalization of retailing. *Journal of Retail and Distribution Management* 18, 4–11.

Wilcox, J., Howell, R., Kuzdall, P. and Britney, R. (1987) Price quantity discounts: Some implications for buyers and sellers. *Journal of Marketing* 35, 60–61.

Winner, R. (1988) *Behavioral Perspectives in Pricing. Issues in Pricing: Theory and Research.* Lexington Books, London.

15 Advertising and Promotions

Ronald W. Ward

Food and Resource Economics Department, Institute of Food and Agricultural Sciences, University of Florida, 1125 McCarty Hall, PO Box 110240, Gainesville, Florida 32611–0240, USA

INTRODUCTION TO COMMODITY PROMOTIONS

You want a glass of milk or orange juice for breakfast. How about a fresh apple or package of fresh almonds? As consumers we turn to many food outlets and make our purchase decisions while giving little thought about the complexity of how the product got there, who determined the product standards, and how the product was priced.

Furthermore, consumers have a reservoir of information about many products. A lot of this information is gained through experience, some through conscious search and some through exposure to advertising and promotions. As one views the process of supplying the market-place, it is useful to keep in mind three fundamental activities that are always present and essential to keeping an ample food supply. Products must be produced in the range of standards and varieties consumers desire, leading to product policy (Chapter 13). Some economic value must be placed on the product (Chapter 14). Finally, there must be a continual flow of information about the product. This chapter deals with issues relating to the last function, providing consumers with information. Our objective is to give insight into commodity advertising and promotion through both theory and examples.

Economics and market information

Products are consumed because of underlying needs and desires, as developed in the theory of consumer behaviour (Chapter 10). Consumers receive satisfaction (or, in economic terms, utility) from the consumption of the goods (Chapter 3). While it is difficult to measure satisfaction, studying purchases of goods, and having knowledge of the array of economic and non-economic factors influencing these purchases, provides a real picture of consumers' relative satisfaction and ranking of goods. Clearly, consumer incomes and price will place limits on how much is consumed at any time. Our ability to consume and store goods have limits, again within definable periods. In the unrealistic setting where all product characteristics are known and understood and all potential consumers have fixed preferences, then there is no need for additional information about the product. Consumers have all the knowledge they need and their preferences cannot be changed.

Such circumstances are obviously unrealistic. Consumers seldom have complete knowledge about a product. Their exposure to the information may be limited and the information about the product changes over time; that is, the attributes of a good may change and thus the information changes. New consumers enter the market while others exit. Entry and exit from certain consumption patterns are a natural part of the life cycle. Young couples buy baby food and are conscious of the attributes of that food product. Older couples are more concerned with health issues related to their diets. Some products may transcend all age groups while others are closely tied to various demographic groups. Certain products may be an essential part of daily consumption while others are not. Some purchases are driven primarily by habit persistence and others through spontaneous decisions (Briz *et al.*, 1995).

Given the range of consumers, their needs and preferences, and their paths of change, then how does market information fit into the scheme? While the information process is diverse and complex, its role can be categorized. Advertising and promotion can provide information about the attributes of the good, thus providing the potential consumer with useful information when making purchasing decisions. Second, the same advertising and promotion may change the underlying preference function for a product. For example, consumers may not be aware of a new style, fad or popular activity. Hence, they have little or no preference for the good or activity. Just as product attributes change, so can consumer preferences change. Advertising can play a role in directing these preference changes. Our preferences reflect needs and attitudes. Some needs are well defined, such as the need to satisfy hunger. We may not have attitudes about particular products. We may have the wrong attitudes and we may change our attitudes. Our attitudes evolve from cultural norms and technological changes. One can simply look at the growth in microwavable foods. This technology produced the product but the attitude toward convenience led to the growth in consumption of these foods. Advertising plays a key role in changing the underlying preferences. Often consumers are not even aware of the extent that they have been influenced by the advertising and promotion. This is particularly true when the information is unsolicited, such as through television and other media. Consumers may be more conscious of the role of information when they have purposely searched out the product attributes before making the purchasing decision (Albion and Ferris, 1981; Forker and Ward, 1993).

Products are not always used for final consumption; rather they may be included in the process of producing another product. Clearly, the role of advertising and promotion is different depending on the potential buyer and use. Buyers dealing in large volumes with well-defined standards will acquire the necessary information in a different way than from a single consumer willing to experiment with new products. Information and the underlying message will differ depending on the primary uses of the product. Simply compare the users of say wheat with those of milk. Ultimately, the same consumer uses both products but the final form changes considerably before reaching that consumer. In the first case, wheat is hardly recognizable as bread whereas milk keeps its product identity at every point in the distribution channel. One would expect to see uniquely different advertising and promotion programmes when confronted with these two products.

Concepts of generic and brand advertising and promotion

All of the above activities lead to consumption of the primary product and in our discussion, the commodity. Advertising and promotions are designed to facilitate and influence this process. Eventually, the informational efforts must have a positive influence on the demand for the commodity. Otherwise, the expenditures would be of no economic benefit to those paying the cost of the programmes. Before turning to the economic theory of advertising, it is important to make two clear distinctions in the types of advertising used. Most daily advertising exposures are for specific *brands*. A brand programme is designed and funded by a specific firm with the intent of benefiting that firm. This does not preclude benefits from the brand effort to others but that is not the usual objective of the brand effort. In contrast, many industries collectively advertise the attributes of the product instead of being brand specific. These types of information programmes are referred to as *generic advertising*. Examples can be found in many countries. Spain's or South Africa's promotion of their citrus are two good examples. In the USA generic promotion of beef, pork, almonds, citrus, dairy, apples and many others provide examples of cooperative advertising.

Forker and Ward (1993) provide a precise definition where:

> ... generic advertising is a cooperative effort among producers of a nearly homogeneous product to disseminate information about the underlying attributes of the product to existing and potential consumers for the purpose of strengthening demand for the commodity.

This definition also provides the basis for distinguishing brand from generic advertising. Brand advertising is not cooperative, at least within the same industry. Efforts are made to establish brand identity and, if successful, the product is differentiated from other suppliers within the same industry. Both programmes disseminate information, but the message is usually substantially different. Generic messages emphasize the product attributes and usually make little reference to specific brands. Brand advertising may note the product attributes but emphasizes the distinguishing attributes of the brand. Finally, although brand advertising may increase

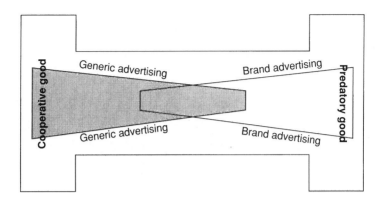

Fig. 15.1. Generic and brand advertising among commodity classifications.

demand for the total industry it is mainly directed at increasing market share for the brand. There are times when brand and generic efforts are complementary. Potentially, brand efforts can expand the total demand and some research shows that to be the case for several commodities (Hall and Foik, 1982; Kinnucan and Fearon, 1986; Sun and Blaylock, 1993).

The relative effects of generic and brand activities depend on the characteristics of the good being advertised. Product differentiation and the acquisition of information are two important dimensions to determining when one or both types of programmes are appropriate. Figure 15.1 provides these classifications, first noting the concepts of *cooperative* versus *predatory* goods. Cooperative goods cannot be differentiated. One producer's product cannot be distinguished from another. Total demand can potentially be increased by advertising but firm shares cannot, since there is no basis to make a distinction among the suppliers. In contrast, with predatory goods, total demand cannot be increased but one firm may gain market share from another. In Fig. 15.1 the end points represent these extremes; most products lie between the extremes. Two wedge-shape boxes are drawn to illustrate the potential for generic and brand advertising. The generic advertising wedge does not cover the full area of the cooperative good. If total demand can be increased and the good is not differentiable, then only generic advertising makes economic sense. As the commodity becomes more differentiable, the role of generic advertising is lessened as implied with the wedge to the right. At some point it is no longer feasible to use generic advertising since the products are too differentiated to emphasize common attributes.

In contrast, for predatory goods brand advertising is needed just to maintain market shares. For goods that are not fully predatory, brand advertising can be used both to expand total demand and shift market shares. At some point, it becomes so difficult to achieve product differentiation that brand advertising becomes less effective to the firm. Industries obviously differ as to where they fit within Fig. 15.1. Market conditions may change and the importance of generic and brand efforts differs over time.

Another classification appropriate to Fig. 15.1 relates to how consumers acquire information. Products lie somewhere in the scale of *search* and *experience* goods. For search goods, potential consumers will pay the cost and search out the product attributes before making the purchasing decision. In contrast, experience goods are those where the consumer will try out the product. One would expect the wedges illustrated in Fig. 15.1 to be larger for experience goods compared with search goods. That is, through advertising consumption may be easily increased because consumers are more willing to experiment with the advertised product. A large number of agricultural commodities lie near the cooperative classification and are experience goods, thus explaining why both brand and generic advertising are found within many commodity groups.

Classifications of commodities and firms

One cannot deal with commodity advertising and promotion without knowing how the product is used. Use determines the audience, the product form and purpose. Every agricultural commodity can be grouped into one of four categories relating to its use. Goods are consumed for food, skins and fibres, aesthetic appeal, and feeds and ingredients. Targeted audiences for advertising and promotion change across these uses. The message content and medium will differ with each use. Some firms may produce a product that is used for ingredients only while another may supply the same product in a packaged consumer form. A good example is the packaging of fresh almonds for direct consumption versus selling the same almonds for use in candy manufacturing. Clearly, the advertising activities would differ between these two types of firms. Furthermore, if most of the industry's products go to ingredients, it is more difficult for a generic programme to be successful, especially if the form of the original good is lost in the manufacturing process. Wheat as used in the production of bread is a good example where the commodity identity is essentially lost.

Within many agricultural sectors some firms who are the first handlers of the raw agricultural product do little to differentiate their goods. Rather than producing brands, they produce products often destined for private labels. Similarly, such firms may supply a product that is later transformed at a different stage in the marketing channel. Other firms in the same industry may have strong brands where clear product differentiation is apparent. Florida's citrus industry provides an excellent example with some processors being simply manufacturers of generic orange juice, while others maintain strong brand identities.

Forker and Ward (1993) provide a list of fundamental commodity characteristics that will influence whether or not an industry can successfully use generic promotion programmes.

Commodity characteristics to be considered:

1. Is the product highly differentiated or is it homogeneous among producers and first handlers? As suggested with Fig. 15.1, the role of generic advertising declines as products become more differentiated.

2. Is the identity of the product maintained as it moves through the marketing channels to the final consumer? The role of both generic and brand advertising of the commodity declines rapidly as the product is transformed. Many agricultural goods maintain their identity throughout the distribution process.

3. Are industry grades and standards sufficient to ensure consistent qualities in reliable packaging? A few negative experiences by the consumer will assuredly negate any advertising effort. Well-established grades and standards are essential to any type of commodity promotion programmes.

4. How much does the potential consumer know about the product? Does the consumer have a good general understanding of the product attributes and uses? The knowledge level and experience with the product will have a major effect on the potential benefits expected to result from both generic and brand efforts.

5. What is the current usage of the product? Are consumers already near a saturation level or is there opportunity for increasing per capita consumption? Many food goods already have a high level of consumption and the potential for large increases in demand may be unreasonable.

6. Does the product have a variety of uses? Generally, the importance of generic advertising is more limited when the commodity has few uses.

7. Are there many substitutes for the commodity? The number of substitutes may point to the need for advertising the specific commodity. Yet the expected gains from advertising and promotions should be lower when there are many substitutes for the advertised good.

8. Is the market for the commodity dynamic in terms of product characteristics and the set of potential consumers? If major changes among the consumer base are likely and the attributes of the good have changed, then the need and potential gains from commodity advertising may increase.

Industry characteristics to be considered:

1. Is the industry monopolized by a few large firms? Generally, the expected benefits and equity from generic commodity programmes decrease when one or a few firms control the industry.

2. Do the producers within the industry have common objectives? If producers are not willing to cooperate then it is virtually impossible to fund and carry out cooperative advertising programmes.

3. Is the industry large enough to underwrite a major advertising campaign? Inadequate funding levels may lead to wasted efforts, especially if there is a threshold effort level needed to reach potential consumers.

4. What is the nature of supply changes in the industry? Are there barriers to entry and can existing firms quickly increase supplies? Generally, the expected industry benefits become less when there can be a quick and substantial supply increase within a relative short period. In contrast, the more inelastic the supply the greater the potential gains back to the industry from advertising.

5. Is the distribution system reliable? The success of any information programme is predicated on the condition that the product reaches the final destination without delays and that distribution uncertainty is at a minimum.

6. Can an effective administrative structure be put in place in order to implement the generic programme?

The above list attempts to set forth a basic set of conditions that every industry must evaluate when considering generic and brand advertising. Significant problems with one or more of these basic conditions can lead to failure with existing or planned programmes.

Types of generic and brand activities

Before moving into the advertising theory and the specific cases, it is useful to lay out a structure for describing how commodity advertising programmes have historically been used. There are five important areas that must be recognized when discussing the types of advertising and promotion activities existing across agricultural commodities: message, media, coverage, timing, and linkage. Table 15.1 provides a summary of each of these areas.

Message reflects the information content to achieve specific industry or firm objectives. Generic messages frequently are designed to change consumer perceptions and expand their knowledge base about the specific attributions of the good. 'Pork, the other white meat' is an example where the USA pork industry through generic advertising has attempted to change the perception of pork from a red to a white meat. Brand advertising messages, while communicating some of the commodity attributes, will still emphasize those characteristics that will give that firm some degree of product differentiation.

Negative attitudes translate into declining demand. In those cases the purpose of the message is to attempt to change the attitudes. Beef has long dealt with perception problems about the negative health effects of red meat. Generic promotion of USA beef continues to address this issue.

Table 15.1 shows a number of media alternatives that are used by commodity groups. Television and radio are the predominate media for most commodities. However, it is impossible to generalize as to what is best simply because the problems and resources available differ across industries. When assessing the appropriate media, one has to look at this on a case by case basis.

Coverage relates to the targeted audience. Some products cut across the interests of all socio-economic groups. For some products such as fluid milk, efforts are often targeted to young potential consumers in order to establish consumption behaviour patterns during the early stages of their development. In contrast, the USA apple industry will target a broad spectrum of consumers since the product can be and is used by all age groups. For some agricultural sectors exports are extremely important. Similarly, some commodities are used primarily for ingredients.

Table 15.1. Types of generic and brand advertising activities.

Message – What is the purpose?
 Changing consumer perception about the product
 Nutritional education
 Changing usage patterns
 Expanding audience
 Responding to specific product problems
 Reminding consumers about existing product attributes
 Providing information about new product attributes
 Attempting to differentiate the product in some form

Media – What methods are used to communicate?
 Television
 Radio
 Magazines and newspapers (print)
 Direct mail
 Coupons
 In-store
 Electronic information (FAX and e-mail)
 Write in
 Samples
 Telephone
 Billboards
 Institutional distribution

Coverage – Who is the audience?
 Selected audiences
 National coverage versus regional
 Domestic and international coverage
 Consumer versus manufacturers' coverage

Timing and allocation – How much is spent and when?
 Setting the budget and allocation to media
 Allocation of budget across time (seasonality)

Linkage – How are programmes coordinated among firms and the industry?
 Separate generic and brand activities
 Joint ventures with complementary industries
 Generic rebates to brand advertisers
 Refunds
 Generic leveraging with retailers
 Generic leveraging with government programmes

Timing and allocation of funds to advertising are usually limited to two primary issues. How much should be spent on the programme and when should it be spent? These allocations are usually determined by experience, habit, media availability and research input. Most agricultural goods have some production and consumption seasonality. Advertising and promotion cannot influence the seasonality in production, but it can potentially change the consumption patterns. One will frequently find commodity groups experimenting with the seasonal allocation of their dollars to try to change the underlying consumption seasonality. This is sometimes very difficult to achieve because of well-established behaviour patterns and for health-related reasons. For example, the high consumption of orange juice in the winter is partially due to the health benefits from having vitamin C during the cold season.

Finally, in Table 15.1, the concept of linkage is introduced. Linkage relates to how firms and the collective groups interrelate. Separate programmes and joint ventures are clear and do not need discussion at this point, whereas, rebates, refunds and leverages need explanation. Rebates among many commodity groups are programmes where firms advertising their specific brands may receive support from generic funds if the brand programmes include certain information content. Figure 15.1 illustrates that there exists a place where both generic and brand advertising may be appropriate.

In contrast to rebates, refunds occur when a producer or handler asks for a direct payback for the initial assessment that was paid. The firm simply does not want to participate in the programme and thus asks for a refund. At one time many USA programmes included some form of refund provisions. Free-rider problems occurred when one or more firms wanted a refund but still benefited from the industry generic efforts. The current USA generic programmes (known as 'checkoff') do not include refund provisions.

Leveraging occurs when generic programme funds are combined with other programmes to generate a total increase in the advertising effort. In the USA, agricultural industries can apply for federal promotion support funds that supplement existing foreign promotion programmes. These industries are able to leverage their industry programme expenditures with federal government support. Similar cases can be found where commodity groups will supplement retailers if they include certain logos and meet other requirements when they advertise. Through leveraging, industry groups are theoretically increasing the total amount of information to the consumer.

ECONOMIC THEORY OF ADVERTISING

In the first section the basic concepts of generic and brand advertising were introduced while eluding to their impacts on demand. Does advertising change demand? Can advertising be included in the demand equation? In this section our purpose is to provide a non-technical explanation of how advertising influences the demand for any commodity.

Consumption characteristics

Consumers purchase products because of the needs for certain sets of characteristics that are satisfied with the particular product. Consumers have an array of needs ranging from hunger to status. These needs are fulfilled through the consumption of characteristics. For example, diet drinks are purchased to satisfy thirst but also to reduce calorie intake. The diet drink provides both characteristics, i.e. fluids and low calories. We purchase flowers because they provide the beauty and fragrance that we desire. We eat an apple because it satisfies our hunger, desire for sweetness and our concern for a healthy food. In each case, a bundle of characteristics is being consumed. These characteristics are derived from the basic good (Nichols, 1985).

Some characteristics are clearly measurable and observable. Product labels often explicitly note the characteristics. For example, the fat content, the number of calories or even the expiration date are commonly reported characteristics. In addition, there are characteristics that are more difficult to measure and evaluate. Take for example the statement that the product is 'new and improved'. It is often difficult to determine exactly what has been changed. Furthermore, we may find the new packaging appealing without even being fully conscious of its impact on our purchasing decision.

There are products with characteristics that are dependent on how others perceive us when we use the product. Wearing cotton or eating caviar fulfils certain fundamental needs. Yet having someone see us eating caviar or wearing the latest brand of cotton blue jeans also fulfils certain needs that are more subjective in nature.

Consumption of goods therefore fulfils our desire for a given bundle of characteristics. The bundle obviously differs with each good and, furthermore, can change over time for some goods. Salt, for example, satisfies our need for a particular seasoning but provides little or no additional appeal. Too much salt can also be related to health issues. Hence, we consume the product but may limit our intake because of other reasons. There probably is no social appeal from the consumption of this product. In contrast, purchase of a well-known brand of mineral water may fulfil many physical and social needs. With each product there exists some theoretical mapping between a set of characteristics and the actual good. In the case of salt the mapping may be defined and not subject to much change. For others, the mapping is fuzzy and subject to considerable evolution over time.

Much of the economic literature expresses these concepts in terms of a 'household production function' (Chang, 1988). Basically, we consume a set of products within the household in order to produce and satisfy a set of characteristics. Our knowledge linking the product and the characteristics comes through experience, through others' reactions and through unsolicited information. Thus, a demand for characteristics can be easily expressed in terms of the commodity or good. From this argument, we conveniently talk about product demand rather than the demand for the characteristics since the product is readily measurable while the characteristics are not.

Information versus persuasion

Where does commodity advertising fit within the scheme linking characteristics and the demand for the commodity? The most important question is whether advertising and promotions have any impact on the demand for

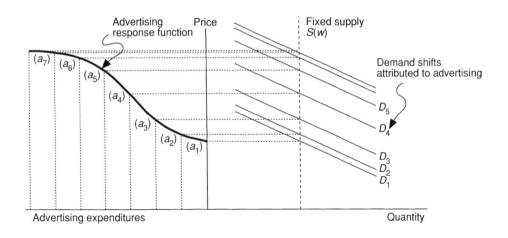

Fig. 15.2. Demand and the advertising response function.

the good. This impact can occur in three distinctly different ways. First, and probably most important, the message content is designed to inform potential consumers and remind previous consumers about the attributes of the good. Factual information about the specific attributes of the product are given to the potential consumer. Second, potential consumers may be influenced to believe that the product has other attributes that were not there before. For example, a celebrity endorsement of a product may convince the consumer that the product must have other desirable attributes that are not as readily apparent. An endorsement may also add believability and trust about specific claims. Furthermore, the same endorsement may help the potential user remember the claims through some form of association.

Advertising may influence the preferences for certain characteristics. This is fundamentally different from telling the consumer about the product attributes. When preferences change so does the demand for the underlying product. Consider the case for USA cotton. One dimension to the marketing of cotton was to let the consumer know about the attributes of cotton relative to other fibres. Simultaneously, one has to convince consumers of the styles and comfort from wearing cotton fibres. That is, the preferences for this fibre were changed. Some of the change in the cotton case may have been due to shifts in styles that have little to do with cotton advertising. Yet some part must be attributed to efforts by the cotton industry to reposition cotton as a fibre of choice.

Demand and advertising response functions

Through the product characteristics, advertising does impact on the demand for a commodity. While the magnitude of the effect will differ by commodity and programme, the expected impact can be illustrated. For convenience, define demand as $D = D(p,a,y)$ and supply as $S = S(w)$. Further, for a reasonably short time (say less than a year) the supply is predetermined by the biological nature of the production process. For a given set of market conditions and no advertising, one would expect the demand for the product to be at D_1 as shown in Fig. 15.2. Note in this conceptual model a positive level of consumption occurs without any advertising.

Now let advertising increase in equal increments as suggested with the left part of Fig. 15.2. An increase from a_1 to a_2 produces an upward demand shift to D_2. Prices increase since the supply (S) is perfectly inelastic. Increases in advertising to a_3, a_4 and a_5 produce larger increases in demand as shown with D_3, D_4 and D_5. For more increases in advertising, the corresponding gains in demand become smaller and eventually almost no additional gains are realized for the extremely large expenditures.

For the fixed supply case, prices are shown to increase with the shifts in demand. Gains are directly proportional to the nature of the advertising response function. In Fig. 15.2 advertising is effective in stimulating the market. If instead, the response function were very flat, then one would conclude that the commodity advertising has little benefit to the industry. There would be little to no shifts in the demand function.

The nature of the response function is closely tied to the characteristics outlined earlier as well as the quality of the advertising programme. If the market is saturated or if the advertising message is poorly developed, one might expect to see the response to be very flat. Clearly, the nature of this response must be dealt with on an industry by industry basis. A key research issue for any commodity group is to determine the nature of the advertising response function.

Figure 15.3 illustrates the situation when supplies can be changed within the current period. This graph is identical to Fig. 15.2 except that two supply responses are drawn. An increase in advertising from a_1 to a_5 first produces an upward shift in the demand curve to D_5. As the supply response is less than perfectly inelastic as illustrated with $S(p,w)$ in Fig. 15.3, then the price increases to p_2 instead of the higher price p_1. Note, however, that more product is being sold. At the pre-advertising level, industry revenues stood at $p_0\ Q_1$ while without the supply change revenues increased to $p_1\ Q_1$ or a gain of $(p_1 - p_0)Q_1$. This gain must then be compared with the additional advertising cost corresponding to a_5. Also, even with the supply response producing a lower price p_2, revenues may be greater. One must compare the two shaded areas in Fig. 15.3 in order to make a judgement about industry revenues.

Figures 15.2 and 15.3 present two important issues always raised when assessing the impact of commodity advertising programmes. What does the advertising response curve look like and does it translate into net increases in industry revenues? The answer clearly depends on the advertising elasticity reflected with the response curve and the demand and supply elasticities.

As implied earlier, advertising that is not all generic may both expand demand and change market shares. Figures 15.2 and 15.3 can be easily

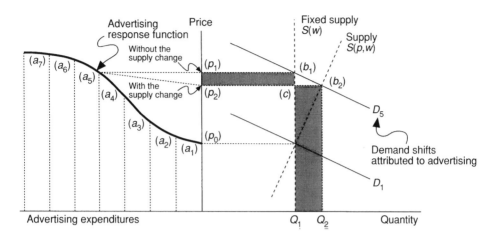

Fig. 15.3. Demand and the advertising response with and without supply changes.

adjusted to illustrate this additional dimension to the advertising response. Using Fig. 15.4, lower coordinates are added to the graph to represent the market share of a major brand advertiser in the industry. Suppose first in this figure that all advertising was generic in form. With no brand advertising a major firm has MS_0 share of the market. If the generic advertising is brand neutral then that firm's share of the market should remain fixed as suggested with the horizontal line (MS_0) immediately under the advertising expenditures. Even when the supply response exists, the firm still maintains the same market share.

Now suppose that some part of the total advertising expenditures includes brand programmes. A combination of generic and brand has expanded total demand as shown before with shifts in D. Furthermore, the brand share has increased from MS_0 to MS_2 in the case illustrated. The market share response is drawn with an upper limit to the market share gains as must always be the case since the maximum share is one.

Assuming no price differential among the brands and given knowledge of both the advertising and market share responses, then it is relatively straightforward to calculate the industry and brand gains. As a rule, brands gain with generic advertising as long as the generic effort is not biased against the brands. If the generic advertising is brand neutral but effective in increasing the market, then the particular brand realizes the increased revenues defined as $(p_1 - p_0)MS_0\,Q_1$. If that firm's market share can be expanded, then that firm realizes additional returns not shared by all firms in the industry with this additional gain being $(p_1 - p_0)Q_1\,(MS_2 - MS_1)$. In contrast, if generic advertising makes it more difficult for brands to achieve some degree of product differentiation, it is possible that the market share curve in Fig. 15.4 is downward sloping instead of that illustrated. In that case, a particular brand would usually oppose the use of generic programmes.

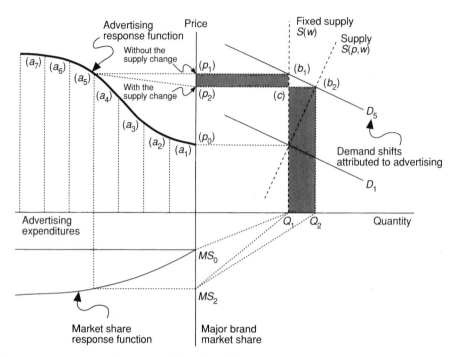

Fig. 15.4. Market share changes and the advertising response curve.

What are the benefits and who gains?

Figures 15.2–15.4 illustrate the final impact of either or both generic and brand programmes. Any gains are predicated on the nature of the advertising response, the market share curves and the slopes of the supply and demand. Each of these is dependent on the particular industry characteristics and how programmes are designed and implemented.

If the parameters implied in Fig. 15.4 are known, then it is possible to estimate the resulting gains to the industry and firms. Theoretical revenues were calculated in the previous section and with the knowledge of all response curves, rates of return to the advertising can be calculated.

The demand changes shown with the above graphs are static in that they do not show the path of change; rather they simply compare two points of equilibrium. Prices are shown to increase with the advertising, yet the model does not reflect the potential reduced search cost to consumers. Also, the benefits from improved nutritional education (if it occurs with the message) are not fully reflected in the model. If nutritional education leads to preference changes favouring the particular advertised good, then the demand shifts capture some of the benefits from expanded education. However, the benefits from nutritional education most likely supersede that of the demand for a single product. The point is that generic advertising in particular may have benefits not fully reflected in the supply–demand configuration for the one good. In that sense, it becomes extremely difficult to calculate the costs and benefits to consumers. Whereas, it is less difficult to measure the benefits accruing back to the industry once the empirical counterparts to Fig. 15.4 are known.

Figure 15.4 shows demand while holding other factors fixed. Obviously other factors are never fixed. Demand D_1 was drawn as if it would remain at that level in the absence of some form of advertising.

Consumption trends for some commodities have declined over time even after accounting for price, income and other effects. The USA beef industry is an example where there has been a continued decline in demand during the 1980s and 1990s. If Fig. 15.4 were redrawn for that industry, one would have to show new demand curves below those at D_1. If advertising can prevent the decline or at least lessen the rate of decline, then the benefits to the programmes must be calculated somewhat differently.

AGGREGATE-LEVEL PROMOTION DECISIONS

Appropriateness of generic programmes

An important distinction in commodity advertising is whether the programmes and funding are from the total industry or from private firms. By definition generic advertising represents efforts to communicate to potential consumers the attributes of the commodity and not the specific characteristics of one or more brands.

This provides the clue as to whether such cooperative efforts at the aggregate industry level are appropriate. A few key words entail the essence of generic advertising: cooperative effort, product attributes, information, potential uses and demand enhancement. Generic advertising and promotions are appropriate if and only if each of these five key factors is in place. First, is whether consumers receive satisfaction. If the good has limited use and low appeal, then any effort through generic advertising will probably be of limited benefit to the industry. Second, is when the information is incomplete or wrong. Information can be incomplete because of little to no exposure to the product or because consumers forget. Generic advertising may precipitate consumer experimentation and may remind consumers in order to encourage repeat use of the product (Forker and Ward, 1993; Sheth, 1974).

A second part of the definition relates to the cooperative effort. Cooperation implies common objectives and the willingness to share in the costs of achieving these common objectives. Industries that are highly diverse in the commodity characteristics and the ways their products are used may experience more problems with the common objectives. Aggregate programmes must have benefits that can be shared by all. Within the USA agricultural system many cases can be shown where the major problem with generic programmes was dealing with an industry where the interest among the producers was too diverse to design programmes that benefited everyone in some equitable way.

Thus, the appropriateness ultimately depends on the needs for information among potential users, and the common interest among those supplying the information. Problems with either will render any planned programme ineffective.

Types of generic programmes

When we view generic advertising programmes there are four key factors that define the types and activities taking place. Generic programmes differ in coverage, funding rules, level of independence and administration. Coverage is in terms of who fits within the group expected to help pay and benefit from the cooperative advertising. For some industries everyone producing and importing the commodity is included. These programmes are considered national in scope since all producers (or handlers) of the commodity are subject to the assessments and must have some voice in the collective policy. Coverage can be regional where programmes and their costs are for producers within well-defined production regions. For example, the funding base for the Florida citrus industry includes only those producers in Florida or volumes of orange juice imported into Florida. In contrast, the USA beef checkoff is a national programme where all beef producers and beef importers must carry their share of the programme cost. Furthermore, coverage can be limited to certain types of product forms within the commodity group. Some programmes may be limited to just fresh forms to the exclusion of processed products. Thus, while some exceptions may exist, as a rule the coverage is defined by region and form (note in Table 15.1 the term coverage implied the targeted audience whereas here it implies who the programmes represent).

Funding policies have evolved over time primarily in terms of setting assessments. In almost every case, a per unit assessment rate based on volume or value is established and then each producer (or handler) is subject to that rate. An extremely important issue is whether or not the assessments are mandatory. As we saw when discussing Fig. 15.1, the 'free-rider' problem has led most USA current programmes to eliminate refund provisions, if they are mandatory.

A third dimension concerns independence from government involvement. In some countries, the government may be the primary (or sole) decision-maker in the design and implementation of generic activities. More typical, however, is where the government has oversight responsibilities while providing little to no input into the design and implementation of the programmes. Most USA generic programmes are of this nature where commodity groups are empowered with the authority to impose assessments for generic advertising while remaining subject to government oversight. There are no exceptions to this when the assessment is mandatory.

Finally, generic programmes differ in how they are administered. With only a few exceptions, most commodity programmes are administered through a Board made up of a cross-section of industry representatives.

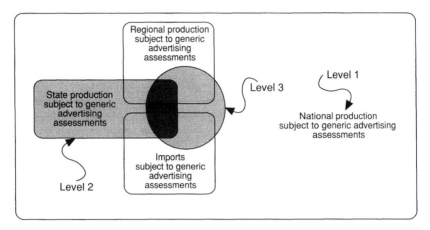

Fig. 15.5. Legal structures for generic advertising programmes.

Each commodity group will have a staff and then the governing Board. Programme policies are usually set by the Boards but the implementation is completed by the staff. Staff size and responsibility obviously differ by commodity depending on the scope and complexity of the programmes.

Underlying legal structure

In those cases where the assessment is mandatory there must be some underlying legal authority for the imposition of the tax. There must be a clearly defined method of recourse and appeal when such assessments are imposed. The USA has the most extensive generic programmes, hence we will draw on the USA case to illustrate the range of legal structures used by commodity groups. In Fig. 15.5 assume that the total production area in the USA is supplying a domestically produced commodity. Within these national boundaries there exist three supply sources that could be subject to the advertising and promotion assessment. For some commodities there exist national research and promotion acts that cover all domestic production and imports. Legally any products supplied within these borders are subject to the assessments. At level 2 in Fig. 15.5 there exist states that maintain authority through state marketing orders that give producers the authority to collect assessments in support of their programmes. Finally, a region may be defined that exceeds one state border but does not encompass all supplies. Under federal marketing orders a region can be defined where product produced within that region is subject to the assessments. Note the overlapping areas in Fig. 15.5.

As of 1990, the USA had approximately ten commodities with national checkoff programmes – dairy, beef, pork, cotton, potatoes, eggs, seafood, wool, honey and watermelons (Forker and Ward, 1993). Each year the list changes with the entry of new commodities seeking national legislative authority. Also, a few have terminated their national programmes. During the same year 13 commodities had advertising programmes funded through the federal marketing orders and 42 using state marketing orders.

Other programmes not reflected in Fig. 15.5 are those that are voluntary and in a few cases where the government subsidizes a domestic programme. In the USA direct government subsidization of a domestic programme is rare unlike in many other developed and developing countries.

Underlying political problems

In addition to organizing and designing programmes to meet certain economic objectives, there exist an array of political problems associated with cooperative advertising activities. Most of the problems can be grouped into two types: (i) those relating to representation and taxation; and (ii) those relating to broader impacts of advertising on society.

Generic programmes require some type of assessment to underwrite the programmes. Politically, those paying for the assessment must have some voice in the policies relating to the use of the funds. Determining how interest groups within the industry are represented and how they are appointed is a political decision that is usually controversial. Who is included and not included in the assessment pool is often difficult to define and who can vote on policies is equally difficult to determine. Of equal importance are setting policies to assure equity in the distribution of benefits. For most commodities the assessment is on the raw commodity without giving any consideration to differences in use. This has on occasion led to challenging the equity aspects of the programmes. Many of the political problems associated with representation and taxation can be solved through setting the correct operating rules, having methods for raising issues in a meaningful way and having useful evaluation programmes (Ward et al., 1983; Strak, 1994).

Politically there are generally three problem areas at the consumer level. First, the advertising and promotions are usually directed to consumers. An important political concern is to assure that the information is factual and not detrimental to good health and food safety. Does it contribute to nutritional education and is it beneficial to developing improved consumption habits? Second, how does advertising and promotion of one good change the total mix of food goods consumed? Does it create waste by simply causing consumers to shift consumption patterns from one commodity to another? Finally, in the USA there are always concerns that public funds should not be used to underwrite domestic programmes. Rather, those potentially benefiting should have to carry the programme costs. The real issue here is how much if any public funds should be used in the promotion of specific commodities? Commodity groups are subsidized in other ways. Hence, should advertising and promotion be included in the list of ways to subsidize an industry?

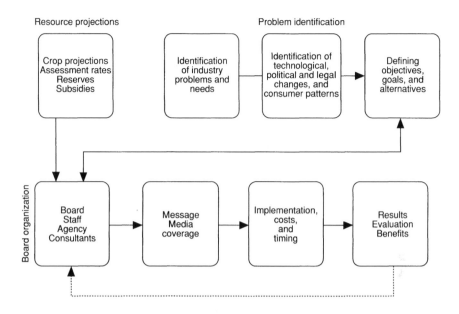

Fig. 15.6. Decision-making structure essential to administering generic programmes.

Funding and management of programmes

Every commodity industry involved in generic advertising and promotion programmes will have its own unique organizational structure. Without exception, however, every programme will have a Board or Commission made up of producers and/or handlers and will have a core staff to plan and implement programmes. Staff size and responsibilities vary depending on the programmes being carried out. As one looks across the many commodity organizations responsible for promotions, three major functions found in each are budgeting, problem identification and programme implementation. In Fig. 15.6 these functions are illustrated.

Resources for generic promotions come from assessments. Available funds for most commodities are projected based on the assessment rate, industry reserves and any external sources of funds. Every active commodity group must project the revenue flow for the crop year. In addition some commodity groups carry an advertising reserve to assure that a pool of funds exits to maintain programme continuity across seasons and for protection against unexpected changes in the crop size. As implied with the first box in Fig. 15.6, the resource projections set the parameters within which a commodity board can operate.

The second major activity of commodity boards is the process of problem identification and setting industry goals and objectives. Most of the identification issues can be grouped into those relating to the industry and those relating to technological and consumer changes. At the industry level, important issues to advertising and promotion activities include product quality and size, problems with structural changes in the industry and equity and issues relating to market development.

Technological, political and legal changes have a direct impact on the types of programmes that are needed and that can be used. The development of new packaging technology, a new variety or an improved form of the commodity are examples that should directly influence advertising and promotion strategies. Removal of trade barriers through political action may open up new markets for the commodity. Hence, new programmes directed to these potential new consumers may be appropriate. Legal issues often set the boundaries within which commodity groups can operate. Certain types of claims and the factual nature of the claims are always subject to legal review. While the list within these categories of technology, political and legal is too long to address in this chapter, the recognition of their importance to setting industry goals, objectives and strategies is essential.

Probably the most fundamental factor influencing the setting of industry goals is the underlying change in consumption behaviour. One must not waste resources on consumer trends that are far greater than what one commodity programme could address. Changes in the USA beef industry provide a good example. Consumers are concerned about fats and cholesterol. Beef advertising most likely cannot influence the general trend in health concerns. Yet, the beef industry can address issues about consumers' perceptions of the health attributes of beef. As implied in Fig. 15.6, determining the resources available and defining the industry goals and strategies set the parameters within which the board must function.

A survey of USA organizations by Lenz *et al.* (1991) provides a data base to give some insight into how USA commodity groups generally implement their programmes. Across all commodity groups in the USA, approximately 32% of commodity programmes used television as the primary medium for delivery of their message. A pronounced trend in the use of television exists where smaller organizations generally allocate a greater share of their funds to nutritional education and public relations and a small per cent to television. Also, as one would expect, the percentage allocations do differ by commodity. One must use the percentages as general guidelines recognizing that such allocations will change as one looks at a specific industry.

Messages obviously differ across commodities. However, for many of the larger commodity groups, a common trend has been to develop slogans that consumers can easily recall. For example, 'orange juice, it's not just for breakfast', 'pork, the other white meat' or 'beef, it's what's for dinner' are all recognizable slogans. In addition, most of the major commodity groups have an identifying logo that accompanies the advertising. For many groups, this logo is a seal of quality and can only be used if the product meets certain industry standards.

Many of the commodity groups have some effort in place to evaluate their programmes. Evaluations differ in the degree of sophistication, ranging from basic recall statistics to full modelling of industry demand as illustrated earlier in Figs 15.2–15.4. A few cases will be discussed later to illustrate the forms of evaluation.

Expected impact

In Figs 15.2–15.4 the impact of commodity advertising was illustrated through shifts in the demand for the commodity. While there are many intermediate ways to evaluate the impact, ultimately if this shift does not occur the programmes will have no impact. Again note that the shift can be calculated relative to the demand before the advertising or what the demand would have been without the programmes. Advertising gains must be measured on a case by case basis. Even so, analysis across several commodities does provide useful insight into the general impact of generic advertising. Most major programmes have shown some positive shift in demand similar to that illustrated in the theory discussion. Demand shifts have usually ranged from 2% to 6%, but again the actual number is unique to each commodity and its programmes. The rates of return to the programmes that have been studied extensively point to relatively large gains where a typical gain of four or more dollars for each dollar of advertising would not be unreasonable. These gains usually represent the average over a specific period and obviously will differ with larger expenditures levels. The typical declining marginal rate of return suggested with the advertising response curve in Fig. 15.2 is usually seen. As a general rule, most programmes have proven successful in terms of the gains in demand and returns to producers.

Calculating the rates of return requires considerable modelling and economic analysis. Also, a historical data base is needed. The evaluation process is time consuming and the results may not be sufficiently timely to make mid-programme adjustments. To compensate, commodity groups often collect data such as consumer awareness and recall to make some judgement about the progress of the advertising and promotion programmes. Indicators such as gross rating points, advertising recall and advertising awareness provide measures of exposure. Without these exposures, demand cannot change as a result of advertising. Yet there is nothing to assure that exposure translates into changing consumption behaviour. Advertising recall provides a necessary but not sufficient condition to realizing economic gains from the commodity promotions. Over a long period there should be some positive correlation between exposure awareness and consumption behaviour if the programmes are working (Chang and Kinnucan, 1992; Kinnucan and Venkateswaran, 1990).

EVALUATION METHODS

Evaluations provide a historical base for planning future programmes. Providing an ongoing review is particularly important to generic programmes because the funding is based on a mandatory tax on producers and/or handlers. Producers have the need and right to know if their funds are being used effectively. Given both the planning and implementation stages to both generic and brand activities, there needs to be both pre- and post-programme evaluation.

Pre-programme evaluation

By definition pre-programme evaluation is an attempt to make some judgement about the potential success of a planned programme. Criteria for judging the potential and actual performance must be defined. For pre-programme evaluation, the criteria will include: (i) historical sales gains and customer loyalty from similar programmes in the same industry and other industries; (ii) measures of consumer awareness; (iii) measures of consumer reaction to the advertised product and to competing products; and (iv) measures of the carryover effect of the message.

It is important that adequate diversity exists among those making the pre-programme evaluations. This process is usually found in all commodity groups and it points out why it is absolutely essential to have a scientific evaluation of the same programmes some time after they have been implemented.

As part of the experience-building base, pre-programme evaluations will include market testing and experimentation. Using focus groups, firms and industries can gain some insight into how a small group of consumers reacts to specific commercials, products and media. Results are from a small group and one must judge how representative the conclusions are to the total market. Unlike focus groups, the programmes can be field tested through using the advertising and promotions in a small test market. Data are collected from the test market and then assessments of the results are made. The two major problems with test markets are not having control over all events during the testing and drawing inferences about the results to the total product market.

Post-programme evaluations

Before programmes are started there should be plans for evaluation, thus assuring that appropriate data are collected as the process evolves. While there are many signals indicating if programmes are succeeding, the changes in demand as initially illustrated in Fig. 15.2 provide the most definitive basis for judging performance. Usually, commodity programmes are national, regional and even international in scope. There is no way to control the events that occur during the programming period. One must monitor the markets as the process evolves. Then the task is one of using the appropriate techniques for separating out the impacts of the advertising and promotions from all other events that may have impacted the demand for the product. It is at this point that statistics and economic modelling techniques of the kind discussed in Chapter 6 become particularly important. Referring back to Figs 15.2–15.4, the basic task is to take actual market data collected over time and then empirically determine if demand has changed as a direct result of the advertising and promotions. Using economic models that are estimated based on the actual market data, the advertising response functions are then estimated. Rather than going into detail on the estimation, programme examples will be shown in the next section.

With this technique and reliable data, one can estimate the rates of return to the various programmes.

In Figs 15.2–15.4 demand was shown to shift outward with more advertising. Having an appropriate measure of the advertising effort is essential to the estimating of the shifts. Many types of advertising intensity measures have been used, including deflated and non-deflated expenditures, advertising awareness, advertising recall, coverage and gross rating points. Expenditures have generally proven to be the most useful if returns to the programmes are to be estimated. Hence, advertising expenditures are generally used as the index of programme intensity when econometric models are included in the evaluation.

SELECTED CASE STUDIES OF INDUSTRIES AND FIRMS

The purpose of this section is to present evidence of the economic impact of selected commodity promotion programmes that have been evaluated over a number of years. Four cases of particular relevance to the issue of whether commodity advertising does or does not have measurable impacts on demand are presented.

National Beef Checkoff Case

After years of debate, the beef industry passed a national checkoff in the mid-1980s. The first national advertising programmes started in 1987 and have continued since then. The beef checkoff is a mandatory programme covering domestic production as well as imports into the USA. An assessment of US$1 is collected each time a live bovine animal is sold. An equivalent rate is applied to imported beef coming into the USA. To date, over US$500 million has been collected. Half of these funds must go to support the national beef checkoff and the other half can remain with the state beef boards. The states, however, can opt to allocate any share of their half to the national efforts.

At the national level, the Beef Board uses three primary demand enhancement efforts: (i) promotions; (ii) consumer and industry information; and (iii) foreign promotions. Promotions generally account for 70% of the checkoff funds with most of the effort being through television advertising. Information programmes are directed to schools, doctors, hospitals and other institutions where diet planning, health education and purchasing coordination may be particularly important. Both the promotions and information efforts are directed to the domestic market. It is the impact of these efforts that has been evaluated.

While advertising programmes differ across commodities, a common issue is how to measure the demands illustrated in Fig. 15.2. Has the national beef checkoff caused a measurable shift in beef demand as depicted with movements from D_1 to D_2 in Fig. 15.2? To address this, an econometric model was developed and estimated using quarterly data extending from 1979 through the second quarter of 1993. Boxed beef and liveweight prices were shown to be related to the supplies of beef, pork and poultry, to income, to seasonal preference shifts, to longer term preference trends, to health concerns and to generic advertising (Ward and Lambert, 1993; Ward, 1994). More specifically, the beef demand model showed a positive and statistically significant impact from the quarterly beef promotion and informational efforts. These impacts are reflected at both the boxed beef and liveweight levels. That is, the gains registered at one point in the market system are, at least, partially transmitted to producers.

Using the latest results from applying the beef checkoff analysis, the following major conclusions are highlighted:

1. Beef consumers have responded to the promotion and informational efforts as reflected with outward shifts in beef demand.
2. There is a one quarter lag in the advertising response. Note also that there is no longer term response as seen in some commodities.
3. For the quarters from 87:1 through 93:2, assessments stood at US$520 million and the revenue gains were near US$3.3 billion.
4. A rate of return to the programmes show that for each US$1 spent on advertising, the USA beef producers gained more than US$5 in returns. This is the average return and will vary from season to season, depending on how much is being spent within a specific period. Also, the beef checkoff response estimates are such that the rate obviously declines with the larger expenditure levels. There exists a point beyond which it is no longer economical to spend additional funds.
5. For the period 91:3–92:2 the beef checkoff increased liveweight cattle prices by US$1.12 per hundredweight (cwt) or approximately 1.73% over what the liveweight price would have been without the beef promotions. From 92:3–93:2, liveweight prices increased by US$1.81 cwt or 2.68% over prices without the programme (Ward, 1994). The important gains listed in (4) above come from multiplying these price increases times the amount of USA beef sold during those quarters.
6. Another way to view the economic impact of the beef case is to compare the shifts in demand relative to a base point with and without the checkoff programmes. First, set the economic conditions to the period of 1979 and calculate the demand for beef as reflected with the estimated liveweight prices. One can show that by the middle of 1993, beef demand declined to 77.3% of the 79:2 level even with the beef checkoff in place. The checkoff did

not reverse the downward trend in demand for beef. However, without the checkoff, beef demand would have declined to 73.7% of the 79:2 level. That is, beef demand would have been approximately 3.6% lower without the demand enhancement efforts.

7. Finally, these programmes need to be put in perspective relative to the other factors causing changes in cattle prices. Specifically, the beef checkoff explains 4.8% of the total variation in cattle prices for the period 87:1–93:2. Clearly, the checkoff's role is quite small compared with all other factors that can affect cattle prices. Similarly, one cannot expect to use such advertising to completely offset the negative price effects from increases in beef, pork or poultry supplies.

The conclusion to the Beef Board is that their programmes do show positive impacts that are measurable. The generic effort for beef at the national level has been effective.

National Dairy Board Case

The National Dairy and Tobacco Adjustment Act of 1983 provided the enabling legislation for the dairy industry to carry out a national checkoff for dairy. Regional programmes were ongoing before the national legislation. Through the dairy act the industry gained the authority to coordinate better their programmes on a national scale. There exists a long history of evaluating these programmes at the state and regional level.

When Congress passed the enabling legislation for the dairy checkoff, they mandated an annual evaluation of the programmes. A dairy model was developed that measured the impact of the dairy advertising programmes on the demand for fluid milk. Econometric models were used, drawing on data from the major USA market order regions and collected over several years. Dairy advertising expressed in real dollars was included in the model and the resulting impact on milk demand was estimated. In addition, the model showed the impact of several demographic variables (Forker and Liu, 1989; Ward and Dixon, 1989; Liu and Forker, 1990):

1. National fluid milk advertising had a statistically significant impact on the demand for fluid milk; that is, the demand curve shifted outward in the manner illustrated in Fig. 15.2.
2. There is a long-term carry-over effect of the fluid milk advertising with the full impact extending over several months.
3. There is evidence that the effectiveness of fluid milk advertising increased after the first year of the national programme. During the first year of the programme, the estimated advertising coefficients were about the same as before the national programmes were started. However, during the second and subsequent years, these same coefficients showed significant improvements. This clearly shows that a coordinated national effort with consistent copy and media produced major improvements to the dairy industry's demand enhancement programmes. In more recent years, these

improvements tended to plateau at a new higher level.

4. From 78:12 through 87:09, the dairy model shows that fluid milk advertising (both regional and national) increased total milk consumption by 4.36%. While the rates will differ over time, the general results are consistent in showing a positive and significant increase in fluid milk consumption that is directly attributed to the milk advertising programmes.

5. During the Post-Act months, the analysis showed that nearly 78% of the Post-Act gain can be attributed to improvements in the advertising effectiveness and not just to the increased expenditures. Such evidence clearly points to the benefits from a coordinated generic programme. Similar percentages are seen for the more recent periods.

6. When the gains from just the national checkoff are calculated, fluid milk consumption is shown to increase by 2.54% over the gains attributed to the regional programmes.

7. Given the actual programmes, the analysis also shows the normal declining marginal response seen in almost every commodity programme studied. For example, 42 pounds of fluid milk sales gains occurred per advertising dollar. If total expenditures were increased by 25%, the marginal gains drop to 35 pounds per dollar. Ultimately, the benefits of the programme depend on the economic value of the gains relative to the additional advertising costs.

These results are cited because they were the first complete analysis of the national dairy programmes. Subsequent studies have confirmed these results and have expanded the models to be more comprehensive, including both the fluid and processed dairy products.

More recent analysis by the Economic Research Service of the USDA (ERS-USDA) continued to confirm the benefits of the national dairy programmes. Sun and Blaylock (1993) noted that increased advertising expenditures raised fluid milk sales by 4.7% or by 7455.4 million pounds, from September 1984 to September 1991. Sales of natural and processed cheese consumed at home rose by 25 million pounds and 290 million pounds in the same period because of increased generic advertising. Conclusions about the economics of the national dairy programme are that it has been effective in stimulating demand for dairy products.

Washington Apple Commission Case

Generic promotion of Washington apples is funded through a state marketing order and has been used for many years to establish the Washington apple industry trademark. Assessments are collected through the handlers and the cost is passed back to the growers. The Apple Commission is responsible for designing and implementing generic programmes for the industry. Econometric models similar to those used for other commodities have been used to measure the economic impact of the apple advertising activities.

Over the years since 1984, the industry used both radio and television media with both spot and national coverage. Apples can be used in both media because of their consumption convenience, potential use across a wide range of consumers and their freshness attributes. Hence, the Apple Commission has long struggled with determining a proper allocation between radio and television. Research has shown that the generic promotion of Washington's apples impacted domestic demand for apples (Ward and Forker, 1991; Ward, 1993).

Given these data and the appropriate models, important conclusions follow:

1. Over the crop years since 1984, apple advertising has produced a positive shift in demand for apples. In the 1991 crop year, prices were 15.31% greater than what would have occurred without the advertising programmes.
2. Given the demand shifts, the apple industry realized a US$7.16 million return per advertising dollar for the full generic advertising effect. On average for each US$1 of advertising, approximately US$7 in returns were realized.
3. The models show the apple industry could nearly double its advertising budget before the returns do not offset the additional cost. What is important is that the industry knows where they are relative to the effectiveness of additional advertising expenditures.
4. Apples have been promoted using both radio and television and the amount spent on each medium changed over time. Given expenditures in excess of US$1 million per month during the normal advertising period, the apple model showed that approximately 70% of the funds should be allocated to television and 30% to radio. While the ratios are interesting, of more importance to the current discussion is that with these type of analyses, one can provide meaningful input into the policy decisions.

Florida Citrus Commission Case

Florida has one of the oldest generic promotion programmes in the country. Programmes have ranged from direct generic promotions to processor and retailer rebates. Television advertising with celebrity endorsements has been used extensively over time. Coupons have been dropped and a variety of informational programmes are ongoing. Furthermore, the industry has strong brands which do their own advertising separate from the generic efforts.

While many economic issues can be tied to a particular season, questions about the relative roles of generic and brand advertising have always been raised. Studies have shown the positive impact from generic advertising of citrus, including the economic gains from dropping coupons.

During the mid-1980s the Florida Citrus Commission had an ongoing advertising rebate programme where eligible processors could receive credit for their own brand promotions. These programmes grew in size, to the point that there was concern about cutting into the resources for generic activities. The obvious issue was to determine if both generic and brand advertising had positive impacts on the total demand for orange juice. Some research showed both generic and brand citrus advertising to influence the total industry. Given that the three largest brands held over 50% of the market, even without empirical evidence it seems intuitive that some net gains in total demand must occur even if market shares are changed. In fact, the models showed that the long-run effects of brand advertising were positive and slightly greater than generic. However, when one considers the allocation of generic funds to brand promotions, the marginal gains from each must be known. Given estimates for both generic and brand, one can draw the following conclusions (Ward and Kilmer, 1989):

1. Both generic and brand advertising programmes have positive impacts on the demand for orange juice. Thus, there may be levels where the use of generic funds to subsidize brand activities is reasonable.
2. However, given the current levels of expenditures where brand was nearly three times that of generic, the marginal gains for the generic promotions exceeded that of brand. Therefore, even though both programme types contributed to the total demand, the industry was probably spending too much on the rebate programmes.

One can cite other important studies for the citrus case. Lee (1981) provides a comprehensive analysis of the grapefruit market, covering the years 1970 through 1978. In that period the Commission spent US$11.83 million on advertising. Lee calculates the gains to be near US$123.41 million, thus giving a rate of return of US$10.44 per advertising dollar.

CONCLUDING REMARKS

Commodity advertising is a very small part of the total advertising effort in the USA and worldwide. There are two conditions that tend to prevail across commodities: (i) there is a need to communicate and convey information about their products; and (ii) the problem set is uniquely different for each industry. What this tells us is that advertising can work but it clearly is dependent on the circumstances of each industry, market and product. One can simply look across markets from drinks to soap and find cases of advertising successes. Yet in each example, the successes (and failure) are conditioned on the product, its potential place in a market, geography, and timing. The same must be true for commodities. If the 'apple is bad' or the 'almonds are old' all the advertising in the world is not going to enhance demand.

All of the commodities do have the underlying attribute of high quality and established grades and standards. For the most part they are repeat purchase goods. Some general conclusions can be drawn about commodity advertising.

1. Most commodities have underlying generic attributes that lend to some degree of product differentiation, whether real or fancied. Hence, the degree of the differentiability will determine the relative importance of generic and

brand advertising. Thus, there clearly can be a place for both types of efforts.

2. The message needed to be communicated clearly differs across the commodities. Each product has its own sets of uses, levels of market penetration and existing perceptions about attributes. Likewise, these conditions are dynamic.

3. As a generalization, most of the major programmes that have been studied do show a measurable positive impact on the demand for the advertised commodity. That is, consumers have responded as reflected through outward shifts in demand.

4. The carry-over effects of these generic programmes differ considerably, ranging from a few months to nearly a year. Part of these differences must be attributed to the types of models used. Nevertheless, there is sufficient evidence to establish the nature of advertising decay across commodities.

5. Brand advertising works. It may enlarge the total demand and/or increase the demand for the specific brand. In either case, the empirical evidence is clear. When products have the potential for some important element for differentiation, brand advertising becomes an important part of the total marketing mix (see Jones, 1986, 1989 for excellent examples across many goods).

6. A substantial part of the measured benefits from generic advertising is passed through to producers in terms of improved prices. That is, those funding the programmes do share in the benefits.

7. Generally, the generic programmes have not been overfunded. When comparing the rates of return to the various programmes, the evidence is that the rates are generally quite large, thus leading to this conclusion.

8. Evaluation can be controversial in terms of methods, inferences, and industry politics. Clearly, advertising decisions must be made within the context of the industry's perceived needs. That is precisely why commodity boards representing a cross-section of the industry are so important.

FURTHER READING

Advertising, and commodity advertising specifically, is taking increasing importance as an instrument for reaching domestic and international markets for a variety of products, including agricultural goods. The literature is developing, as suggested by the references in this chapter. However, there are several books and articles that are suggested for further reading among those listed in the references. These suggestions range from more general topics in advertising to the specifics dealing with commodities.

Albion, M. and Ferris, P. (1981) *The Advertising Controversy: Evidence on the Economic Effects of Advertising.* Auburn House Publishing, Boston, MA.

Bain, J.S. (1972) *Essays on Price Theory and Industrial Organization.* Little, Brown & Co., Boston, MA.

Dorfman, R. and Steiner, P. (1954) Optimal advertising and optimal quality. *American Economic Review* 44(5), 826–836.

Farm Foundation (1985) *Research on Effectiveness of Agricultural Commodity*

Promotion. Proceedings from the Seminar held in Arlington, Virginia, April 1985. Farm Foundation, Chicago, IL.

Forker, O. and Ward,. R.W. (1993) *Commodity Advertising. The Economics and Measurement of Generic Programs.* Lexington Press, New York.

Henneberry, S.R., Ackerman, K. and Eshleman, T. (1992) US overseas market promotion: An overview of non-price programs and expenditures. *Agribusiness* 8(1), 57–58.

Jones, P.J. (1986) *What's in a Name? Advertising and the Concepts of Brands.* Lexington Press, New York.

Jones, P.J. (1989) *Does it Pay to Advertise? Cases Illustrating Successful Brand Advertising.* Lexington Press, New York.

Kinnucan, H., Thompson, S.R. and Chang, H.S. (1992) *Commodity Advertising and Promotion.* Iowa State Press, Ames, IA.

Nichols, L.M. (1985) Advertising and economic welfare. *American Economic Review* 75, 213–218.

Strak, J. (1994) *Levies on Farm Products: Who Pays and Who Gains?* E 74. Euro PA & Associates, Northborough, Cambridge.

Thraen, S.G. and Hahn, D. (eds) (1989) *Advertising, Promotion and Consumer Use of Dairy Products: Insights from Economic Research.* Ohio State University Press, Columbus, OH.

Ward, R.W., Chang, J. and Thompson, S. (1985) Commodity advertising: Theoretical issues relating to generic and brand promotion. *Agribusiness* 1(4), 269–276.

REFERENCES

Albion, M. and Ferris, P. (1981) *The Advertising Controversy: Evidence on the Economic Effects of Advertising.* Auburn House Publishing, Boston, MA.

Briz, J., Ward, R.W. and de Felipe, I. (1995) Habit Persistence and the Demand for Fluid Milk in Spain. Unpublished paper in review. Gainesville, FL.

Chang, H.S. and Kinnucan, H. (1992) Measuring exposure to advertising: A look at gross rating points. *Agribusiness* 8(5), 413–423.

Chang, J. (1988) A theoretical model of generic and brand advertising. PhD Dissertation, University of Florida.

Forker, O.D. and Liu, D.J. (1989) *Generic Dairy Promotion Economic Research: Past, Present and Future.* Agricultural Economics Staff Paper 89-34. Department of Agricultural Economics, Cornell University, Ithaca, NY.

Forker, O.D. and Ward, R.W. (1993) *Commodity Advertising. The Economics and Measurement of Generic Programs.* Lexington Press, New York.

Hall, L.L. and Foik, I.M. (1982) Generic versus brand advertising for manufactured milk products – The case of yogurt. *North Central Journal of Agricultural Economics* 5(1), 19–24.

Jones, P.J. (1986) *What's in a Name? Advertising and the Concepts of Brands.* Lexington Press, New York.

Jones, P.J. (1989) *Does it Pay to Advertise? Cases Illustrating Successful Brand Advertising.* Lexington Press, New York.

Kinnucan, H. and Fearon, D. (1986) Effects of generic and brand advertising of cheese in New York with implications for allocations of funds. *North Central Journal of Agricultural Economics* 8(1), 93–107.

Kinnucan, H. and Venkateswaran, M. (1990) Effects of generic advertising on perceptions and behaviour: The case of catfish. *Southern Journal of Agricultural Economics* 22(2), 137–151.

Lee, J. (1981) Generic advertising, FOB price promotion, and FOB revenue: A case study of the Florida grapefruit juice industry. *Southern Journal of Agricultural Economics* 13(2), 69–78.

Lenz, J., Forker, O. and Hurst, S. (1991) *US Commodity Promotion Organizations: Objectives, Activities, and Evaluation Methods.* Agricultural Economics Research 91–4. Department of Agricultural Economics, Cornell University, Ithaca, NY.

Liu, D.J. and Forker, O.D. (1990) Optimal control of generic fluid milk advertising expenditures. *American Journal of Agricultural Economics* 72(4), 1048–1055.

Nichols, L.M. (1985) Advertising and economic welfare. *American Economic Review* 75, 213–218.

Sheth, J.N. (ed.) (1974) *Models of Buyer Behaviour: Conceptual, Quantitative, and Empirical.* Harper & Row, New York.

Strak, J. (1994) *Levies on Farm Products: Who Pays and Who Gains?* R 74. Euro PA & Associates, Northborough, Cambridge.

Sun, T. and Blaylock, J. (1993) An Evaluation of Fluid Milk and Cheese Advertising. Technical Bulletin No. 1815. USDA, ERS, Washington, DC.

Ward, R.W. (1993) *Washington Apple Advertising: An Evaluation Update.* WAC93#1. University of Florida, Gainesville, FL.

Ward, R.W. (1994) *Economic Returns from the Beef Checkoff.* UF#NCA94.1. University of Florida and the National Cattlemen's Association, Gainesville, FL.

Ward, R.W. and Dixon, B.L. (1989) Effectiveness of fluid milk advertising since the dairy and tobacco Adjustment Act of 1983. *American Journal of Agricultural Economics* 71(3), 730–740.

Ward, R. and Forker, O. (1991) *Washington Apple Advertising: An Economic Model of Its Impact.* Washington Apple Commission, WAC91#1. University of Florida, Gainesville, FL.

Ward, R.W. and Kilmer, R.L. (1989) *The Citrus Industry: A Domestic and International Perspective.* Iowa State Press, Ames, IA.

Ward, R. and Lambert, C. (1993) Generic promotion of beef: Measuring the impact of the US beef checkoff. *Journal of Agricultural Economics* 44(3), 456–465.

Ward, R.W., Thompson, S. and Armbruster, W. (1983) Advertising, promotion, and research. In: Armbruster, W., Henderson, D. and Knutson, R. (eds), *Federal Marketing Programs in Agriculture: Issues and Options.* Interstate Printers and Publishers, Inc., Danville, IL, pp. 91–120.

16 Distribution

Matthew T.G. Meulenberg

Department of Marketing and Marketing Research, Wageningen Agricultural University, Hollandseweg 1, 6706 KN Wageningen, The Netherlands

INTRODUCTION

Distribution is a basic activity in the marketing of goods and services. It has been a core element of marketing theory right from the beginning. Indeed, one of the first scientific papers on marketing was entitled 'Some problems in market distribution' (Shaw, 1912). Distribution has become a marketing activity which is integrated in a total marketing plan. It is concerned, among other things, with the choice of a marketing channel, logistical planning, and relationships with clients. In this chapter the distribution of agricultural and food products, both basic characteristics and evolution, is analysed. Attention will be paid to the role of wholesale and retail companies in agro-food marketing.

THE ROLE OF DISTRIBUTION IN THE MARKETING OF AGRICULTURAL AND FOOD PRODUCTS

General characteristics of distribution

Distribution, as a marketing function, is concerned with adapting supply to demand as far as time and place are concerned. In marketing commodities, such as wheat, distribution can be considered the central marketing activity, whilst in the marketing of branded food products, distribution is a marketing instrument integrated within a marketing policy.

Narrowing the gap between the production and consumption of agricultural products in terms of time and place is essential in agro-food marketing. In The Netherlands, for example, in 1988, 28% of total food expenditure accrued to the primary sector, 27% to processing and 45% to the distribution sector (van Bruchem, 1992). Distribution is particularly important in agro-food marketing for several reasons. First, agricultural products are grown in specific climatological zones or on special types of soil, while food consumption is concentrated in urban areas. Second, products are perishable and therefore require special transport and storage. Third, seasonal production has to correspond with a consumption pattern that extends over a long period of time. Finally, because there are many small product suppliers and consumers in agricultural and food markets, considerable effort goes into collection, regrouping and dispersing products. The share distribution assumes in total food expenditure is increasing, partly because of increasing consumer demand for services and a growing quality consciousness amongst consumers. In The Netherlands, for instance, the share of distribution in total consumers' food expenditure increased from 35% in 1961 to 45% in 1988 (Van Bruchem, 1992). Distribution of agricultural products can be described according to Alderson's 'sorting principle'. Alderson (1965) distinguished four successive types of sorting in marketing operations: (i) sorting out – breaking down a heterogeneous collection into several homogeneous groups; (ii) accumulation – building up larger homogeneous collections; (iii) allocation – breaking down homogeneous groups into smaller, homogeneous groups; and (iv) assorting – building up of heterogeneous collections which suit the needs of specific customers. The distribution of fresh apples is a case in point. Heterogeneous lots, harvested at individual farms, are broken down – often by the fruit grower himself – into homogeneous lots on the basis of grading and sorting schemes, and subsequently small homogeneous lots of the same standard are assembled by wholesalers into larger, homogeneous lots. Wholesalers transmit these homogeneous lots to other intermediaries and retailers, who break down homogeneous lots and build up heterogeneous collections of apples and other fresh fruit for the eventual consumer. Fulfilling these distribution functions creates time, place and possession utilities.

Differences in distribution in relation to type of product and stage of market development

The distribution of agricultural products differs according to the type of product and the stage of market development concerned. Methods of distribution are related to differences in perishability and exclusiveness, as well as the processing required by some agricultural products. Fresh produce such as fruit, vegetables and fresh meat need rapid and refrigerated transport in order to preserve product quality. In marketing commodities such as wheat and potatoes, storage is of central importance. Milk is processed into a large variety of dairy products, initiating a set of different marketing channels between producer and the eventual consumer.

Marketing sophistication also has a strong impact on the role of distribution. In less developed economies, for example, the distribution of agricultural products is primarily concerned with market transparency and the efficient performance of physical functions. The assembling of products by small wholesalers is often important in such a situation. In sophisticated food marketing operations, such as the marketing of branded products by large dairy companies (e.g. Danone and Yoplait in France, Land O'Lakes in the USA or Südmilch in Germany), distribution strategy (choice of marketing channel) and physical distribution (coordinated planning of transport and storage) are extremely important elements in the marketing operation.

Companies specializing in distribution

Whilst every company performs a certain number of distribution functions, some companies, such as wholesalers, retailers and other intermediaries specialize in distribution. Wholesalers play a central role in many agricultural markets. In wheat marketing, grain merchants, such as Cargill in the USA, are global players. In flower marketing, wholesale companies such as the Dutch company Zurel operate internationally. Cocoa trading companies including Lonray, Inc. and Gill and Duffus, Inc. in the USA, and Kakao-Einkaufsgesellschaft (KEG) in Germany, for example, also operate globally. There are many agricultural wholesalers, however, who operate at a national or regional level.

Even more important in food distribution in Western agriculture is the way in which food retailers have specialized in distribution. Davis (1966) argues that in London at the end of the sixteenth and the beginning of the seventeenth century '... shopkeeping, as distinct from random trading and the keeping of a miscellaneous warehouse, was becoming an important occupation'. But, she argues, that at the same time,

> ... quite different changes were taking place on the other side of distribution, the selling of food. As a commodity, food was in a class by itself, for its trade was supervised with anxious care by the civic authorities, who dreaded, with almost superstitious intensity, any departure from traditional methods.
>
> (Davis, 1966)

Even in the nineteenth century food was mainly retailed in the marketplace: 'Fruit and vegetables too ... had made little headway into fixed shops, ...' (Davis, 1966). The chain store emerged in the second part of the nineteenth century: 'Most authorities trace the beginnings of the chain system of retailing to the origin of The Atlantic and Pacific Tea Company in 1858' (Duddy and Revzan, 1953);

> Thomas Lipton started a one-man grocery shop in Glasgow in 1872; within eighteen years he had seventy branches in London, and eight years later still they had shot up to 245 all over the kingdom.
>
> (Davis, 1966)

Another breakthrough was the emergence of the supermarket in the 1930s.

> The first period, prior to 1930, was characterized by few units, located principally in Los Angeles. In 1929 there were only 25 units in operation in that locality, and practically all were operated by independent proprietors. ... The

second period, 1930 to 1935, was characterized by a mushroom growth, influenced in no small measure by the depression and by attempts to overcome corporate chain competition ... One source estimates the growth as follows: 1934: 400, 1939: 4982, 1945: 9575.

(Duddy and Revzan, 1953)

Whilst they were of limited size and locally/regionally oriented until the 1950s, food retail companies subsequently became big business and started to operate at a national and international level. Large food chains play an important role in food distribution: in 1993 Kroger, for example, had a 5.7% share of the market in the USA; Metro/Asko in Germany had a market share of 4.1% in Europe; Leclerc in France had a 2.6% share of the European market; and Sainsbury in the UK had a 1.9% share in Europe (Heijbroek *et al.*, 1994). Retail chains have become important players in food distribution because of their large scale and well-planned marketing strategies. They have a strong bargaining power, require special services and product qualities from suppliers and are keen on low purchasing prices and discounts.

Wholesaling and retailing are analysed in more detail in the section on companies specializing in the distribution of food and agricultural products.

OBJECTIVES, FUNCTIONS AND STRATEGIES OF DISTRIBUTION

Distribution serves routinized consumption by making products available to the consumer in the right form and at the right time and place: consumers are provided with milk, bread and potatoes daily. In sophisticated markets distribution should also stimulate demand by making products available at a specific place and time. For instance, consumers will purchase a specific variety of fruit, such as avocados for example, or a specific dairy dessert only if their supermarkets carry the product; impulse buying of flowers will be enhanced if supermarkets carry these products. Creating demand by choosing appropriate marketing channels has become important for food producers.

Basic distribution objectives

The basic objectives of distribution in today's food markets can be summarized as follows:

1. Maximizing access to target groups. Making products available at the right time and place does not only preserve loyalty amongst existing customers, but also generates new clients.
2. Minimizing distribution costs. Because distribution costs make up a large proportion of total product costs they are extremely important as far as profitability and competitiveness are concerned. In this context the relationship between the costs of performing different distribution functions (transport and storage) within a company and in the channel will have to be considered: a total cost concept relevant for the marketing channel as a whole is in order.

3. Sufficient bargaining power. A company must have bargaining power *vis-à-vis* its partners in the marketing channel in order to receive a 'fair' share of the consumers' dollar.

These distribution objectives lead us to the following strategic questions:

1. Which marketing channels should be chosen and which marketing policies should be followed *vis-à-vis* these channels to get maximum market coverage? This question is extremely relevant for food companies. Many food producers, such as dairy companies, aim at intensive distribution and try to get big retail chains as their customer by outperforming competitive suppliers in one or more of the marketing instruments – product quality, logistical service, or price. Agribusiness companies which supply farmers with the means of production and services also face the strategic question of channel choice. For instance, compound feed mills compete for the clientele of efficient large farms and are less interested in small ones, which might fade away in the future.
2. Which logistical plan realizes the desired customer service at the lowest cost? Food companies use logistical planning as a competitive weapon. The marketing of Chiquita bananas provides an example of a product which has a strong market position because of effective and efficient logistics, amongst other things.
3. Which type of relationship should a company develop with its partners in order to prevent one participant in the marketing channel having excessive market power? Cooperatives are institutions partly set up to strengthen the bargaining power of farmers in the marketing channel.

Distribution functions

Distribution objectives are attained by performance of the *exchange functions*, buying and selling, the *physical functions*, storage and transport, and the *facilitating functions*, market information, grading/sorting and credit delivery. Performance of these marketing functions is facilitated in agricultural markets by a good infrastructure of roads, railway systems, communication systems, and markets.

Exchange functions

Buying, selling and price formation are important to every exchange process. These functions are important in the marketing of agricultural commodities, such as wheat, corn, rubber and tobacco. Special market institutions, including auctions and futures markets, have been set up in order to perform exchange functions in the marketing channel more effectively and efficiently. In marketing differentiated food products, such as branded products, exchange functions are performed by the salesmen employed by food companies or by wholesalers in direct contact with the buying agents from retail companies.

Physical functions

Storage and transport are core elements in distribution processes. In this section they will be discussed separately.

Storage   Whether or not a company in the marketing channel of an agricultural commodity will hold stocks or not depends on its relative cost advantages and on strategic considerations. In comparison to farmers, wholesalers or specialized warehouse companies may realize economies of scale in holding stocks. By holding stocks a wheat merchant can serve his clients better, particularly if there is a sudden change in market demand.

Stocks of raw materials/final products enhance and ensure a smooth production process and a high level of customer service. However, the Just-In-Time (JIT) concept argues that stocks are an expensive nuisance and they hide problems in the planning of production and logistics. Companies should avoid stocks and produce the necessary quantity of products and deliver them at exactly the right time and place. Holding stocks for a long period may have a negative influence on the quality of products delivered to consumers. However, apart from seasonality in production, there are various other reasons for keeping stocks of food and agricultural products. These include a stochastic consumer demand and a stochastic delivery time in the distribution process.

A need for efficient and effective storage has stimulated the development of inventory models. The Economic Order Quantity (EOQ) model, for example, is a simple model that minimizes inventory costs. This model assumes a deterministic constant product demand, a deterministic lead time, no back ordering costs, fixed inventory costs per unit per time period, arrival of the order quantity at a particular point in time, fixed ordering costs per order, order quantity employed as the defining parameter, no side constraints to the Economic Order Quantity and no interaction between inventory costs, transportation costs and/or production costs.

The model is specified as:

$$TC = \left(\frac{Q}{2}\right)c_s + \frac{c_0}{\left(\frac{Q}{D}\right)} \tag{16.1}$$

where: TC = total inventory costs per time period; Q = order quantity; c_s = carrying costs per unit per time period; c_0 = ordering costs per order; D = product demand per time period.

The first term of Eqn 16.1 specifies storage costs per time period, since $Q/2$ is equal to the average stock per time period. The second term of Eqn 16.1 specifies the ordering costs per time period since Q/D is the time in which the order expires. The Economic Order Quantity minimizing total inventory costs, Q_0, can be derived by equating dTC/dQ, the first derivative of Eqn 16.1 with respect to Q, to zero. This will be a minimum, if the second derivative of TC with respect to Q is positive:

$$\frac{dTC}{dQ} = \frac{1}{2} c_s - \frac{c_0}{\dfrac{Q^2}{D}} = 0 \tag{16.2}$$

$$\frac{d^2TC}{dQ^2} = \frac{2c_0 D}{Q^3} > 0 \tag{16.3}$$

$$Q_0 = \sqrt{\frac{2c_0 D}{c_s}} \tag{16.4}$$

Note that in the case where $Q = Q_0$, total carrying costs per time period, $(Q/2) c_s$, and total ordering costs per time period, $C_0/(Q/D)$, are equal:

$$TC = \sqrt{\frac{c_s c_0 D}{2}} + \sqrt{\frac{c_s c_0 D}{2}} = \sqrt{2 c_s c_0 D} \tag{16.5}$$

The following example illustrates the procedure. Carrying costs per unit per month = US$0.5 per month; ordering costs per order = US$5; product demand per month = 500 items per month:

$$Q_0 = \sqrt{\frac{2c_0 D}{c_s}} = \sqrt{\frac{(2 \times 5 \times 500)}{0.5}} = 100 \text{ items} \tag{16.6}$$

The assumptions made in the EOQ model are often not realistic and have to be mitigated, and a number of extensions of the EOQ model have been developed which take better account of real life situations (see, for example, Tersine, 1988; Anderson *et al.*, 1991).

Transport   The transport decisions made by a marketing organization can be divided into three categories: (i) choice of transport mode; (ii) whether or not transport should be contracted out; and (iii) the method used in planning transport. Decisions on these issues are often interrelated.

The choice of a transport mode is based on a trade-off between desired customer service and transportation costs. A comparison of different transport modes on a number of characteristics is given by Bowersox *et al.* (1986) (Table 16.1).

Transport by truck is the rule for fresh agricultural products, because its performance in terms of availability, speed (flexibility) and frequency is comparatively good. Many commodities such as wheat, corn and soybeans are transported by ship and rail, which are particularly competitive in terms of capability and costs.

The choice between own transport and contracting transport depends on the customer service and transport costs realized by both alternatives. Own transport may require substantial investment on the part of a food company, contracting out does not require this. A freight company can operate at full capacity by assembling freight from various suppliers, own transport of a food company cannot do this. Generating backload will be more difficult for a food company with its own transport than for a freight company. At the same time, a freight company that operates internationally may

Table 16.1. Ranking of five basic transportation modes on five operating characteristics (source: Bowersox *et al.*, 1986).

Operating characteristic	Transportation mode				
	Rail	Highway	Water	Pipeline	Air
Speed	3	2	4	5	1
Availability	2	1	4	5	3
Dependability	3	2	4	1	5
Capability	2	3	1	5	4
Frequency	4	2	5	1	3

be in a better position to innovate methods of transportation than a medium-sized food company with its own transport. An advantage of own transport for a food company is that it makes it possible to take quick decisions on transport, which is beneficial both in terms of customer service and product quality. Own transport also offers food marketers more opportunities for making personal contacts with clients.

There is substantial difference between countries in the importance of contracting out logistical services to third parties. Cooper *et al.* (1994) quote a study from 1989 reporting that specialist distribution contractors are prominent in UK grocery multiples (44%), of some importance (17%) in German and French multiples, but are practically non-existent in Spain and Italy.

Transport planning is a core element in logistical management. Planning procedures have become more sophisticated because of advances in Information Technology (IT) and in decision methodology. A well-known planning model is the transportation model: how to allocate supply locations S_i ($i = 1, \ldots m$) to demand locations D_j ($j = 1, \ldots n$) in order to minimize transport costs. A dairy company (mixed feed company) for example has m plants or warehouses which serve n shops/distribution centres (farms). Let us assume that the capacities of the supply locations S_i, $Q_{i,0}$ and the requirements of the demand locations D_j, $Q_{0,j}$ per time period are given. Also it is assumed that the total capacity of supply locations $\Sigma_i Q_{i,0}$ equals total requirements of demand locations $\Sigma_j Q_{0,j}$, that transportation costs per unit from S_i to D_j, $c_{i,j}$, are fixed and independent of the quantity supplied and that supply locations S_i and S_k ($i,k = 1, \ldots m$) can substitute each other. The question is how to assign transport routes to supply locations, $S_1, \ldots S_m$ such that requirements $Q_{0,j}$ of demand locations $D_1, \ldots D_n$ are satisfied and total transportation costs $\Sigma_{i,j} Q_{i,j} c_{i,j}$ are minimized. This decision problem can be solved by linear programming, more specifically by the transportation problem (see, for example, Anderson *et al.*, 1991).

Other well-known transportation models are network models. One problem involves determining the shortest route from a supply location through a network of roads to a client. Network models and dynamic programming are used to solve such problems (see, for example, Anderson *et al.*, 1991).

Another set of transportation problems are routing problems. For instance, how to combine customers when planning routes for a number of vehicles that supply these customers from one location, such that: (i) total transportation costs are minimized; (ii) the requirements of the demand locations are satisfied; and (iii) the capacity constraints of the vehicles are not violated. The 'Savings' method is used to solve this problem (Bowersox *et al.*, 1986). A case in point is a mixed feed plant planning its feed transport to a fixed number of poultry farms.

Facilitating functions

Important facilitating functions in distribution are standardization, information and credit delivery. In developing economies in particular, poor performance of facilitating functions often seems to be the bottleneck in the distribution of farm products.

Standardization Grading and sorting schemes have been introduced for many agricultural products. For instance, in the EU for fresh fruit and vegetables three classes, E(extra), I and II are distinguished on the basis of product quality, sorting, packaging and indication. A lower class III has been barred from normal marketing channels since 1 May 1988 (LEI-DLO, 1992). The use of grading and sorting schemes makes market supply more transparent and decreases the need for a physical inspection of goods. Well-graded products supplied by different farmers can be assembled in one lot, and this is particularly advantageous for logistical efficiency.

Agricultural marketers, both wholesalers and food companies, increasingly differentiate products beyond what is possible by grading and sorting schemes, by specific product attributes, packaging, or (inter)national brands. Standardization of packaging, especially where master cartons are concerned, contributes to logistical efficiency. Great efforts have been made in this field, but much has to be done yet. The increasing use of containers in distributing food and agricultural products draws attention to the importance of standardization in this field, such as the standardization of containers in the international pot plant trade.

Market information Market information tells producers and traders where and when there is a demand for a specific product. As such it is an essential ingredient in every distribution operation. In less-developed economies this information is scarce and setting up market information services is often a first step towards effective and efficient distribution.

Commodity exchanges, wholesale markets and auctions supply information on quantities sold and their prices. Various industrial bodies, such as marketing boards, statistical offices and individual companies collect and disseminate market information. Strong competition, rapid innovation and concentration in the food industry have stimulated the need for more precise market information. Advances in Information Technology (IT) can serve distribution in this respect. The Universal Product Code (UPC) was introduced in the USA by the supermarket industry in 1974 and Europe followed suit with the European Article Numbering Association (EAN). Scanning bar codes has led to speeding up checkouts in supermarkets and facilitated the registration of inventory level. Having begun with packaged dry groceries, bar codes are now invading the fresh produce market too.

Electronic communication in food distribution is increasing. In the Videotex system a customer, such as a supermarket organization for example, has the opportunity of entering by terminal or PC the supplier's computer-based information system. The Electronic Data Interchange (EDI) communication is organized by automated electronic exchange of structured information between computers of different organizations. The important method of standardization in EDI is Electronic Data Interchange For Administration, Commerce and Transport (EDIFACT) and for agriculture, Agricultural Data Interchange Syntax (ADIS). Better and more rapid information exchange between producer and customer has decreased the costs of order entry and processing and improved customer service because of the continuous information available on inventory levels. A full discussion of the role of market information systems will be found in Chapter 9.

Credit delivery Credit delivery advances a smooth product flow through the marketing channel. By credit delivery, suppliers offer clients the opportunity of buying a product when they need it instead of postponing their purchase. Farmers may be better able to produce at the right time if they have received credit from wholesalers. The reverse side of the coin is the possible dependence of farmers on a wholesaler. Cooperatives have been set up and one of their objectives was to avoid this credit trap in marketing of agricultural products.

Final remark

The performance of distribution functions has improved because of better planning methods and more advanced distribution technology. Logistical management, the integrated planning of purchasing of inputs, of materials management in the factory and of the physical distribution of final products, is becoming increasingly important (see, for example, Bowersox *et al.*, 1986).

Distribution strategy

The core decision in a distribution strategy is the choice of marketing channel. Access to the target group, distribution efficiency and channel power are important decision criteria in this context. Basic strategies in choosing a marketing channel are:

1. Intensive distribution; sell your product through as many outlets as possible. This strategy is attractive for those food products that are regularly purchased (routinized buying), such as milk.

2. Selective distribution; choose outlets which offer adequate service with the product and whose image fits to the image of the respective product.

This distribution strategy is relevant for shopping goods and specialty goods, such as quality wines and luxuries, which are sold in specialty shops. This aspect has become more important since market segmentation became a popular strategy in food marketing.

3. Exclusive distribution; one retailer has the exclusive right to sell a product in a specific area. Whilst this strategy might be attractive for some durable products such as stylish furniture, it is not that important in food distribution.

The distribution strategy of food producers *vis-à-vis* big retail chains is evolving towards relationship marketing in which the choice of a specific marketing channel is augmented with additional marketing activities such as logistical services, joint promotion programmes and sometimes even special product development. In this way distribution policy is integrated into the total marketing policy of food and agribusiness companies.

COMPANIES SPECIALIZING IN THE DISTRIBUTION OF FOOD AND AGRICULTURAL PRODUCTS

Distribution functions are either performed by farmers and food companies themselves or are transferred to companies, such as wholesale and retail companies, which specialize in carrying out distribution functions. Having described wholesale and retailing generally in the section on companies specializing on distribution, we will now elaborate their function in the distribution of food and agricultural products.

Wholesaling

Wholesale companies assemble, regroup and dispatch farm products to the processing industry or to retailers. They are more effective distributors than farmers because of their intimate market knowledge and because they handle large product volumes. However, a wholesaler is not always more efficient in moving goods through marketing channels than farmers or the food industry. In large farms there is less need for wholesalers to assemble farmers' products. For example, there is less need for wholesalers to assemble eggs in a market supplied by poultry farms that have 100,000 laying hens per farm than in a market supplied by mixed farms with a flock of 300 laying hens. Wholesale companies can specialize in one product group, region, or in one or more specific functions.

Various market developments influence the wholesale marketing of food and agricultural products which include: (i) concentration in food industry, retail trade and agriculture; (ii) advances in communication technology and logistical procedures; and (iii) a continuous search for better product quality and more service to the final consumer. As a result various food producers and retail chains abstain from using the services of wholesale companies.

Wholesale companies have responded in different ways to these threats and opportunities. Some wholesalers contract production from farmers or engage in production themselves (backward integration). Others have developed special relationships with retailers, as in wholesaler-sponsored voluntary chains, for example where wholesalers cooperate with independent retailers on the basis of a specific retail formula (forward integration). Another reaction of wholesalers to market developments has been to specialize in specific distribution functions. Cash and carry wholesalers specialize in keeping a broad assortment but avoid product delivery, credit delivery and giving much advice to clients. They compete on the basis of low prices. Makro, the international wholesale chain, is an example of this. In The Netherlands and in Germany cash and carry wholesalers also operate in the market for ornamentals.

Truck jobbers specialize in transport, carry a limited assortment and have low overheads. They are low-cost traders characterized by their low prices and they operate in areas such as perishables (fruit, vegetables and flowers). Brokers, who act as intermediaries between producers or country shippers on the one hand and wholesalers/retailers on the other hand, specialize in the functions of exchange and market information. Agents, who sell on behalf of a specific producer, concentrate on selling without taking market risk. Some wholesalers specialize in specific products, like exotic fruits, special cheeses or wine, on specific regions such as Eastern Europe, or in special types of clients such as hotels/restaurants, for example. Others react to current market developments by trying to become/remain strong, full-service wholesalers making use of their superior market knowledge, international relationships with suppliers and buyers, and international logistical networks.

Retailing

The strategy and structure of food retailing will be analysed along two dimensions: retail operation and retail organization.

Retail operation

Food retailing began with selling at markets or by visiting consumers and selling on their doorstep. Today, the fixed outlet has become the dominant type of retail outlet. From a mixture of selling and handicraft at the beginning of this century, food retailing has become a full-scale marketing operation dominated by large food chains. Food retailing can best be understood by analysing the role played by the basic marketing instruments – product (service), price, promotion and distribution (service) – in retailing.

Product Food retailers carry food assortments and offer services which correspond to the needs of their target group. For instance, supermarkets carry a complete assortment of everyday food products and an assortment of frequently purchased durables in order to serve the general food consumer.

Fig. 16.1. Location of specific types of food retailing on the assortment dimensions 'Broad – Narrow' and 'Deep – Shallow'.

Specialty shops carry a narrow and deep assortment of food products, focusing on the specific needs and wants of consumers.

Basically, the marketing policies of food retailers and the types of retail shops that result can be classified on the basis of the product assortment. The two assortment dimensions are 'Broad – Narrow' and 'Deep – Shallow'. The first dimension represents the need categories of consumers to be satisfied, the second dimension represents the degree of consumer satisfaction in terms of the quality and variety within that need category (Fig. 16.1).

Intermediate positions on these two dimensions result in a great diversity of food shops. For instance, some supermarkets carry a deep assortment in a limited number of departments, such as bread and other produce. Food shopping is not only a necessity but also fun. As a result the atmosphere in the shop has become important in food retailing.

Price Food retailers try to build an attractive 'price image'. Both the general price level (the strategic component of pricing) and special offers (the tactical component of pricing) are important. Roughly speaking three categories of retail price strategy can be distinguished: (i) low prices, such as discounters and market stalls; (ii) high prices, such as specialty shops; and (iii) food shops, which do not see price level as basic to their marketing strategy, but go rather for competitive prices in terms of value for money.

Price and product policies are correlated. 'Low price' strategies are based on low purchasing prices, quick inventory turnover and limited service. Discounters and market stalls carry a shallow product assortment and offer limited service. Specialty shops carry a wide assortment of high quality products, offer considerable service, and charge high prices. Retail companies using competitive prices reflecting value for money will use sophisticated price policies in order to create a good price image.

It is important for a retailer to know whether some product prices are perceived by consumers as indicators of the general price level of the shop. According to Corstjens and Corstjens (1995, pp. 153, 154): 'Consumers form their impression of a retailer's relative price position in five main ways: (1) Direct price comparisons ... (2) Promotional activity ... (3) Store presentation ... (4) Direct communication ... (5) Positioning of own brands'. Weekly specials are a tactical instrument used by practically every food retailer. They reinforce 'low price' strategies of discounters, support the price image of supermarkets charging 'average prices' and mitigate the 'high price' image of specialty shops.

Promotion   Promotion and information by food retailers are very important. Weekly specials are supposed to attract consumers. They also underpin the price image of retail companies. Food stores and supermarkets use daily papers and folders as a medium to promote low prices and weekly specials. National supermarket chains, selling substantial product quantities under own brand, also use national TV advertising in order to build a strong retail image. Progress in Electronic Data Interchange (EDI) will advance the use of electronic shopping. However, up till now communication by computer screen in combination with home delivery has not made a great deal of progress in food retailing.

Distribution   Distribution as an element of the marketing mix of retail companies includes: (i) *time elements*, like opening hours and mail order/electronic shopping; and (ii) *place of shopping* such as store location and doorstep delivery. It also includes a great many service elements, such as parking facilities, service at check-outs and the handling of complaints. Some food retailers base their strategy on a specific way of distribution, such as home delivery (milkman, mail order houses) or store location (snack bars at railway stations).

Retail organization

Various organizations exist in food retailing. While there are still a large number of small independents, many food retail outlets are members of large chains.

Food retail chains   In the second part of the nineteenth century the first food chains were set up: in the UK consumer cooperatives appeared in 1856 and multiples such as the chain set up by Lipton emerged after 1872 (Davis, 1966); in the USA the Great Atlantic and Pacific Tea Company was established in 1858 (Duddy and Revzan, 1953); Delhaize was formed in 1866 in Belgium; and Kaisers Kaffeegeschäft in Germany in 1890 (Muiswinkel, 1961). Retail chains have attractive features including: low transaction costs and the discounts that result from purchasing large quantities, greater opportunities for labour specialization, better logistical planning, economies of scale in advertising and more effective control mechanisms, including comparing the results of different outlets. In fact, food chains have become big business operating at an international scale (Table 16.2).

Table 16.2. Turnover of the two largest food retail companies in selected countries in 1993 (source: Heijbroek, 1994).

Companies	Turnover (US$ billion)	Country
Kroger	22.4	USA
American Stores	18.8	USA
Ito-Yokado		Japan
Total	27.4	
Food	9.3	
Daiei		Japan
Total	22.7	
Food	6.8	
Metro/Asko	33.9	Germany
Rewe	22.7	Germany
Leclerc	20.1	France
Intermarché	20.1	France
Sainsbury	15.9	UK
Tesco	12.9	UK

Food chains have strengthened their position in food marketing channels by:

1. *Marketing policies* focusing on well-defined target groups, for instance quality conscious, environmentally concerned, price conscious or 'modern' consumers. Various characteristics of big retail chains are helpful in planning and implementing corporate marketing policies, such as marketing expertise, national/international coverage, electronic monitoring of product sales and of stocks, and own brand as a quality guarantee. Sainsbury, the food retail leader in the UK, for example, obtains 50% (Heijbroek *et al.*, 1994) to 60% of its turnover on its own brands[1].

2. *Efficiency improvement*, in particular by logistical planning and efficient use of shelf space. Improved planning methods, the use of bar codes and electronic data interchange and efficient distribution centres increase retail efficiency.

3. *Strong bargaining* power *vis-à-vis* food producers, because food retail chains command substantial market shares and because there is an overcapacity in Western food processing and agricultural production.

Many food chains have expanded into conglomerates including a diversified group of food chains and other food operations. For example, Metro/Asko and Rewe, big German food chains, include both supermarkets, discount stores and department stores in their organization, and Ahold, a large food retail holding in The Netherlands, has, apart from a number of supermarket chains, among others, also chains of liquor stores and drugstores under its control.

Another recent development in Europe is the emergence of alliances of food chains. Examples are Deuro Buying AG, Eurogroup, ERA (European

Retail Alliance), AMS (Associated Marketing Services), Bigs, Gedelfi and Interspar (Patt, 1993). These alliances are set up partly to secure purchasing advantages.

Voluntary chains   Wholesaler-sponsored voluntary chains are the centrally managed retail organization of one or more wholesalers and a large number of retailers. They are supposed to combine the advantages of corporate food chains (low purchase prices, uniform corporate marketing policy and management) with those of independents (strongly motivated, well informed about and responsive to local conditions). The first voluntary chain, 'Red and White', was set up in Buffalo, USA, by the wholesaler Flickinger in 1922. In Europe various voluntary chains have been established, which have spread out internationally, such as Spar which was founded in 1932 by the Dutch wholesaler van Well. In many countries voluntary chains command substantial market shares. During the last 30 years, however, they have had to compete with corporate food chains. The relationship between wholesaler and retailer appeared in some voluntary chains to be loose and did not encourage the development of consistent marketing policies and the effective management of the organization.

Franchising   Franchising in retailing implies a contractual relationship between a franchiser (producer or wholesaler) and franchisee (independent retailer). A franchiser authorizes the franchisee to become a member of a well-defined retailing system, which is characterized by specific marketing strategies and specific business planning, for example. A well-known franchise organization is McDonald's restaurants. McDonald's franchises local businessmen. According to Stern and El-Ansary (1992), McDonald's owns 30% of its USA outlets; the remainder are franchisees. A franchiser serves franchisees with his franchise concept, and with his marketing strategy in particular. Franchisees implement the franchise strategy in their outlets. They pay royalties, fees and initial charges to the franchiser for the services offered. The synergy of the franchise operation seems obvious: the independent retailer profits from the capacities, expertise and image of the franchiser and the franchiser can expand his business without having to invest heavily in store locations.

The relationship between wholesaler and retailer in a franchise system is often more specific, especially where market planning and implementation are concerned, than in a voluntary chain. As a result some wholesalers try to enlarge the competitive strength of voluntary chains by franchising. For instance, '... Wetterau (a major grocery wholesaler) authorizes IGA, Foodland and Red and White stores in the market it serves' (Stern and El-Ansary, 1992). Some specialized butchers and greengrocers have joint franchise systems in order to fight competition from corporate food chains. A number of corporate food chains have franchised the concept of their own outlets to a limited number of independents.

Cooperative structures   Consumers' cooperatives in retailing were set up in the UK by the Rochdale pioneers in 1844 (Davis, 1966) and have become important particularly in some Northern European countries. They were set up to confront the power of retailers who charged high retail prices but have lost their market share in the second half of this century. Retailer-sponsored cooperatives have been established in many countries, such as Topco in the USA (Stern and El-Ansary, 1992); Leclerc and Intermarché in France (Heijbroek *et al.*, 1994). While retailer-sponsored cooperatives differ from voluntary chains in their organizational set-up – backward versus forward integration – they have similar operational structures and marketing objectives.

Consequences for agricultural marketing

Concentration and market orientation make food retail companies more powerful in the marketing channel of food and agricultural products. It is often argued that retail chains are the captains of food marketing channels. However, various food companies, for example Danone, Heinz, Kraft, Nestlé and Unilever, still have substantial channel power on the basis of strong brands, international market coverage and innovative capacities. Even today many producers and wholesalers of agricultural and food products can make themselves attractive, sometimes even indispensable, partners for food chains by offering high product quality, excellent logistical services and competitive prices.

MARKETING CHANNELS AND MARKETING CHANNEL STRATEGY

The distribution decisions of companies, involved in the production and marketing of a product, crystallize out in the marketing channel of the product concerned. The shortest marketing channel is direct contact between producer and final consumer. Often a marketing channel consists of a number of successive intermediaries (agents, wholesalers and retailers) between producer and final consumer. Different approaches to the description and analysis of marketing channels will be discussed.

Description of marketing channels

Marketing channels can be described on the basis of:

1. Who is performing which marketing function in the channel? For example, are potatoes stored by farmers or by wholesalers? Does a fruit grower sell products himself or does he transfer this function to a marketing cooperative? Such a description of a marketing channel on the basis of the functions performed in the channel can make use of the classification: exchange functions (buying, selling, price formation), physical functions (storage, transport) and facilitating functions (sorting/grading, information, credit delivery).
2. The length of the marketing channel. A marketing channel can be described on the basis of the number and types of successive intermediaries in the marketing channel from producer to consumer.

3. The number of channel constructs. Bucklin (1970) has suggested three critical dimensions in the description of a marketing channel: flows, degree of aggregation of channel work into agencies and the number of levels of such agencies (see Mallen, 1977, for a similar approach).

Factors influencing the structure and evolution of agricultural marketing channels

It has been argued that distribution decisions are determined by criteria related to efficiency (costs), effectiveness (sales, margins) and bargaining power (share of the consumer dollar). These criteria are important in theories which try to explain the structure of marketing channels.

Channel theories which focus on channel efficiency

Alderson (1954) suggested that a middleman will emerge between producer and consumer if this will result in a smaller number of transactions: in the case of m producers and n consumers, a middleman will emerge in the marketing channel if $m \times n > (m + n)$. The practical value of this theory seems limited since the assumptions implied, such as no economies of scale in costs per transaction, every producer contacts every consumer, the product is homogeneous and market entry is free, are unrealistic in many circumstances.

Various authors argue that companies transfer marketing functions to other institutions in the marketing channel if this transfer will diminish marketing costs. In this context Stigler (1951) speaks of vertical specialization and Mallen (1973) about 'spin off'. Etgar and Zusman (1982) have modelled the emergence of middlemen in the marketing channel from the point of view that a middleman buys and sells market information. Bucklin (1965) argues that postponement and speculation are two factors that determine the structure of a marketing channel: '... postponement ... may be seen as a device for individual institutions to shift the risk of owning goods to another' and

> The principle of speculation holds that changes in form, and the movement of goods to forward inventories, should be made at the earliest possible time in the marketing flow in order to reduce the costs of the marketing system.
>
> (Bucklin, 1965)

Bucklin argues that

> The minimum cost and type of channel are determined by balancing the costs of alternative delivery times against the cost of using an intermediate, speculative inventory. The appearance of such an inventory in the channel occurs whenever its additional costs are more than offset by net savings in postponement to the buyer and seller
>
> (Bucklin, 1965).

Channel theories which focus on effectiveness and bargaining power

Channel structure is not only determined by efficiency criteria but also by criteria of effectiveness and bargaining power. Does a marketing channel generate sufficient sales and an adequate gross margin? Two examples may illustrate this point:

1. Aspinwall's (1958) 'Characteristics of Goods Theory' explains channel structure on the basis of product characteristics. Aspinwall proposed five criteria for classifying goods: (i) replacement rate ('The rate at which a good is purchased and consumed by users in order to provide the satisfaction a consumer expects from the product'); (ii) gross margin ('The money sum which is the difference between the laid in cost and the final realized sales price'); (iii) adjustment ('The services applied to goods in order to meet the exact needs of the consumer'); (iv) time of consumption ('The measured time of consumption during which the good gives up the utility desired'); and (v) searching time ('The measure of average time and distance from the retail store') (Aspinwall, 1958). Goods that score high on the first criterion and low on the other four, so-called 'red goods', will be marketed by intensive distribution and consequently their marketing channels will include one or more middlemen. Goods, having opposite ratings on the five criteria, so-called 'yellow goods', will often be sold directly to the customer, without the interference of a middleman. Many agricultural/food products are 'red goods' and as a result will be distributed indirectly. In fact, retailers and wholesalers play an important role in agricultural marketing channels. However, developments in food retailing, such as concentration and increasing sales of own brand have fostered direct sales from food producer to retailer that exclude wholesalers. It demonstrates the partial character of analysing marketing channels on the basis of Aspinwall's approach.

2. Stern (1969) and others have applied behavioural concepts such as power, cooperation and conflict to the analysis of marketing channels. Drawing on the work of French and Raven (1959) they distinguish the following sources of power: 'rewards, coercion, expertise, reference, and legitimacy' or, more concisely, coercive and non-coercive sources of power (Hunt and Nevin, 1974). Analyses have been made, for example, of the use of power within a franchise channel of distribution (see Frazier and Summers, 1986) and of channel domination and countervailing power in distributive channels (Etgar, 1976).

Marketing channels as a system

Marketing channels can develop into a system of producers plus middlemen pursuing a coordinated marketing policy *vis-à-vis* final consumers. Such a development is also relevant for agricultural markets. Coordination of marketing policy in the channel may concern total marketing policy for a product or specific marketing elements only, such as product quality or logistical operations. Various economic and marketing theories have contributed to our understanding of marketing channels as a vertical system. Transaction costs theory and the analysis of vertical marketing systems have proved to be particularly useful.

Transaction costs theory as developed by Williamson (1975, 1985) seems relevant to the analysis of agricultural marketing channels. According to this theory, dimensions of transactions, such as asset specificity, uncertainty/complexity and frequency, influence the governance structure of transactions, and as a result the structure of a marketing channel. If asset specificity and uncertainty are high, for instance, vertical integration is attractive, but if both are low spot contracts might be preferred (Douma and Schreuder, 1991).

In marketing agricultural commodities such as wheat, asset specificity and uncertainty/complexity seem low and transactions between actors in the channel are frequent. As a result spot contracts dominate the conventional marketing channels through which agricultural commodities move. However, marketing chicken meat as a branded product, for example, may require coordination of policies in the marketing channel by contract or vertical integration.

Marketing theory distinguishes between: (i) administered vertical marketing systems in which marketing is coordinated between channel companies by deliberation without contracts; (ii) contractual vertical marketing systems, in which marketing operations are coordinated by contracts; and (iii) corporate marketing systems, which are coordinated by the vertical integration of companies. All three types of vertical marketing systems occur in marketing agricultural and food products. Some retail chains have worked out programmes for product quality improvement, which are based on coordination of production and marketing planning between retailer and food producer (and/or farmers) without binding contracts. Marketing channels for fruit and vegetables for the canning industry and marketing channels for pigs and chickens are often coordinated by contractual relationships concerning product quality, delivery time and price. Examples of corporate vertical marketing systems are dairy cooperatives which have integrated cheese wholesale companies (forward integration), and food retail companies which have integrated production plants, such as bakeries (backward integration).

CONCLUDING REMARKS

This chapter has shown that distribution is a dynamic field of agro-food marketing. Recent developments in the environment of the agro-food system suggest that distribution will remain a dynamic marketing subject in the future. Amongst others, the following developments seem important in this respect:

1. Societal concern about sustainability will stimulate green logistics, i.e. focusing on energy savings and on reduction of pollution and waste. This concern and overcrowded highways, together with other things, will enhance the use of combined transport, such as road–rail services.
2. Government deregulation and the abolishment of international trade barriers within the context of the World Trade Organization will internationalize

agricultural markets and logistical networks even more.

3. The growth rate of gross domestic product of OECD countries and of Third World countries, estimated to be 2.75% and 6.25% respectively in 1995 (Centraal Planbureau, 1994) will stimulate international trade of agricultural and food products.

4. Improvements in IT, amongst others the advance of the electronic highway, will improve communication both between businesses and between households and businesses. Consequently the flows of physical goods and the flow of information can better be uncoupled, which is advantageous for logistical efficiency.

So, it looks as if distribution, the mother of agricultural marketing, will remain the bright and breezy lady of agricultural marketing in the future as well.

NOTES

1.   *The Economist* (1995) Change at the check-out. A survey of retailing. March 4, p. 8.

FURTHER READING

Cooper, J., Browne, M. and Peters, M. (1994) *European Logistics, Markets, Management and Strategy*, 2nd edn. Blackwell Publishers, Oxford.

Corstjens, J. and Corstjens, M. (1995) *Store Wars, the Battle for Mindspace and Shelfspace*. J. Wiley & Sons, New York.

Davies, G.J. and Brooks, J.M. (1989) *Positioning Strategy in Retailing*. Paul Chapman Publishing, London.

Gosh, A. and McLafferty, S.L. (1987) *Location Strategies for Retail and Service Firms*. Lexington Books, Lexington, MA.

Stern, L.W. and El-Ansary, A.I. (1992) *Marketing Channels*, 4th edn. Prentice-Hall International Editions, Prentice-Hall, Inc., Englewood Cliffs, NJ.

Tersine, R.J. (1988) *Principles of Inventory and Materials Management*, 3rd edn. North-Holland, New York.

REFERENCES

Alderson, W. (1954) Factors governing the development of marketing channels. In: Clewett, R.M. (ed.), *Marketing Channels for Manufactured Products*. Richard D. Irwin, Inc., Homewood, IL, pp. 5–34.

Alderson, W. (1965) *Dynamic Marketing Behavior: A Functionalist Theory of Marketing*. Richard D. Irwin, Inc., Homewood, IL.

Anderson, D.R., Sweeney, D.J. and Williams, T.A. (1991) *An Introduction to Management Science*, 6th edn. West Publishing Company, St Paul, MI.

Aspinwall, L.V. (1958) The characteristics of goods and parallel systems theories. In: Kelley, E.J. and Lazer, W. (eds), *Managerial Marketing*. Richard D. Irwin, Inc., Homewood, IL, pp. 434–450.

Bowersox, D.J., Closs, D.J. and Helferich, O.K. (1986) *Logistical Management*, 3rd edn. Macmillan Publishing Company, New York.

van Bruchem, C. (ed.) (1992) *Landbouweconomisch Bericht 1992*. LEI-DLO, 's Gravenhage.

Bucklin, L.P. (1965) Postponement, speculation, and the structure of distribution channels. *Journal of Marketing Research* 2, 26–31.

Bucklin, L.P. (1970) The classification of channel structures. In: Bucklin, L.P. (ed.), *Vertical Marketing Systems.* Scott, Foresman and Company, Glenview, IL, pp. 16–32.

Centraal Planbureau (1994) *Macro Economische Verkenning 1995.* Sdu Uitgeverij Plantijnstraat, Den Haag.

Cooper, J., Browne, M. and Peters, M. (1994) *European Logistics. Markets, Management and Strategy,* 2nd edn. Blackwell Publishers, Oxford.

Corstjens, J. and Corstjens, M. (1995) *Store Wars, the Battle for Mindspace and Shelfspace.* J. Wiley & Sons, New York.

Davis, D. (1966) *A History of Shopping.* Routledge & Kegan Paul Ltd, London.

Douma, S. and Schreuder, H. (1991) *Economic Approaches to Organizations.* Prentice-Hall International (UK) Ltd., Hemel Hempstead.

Duddy, E.A. and Revzan, D.A. (1953) *Marketing, an Institutional Approach,* 2nd edn. McGraw-Hill Book Company, Inc., New York.

Etgar, M. (1976) Channel domination and countervailing power in distribution channels. *Journal of Marketing Research* 13, 254–262.

Etgar, M. and Zusman, P. (1982) The marketing intermediary as an information seller: A new approach. *Journal of Business* 55(4), 505–515.

Frazier, G.L. and Summers, J.O. (1986) Perceptions of interfirm power and its use within a franchise channel of distribution. *Journal of Marketing Research* 23, 169–176.

French, J.R.P. and Raven, B. (1959) The bases of social power. In: Cartwright, D. (ed.), *Studies in Social Power.* University of Michigan Press, Ann Arbor, MI, pp. 150–167.

Heijbroek, A.M.A., van Noort, W.M.H. and van Potten, A.J. (1994) *The Retail Food Market. Structure, Trends and Strategies.* Rabobank Nederland, Utrecht.

Hunt, S.D. and Nevin, J.R. (1974) Power in a channel of distribution: Sources and consequences. *Journal of Marketing Research* 11, 186–193.

LEI-DLO (1992) *Tuinbouwcijfers 1992.* LEI-DLO, 's Gravenhage.

Mallen, B.E. (1973) Functional spin-off: A key to anticipating change in distribution structure. *Journal of Marketing* 37, 18–25.

Mallen, B.E. (1977) *Principles of Marketing Channel Management.* Lexington Books, Lexington, MA.

Muiswinkel, F.L. (1961) *De Handelsonderneming,* 2nd edn. N.V. Noord-Hollandsche Uitgeversmaatschappij, Amsterdam.

Patt, P.J. (1993) Internationalisierung im Einzelhandel. In: Irrgang, W. (ed.), *Vertikales Marketing im Wandel.* Verlag Franz Vahlen, München, pp. 81–95.

Shaw, A. (1912) Some problems in market distribution. *Quarterly Journal of Economics* 26, 706–765.

Stern, L.W. (1969) *Distribution Channels: Behavioral Dimensions.* Houghton Mifflin Company, Boston.

Stern, L.W. and El-Ansary, A.I. (1992) *Marketing Channels,* 4th edn. Prentice-Hall International Editions, Prentice-Hall, Inc., Englewood Cliffs, NJ.

Stigler, G.J. (1951) The division of labor is limited by the extent of the market. *Journal of Political Economy* 54, 185–193.

Tersine, R.J. (1988) *Principles of Inventory and Materials Management,* 3rd edn. North-Holland, New York.

Williamson, O.E. (1975) *Markets and Hierarchies: Analysis and Antitrust Implications.* The Free Press, New York.

Williamson, O.E. (1985) *The Economic Institutions of Capitalism.* The Free Press, New York.

17 Control of Marketing Programmes

Oswin Maurer

Department of Agricultural Economics, University of Kiel, Olshausentrasse 40, 24118 Kiel, Germany

INTRODUCTION

According to the marketing literature, the major tasks of a marketing manager's job are to plan, to implement and to control marketing activities. Just as considerable care must be taken in implementing marketing strategies, equal care must be given to establish an effective evaluation and control system. Looking at real world business situations, two managers facing similar market conditions and possessing equal marketing resources could generate dramatically different performance results. It can be argued that one manager employs accurate tools to monitor and to control the performance of the marketing strategy whereas the other one does not. Therefore, the discerning marketing manager evaluates and controls the effectiveness and efficiency of the marketing strategy implemented to isolate problems, to recognize opportunities and to alter the marketing strategy as competitive conditions change (Hutt and Speh, 1992).

BASIC CONSIDERATIONS IN MARKETING CONTROL

Managing a firm's marketing strategy can be compared with the coaching of a sports team. The excitement and challenge is to be found in the formulation of strategy. Like a coach, the business marketer also applies managerial talent creatively when developing and implementing marketing strategies that respond to customer needs and capitalizes on competitive weaknesses of other firms. Nevertheless, formulating and implementing a marketing strategy is only one part of the necessary management tasks to be fulfilled. The other part deals with the evaluation of marketing performance, with marketing control systems which enable a firm to check actual against planned performance and to take corrective action where needed.

© CAB INTERNATIONAL 1997. *Agro-food Marketing*
(eds D.I. Padberg, C. Ritson and L.M. Albisu)

The process of control

Evaluation and control is the process in which marketing activities and performance results are monitored so that actual performance can be compared with desired performance, and managers can use the resulting information to take corrective action and resolve problems. Although control is commonly considered to be the final element in the marketing process, it should allow us to pinpoint weaknesses in previously implemented marketing plans and thus stimulate the entire process from planning to implementation to begin again.

Controlling activities and behaviour

According to Jaworski (1988) marketing control systems should focus on two aspects: control of activities and control of personnel. Control of activities, i.e. strategies, programmes, plans, tactics, etc., helps to ensure that marketing activities produce the desired results. For example, increasing the advertising expenditures for a brand losing market share could be termed a control action. Control of personnel refers to attempts by management to influence the behaviour and activities of marketing personnel to achieve desired results. Therefore, an appropriate control device is designed to affect individual action which, in turn, is expected to influence performance.

Most control systems are output oriented, meaning that they are focused on evaluating the results of activities. By employing this system, desired output levels are specified, the implementation process of the marketing strategy is monitored, and corrective action is taken where substantial deviations from expected performance targets appear. This output-oriented view of marketing control is limited in scope and application. Since informal controls, as mechanisms that influence individual or group behaviour, may have more impact on performance than management controls, equal attention needs to be given to personnel control. Furthermore, if attention is restricted to output, problems can occur, since the positive effects expected from a tight, formal system of control might be offset by informal controls. For example, sales representatives might establish norms of low productivity or a market researcher may provide just the information which supports management's a priori view (Jaworski, 1988). Therefore, marketing control systems should be broadly based enough to account for the control of marketing personnel, too.

Controlling results and guiding strategy implementation

Information generated by the marketing control system is fundamental when revising current marketing strategies, formulating new strategies and allocating funds (Hutt and Speh, 1992). Both the long-term and the short-term objectives, as set out in the strategic marketing plan, and the annual marketing plan respectively, need to be evaluated so that control systems can be put into place. The requirements for an effective control system are strict and data on the appropriate performance measures has to be collected, processed and to be made available to decision-makers on a continuous basis. Besides this, it must also provide the necessary tools to monitor the quality of strategy implementation. Often poor marketing performance is caused by implementation problems which may lead to the rejection of a sound strategy. If an effective marketing strategy is based on a thoughtfully designed and well-applied control system, many implementation problems can be avoided. In this context, it is most important that a balance between a too rigid control and *laissez-faire* policy is achieved when evaluation and control systems are established. A too bureaucratic control system can be as counter-productive as a no-controls system, whereas a balanced system is able to provide timely feedback and focuses on the critical success factors.

THE CONTROL PROCESS AS A FIVE-STEP FEEDBACK MODEL

Control follows planning – and it may be the first step in initializing a new planning process. It ensures that the corporation achieves what it set out to accomplish. Just as planning involves the setting up of objectives along with the strategies and programmes necessary to accomplish them, the control process compares actual performance with desired results and provides the feedback necessary for management to evaluate results and take corrective action by revising strategies, policies, programmes, budgets and procedures. This process can be viewed as a five-step feedback model (Wheelan and Hunger, 1991).

Determine what to measure

It has to be specified what implementation processes and results will be monitored and evaluated. Both must be capable of being measured in an objective and consistent manner, whereas emphasis should be put on the most significant elements in the process – on the critical success factors.

Because marketing strategies are never timeless, the marketing control process serves as the mechanism for achieving strategic adaptation to environmental change and operational adaptation to productivity needs. The process consists of two complementary activities: strategic control and operations control. Strategic control evaluates the direction of the firm as stated in its objectives, strategies and capacity to perform in the context of changing environments and competitive actions. On the other side, operations control assesses how well the organization performs the marketing activities with respect to planned results. Operations control implicitly assumes that the organization is heading in the right direction and only the ability to perform specific tasks needs to be improved (Kerin and Peterson, 1993). The distinction between the two types of control is important, because undesirable results may be identical; but, performance evaluation and the type of response needed will differ.

Establish standards of performance

Since standards used to measure marketing performance are detailed expressions of marketing objectives, they are measures of acceptable performance results. They must be stated in meaningful terms so that when they are compared to actual performance, the comparison will yield useful information for decision-making. The standards set usually include a tolerance range within which deviations accepted as satisfactory are defined. Standards should be set not only for final output, but also for intermediate stages of the process.

Establishing performance standards may be done by using comparative standards or ideal standards. Comparative standards are set on competitor performance, industry averages or historical performance levels. Ideal standards are established on what is thought to be perfection, and they are subjective by nature. This type of measurement criterion is commonly used if comparative standards are not easily available.

Measure actual performance

Measurements must be made at predetermined times by using measures which are appropriate for comparing results with the standards set. Essentially, there are two classes of measurement: effectiveness and efficiency. The differences between both terms were appropriately summarized by Drucker (1974): 'Effectiveness is the foundation of success – efficiency is the minimum condition for survival after success has been achieved. Efficiency is concerned with doing things right. Effectiveness is doing the right things'.

Effectiveness measures – 'are we doing the right things?' – relate outputs to the objectives of an organization. They are based on a comparison of performance to the goals formulated in the marketing strategy and measure what current operations of a marketing department contribute to achieving the goals and strategies (Bonoma and Clark, 1988). This type of performance measurement deals with factors such as market share, sales results, etc., but also considers qualitative factors such as customer service satisfaction, customer attitudes, or perceptions of product quality.

Efficiency measures – 'are we doing the things right?' – are essentially marketing productivity ratios of results over expenses. They record the relation between inputs and outputs, maximizing the outputs relative to the inputs. Efficiency measures include cost and profit factors as well as marketing expenditures and revenues, cost–profit breakdowns by product, market segment, distribution channel or sales representative. Bonoma and Clark (1988) point out that marketing productivity analysis traditionally focuses on the measurement of efficiency – the 'how well is marketing done?' aspect – and often disregards effectiveness as a performance measurement.

In summary, the brief discussion on effectiveness and efficiency should highlight that the assessment of marketing performance should not be limited to a review of a few single factors as market share, sales or profit.

Within a profound control system a variety of efficiency, effectiveness and environmental factors have to be considered. These include aspects of corporate culture, management skills and marketing structures, customer expectations and attitudes as well as other competitive factors external to the corporation. In this respect, perhaps the most advanced tool currently available to assess marketing performance is the marketing audit. As pointed out by Bonoma and Clark (1988) '... audits can provide the capstone to a marketing control system by providing an independent perspective on the system ...', by concentrating '... on illuminating the context within which marketing implementation occurs, such as strategy, environment, and information systems capability. The context provides meaning that simple profit/expense ratios cannot capture'.

Evaluate performance

The process of evaluation should begin as soon as possible after a firm's marketing plans have been set into operation. Evaluation logically follows planning and implementation and commonly a circular relationship exists between these steps. Since performance measurement reveals problems that have occurred in long- and short-term strategy implementation, performance evaluation should help to determine why these problems have occurred.

Proper implementation of strategic and operations control requires awareness of several pertinent considerations. One of these considerations, the dynamic tension which exists between effectiveness and efficiency, has already been discussed. Furthermore, most relevant considerations in marketing control are the qualitative difference between data and information (Kerin and Peterson, 1993). Effective control requires distinguishing between problems and symptoms. In this context, managers have to fulfil a diagnostic role by developing causal relationships between occurrences. For example, if there is evidence of sales decline or poor profit margins, managers have to 'look behind' the numbers to identify the underlying causes and then attempt to remedy them. On the other hand, the qualitative difference between data and information may affect the decision-making process negatively. Data are essentially reports, i.e. of activities, events, or performance. In contrast, information has to be viewed as a classification of activities, events, or performance designed to be interpretable and useful for future decision-making.

Basically, two causes of marketing strategy implementation problems exist: an unsatisfactory marketing effort, and changes in the external environment which can affect the long-run performance of a firm, like alterations in primary demand, legislative changes, new marketing channels or increased competitive activity. If the marketing effort is evaluated to be unsatisfactory the focus has to be on finding ways to improve the productivity of marketing operations. If changes in the external environment are revealed as major causes for undesired performance this commonly calls for a modification of the objectives originally set in the marketing strategy. An improper assessment of the original cause of the problems identified may call for the wrong type of corrective action. Modifications to improper operational performance are called a steering control system, whereas modifications to strategic objectives are described as an adaptive control system (Assael, 1990). Being a typical reactive system, a steering control system allows the identification of operational problems already throughout the implementation period of a marketing strategy. It supports corrective action to be taken as long as performance results are still within the determined tolerance range of the standards set. In contrast, an adaptive control system is a proactive system, by anticipating changes in the competitive environment and by modifying objectives and strategies to adopt to the changed situation. Therefore, adaptive control is a fundamental aspect of strategic control, because marketing strategies may sometimes reflect objectives which have already become obsolete during time.

Take corrective action

If actual marketing performance results do not meet the standards set, or fall outside the depicted tolerance range, action has to be taken to correct the deviation. The evaluation of why performance has not met the standards provides the information needed to begin correcting the situation. However, before starting action, it has to be determined whether the deviation is only a chance fluctuation, the processes have been carried out incorrectly, or, whether the processes used are inappropriate to the achievement of the standards. Although it is essential that appropriate action is taken as soon as possible, hasty reaction may lead to detrimental results. Thus, it is important to achieve a balance between satisfying short-term and long-term performance requirements. Corrective action should not only be targeted at how to rectify the deviation, but how to prevent its occurrence again.

Corrections of marketing performance at the operational level fall into three categories (Bovée and Thill, 1992): (i) changing the performance standards; (ii) redirecting resources; and (iii) stimulating improved performance, whereas in many cases a combination of these actions is used. To change performance standards may be necessary in a situation when it becomes apparent that the standards set are not realistic any more. This may be due to changes in the external environment, as for example the introduction of a new product by a competitor, or to an inaccurate forecast of future developments during the planning period. To redirect resources from one product or from one marketing element to another is the second way to correct performance. The evaluation of performance results for the most recent operating period will indicate how each individual marketing strategy element contributed to the achievement of desired objectives. Performance below or above expected levels clearly signals where funds should be reallocated. Stimulating improved performance is the third way to correct performance problems. In this case, any element of the marketing mix is examined to find opportunities for improvement. These efforts to stimulate improved performance can take on many different forms such as the introduction of sales contests, the relaunch of an advertising strategy, or the improvement of products.

Apart from the situation that actual performance is below the desired standards, performance levels that exceed the standards may also call for corrective action. Commonly, exceeding the standards set is a most desirable situation, but in marketing it may also lead to some additional problems. Sales exceeding forecasts simply means that the forecast was too low. This may result in lost opportunities, which comes also at some cost by causing problems in procurement, manufacturing, sales and distribution.

LEVELS OF MARKETING CONTROL[1]

Marketing control is a universal process which serves as the mechanism for achieving strategic adaptation to environmental change as well as operational adaptation to productivity needs (Jaworski, 1988). It therefore allows the examination of performance results at the strategic, tactical and operational level of marketing. According to this, three different types of control can be distinguished (Kotler *et al.*, 1989): (i) annual plan control; (ii) profitability control; and (iii) strategic control. Hutt and Speh (1992) add an additional dimension to these three control types – strategic component control – by categorizing evaluation and control efforts according to the individual elements of the marketing mix. For practical reasons the four-step approach which is delineated in Table 17.1 seems to be the most appropriate one. It also takes into account how well resources are being utilized in each element of the marketing strategy.

Annual plan control

Annual plan control checks ongoing performance of marketing personnel against the objectives set in the annual marketing plan. These objectives are the benchmarks against which actual results are compared. Since the purpose is to ensure that a company achieves the goals established in the annual marketing plan, the primary responsibility for this type of control is with top and middle management. Typical performance standards utilized in annual plan control are sales volume, profit and market share. Specific control tools commonly used are sales analysis, market-share analysis and marketing expense-to-sales analysis which are discussed in the following.

Sales analysis

Sales analysis consists of a detailed breakdown of a company's sales records to compare and evaluate actual sales with sales goals. Therefore, it comprises a detailed study of sales data for the purpose of appraising the appropriateness of a marketing plan. Sales data commonly includes information about product class and attributes, price and financial arrangements, sales

Table 17.1. Types of marketing control (source: adapted from Kotler *et al.*, 1989, and Hutt and Speh, 1992).

Type	Purpose	Control devices
Annual plan control	To examine whether the planned results are being achieved	Sales analysis Market-share analysis Expense-to-sales ratios
Profitability control	To examine where the company is making or losing money	Profitability by product, territory, market segment, trade channel, order size
Strategic component control	To examine how well resources are utilized in each element of the marketing strategy	Expense ratios Advertising effectiveness measures Market potential Contribution margin analysis
Strategic control	To examine whether the company is pursuing its best opportunities with respect to markets, products and channels	Marketing audit

territories, customer types, order size and so on. By themselves, raw sales data just paint an incomplete picture. It has to be considered that expected sales may not be realized because of short-term activities as price reductions or special promotional activities of competitors. Therefore it is necessary to untangle changes in sales performance which reflect fundamental changes in the customer basis from those which are caused by tactical actions. In order to relate the results of specific marketing activities to their principal causes, appropriate tools such as sales variance analysis and micro-sales analysis have to be used (Kotler *et al.*, 1989).

Sales variance analysis looks for exceptions or variations from planned sales levels and measures the relative contribution of different factors to a deviation in sales performance. It can help to reveal the principal factors causing a gap in sales performance, as shown in the following example: suppose the annual sales goal for a specific product set in the marketing plan was 20,000 units at US$1 per unit, or US$20,000. Actually, the firm achieved sales of 18,000 units at US$0.95, or US$17,100. The performance variance is US$2,900, a gap of 14.5% to planned sales. As both the expected sales price and the sales volume were not achieved, the question arises, to which extent each of the two factors contributed to the total variance in sales?

Total variance:
$(US\$1 \times 20{,}000) - (US\$0.95 \times 18{,}000) = (US\$20{,}000 - US\$17{,}100) = US\$2900$
Variance due to lower price:
$(US\$1 - US\$0.95) \times 18{,}000 = (US\$0.05 \times 18{,}000) = US\$900$
Variance due to lower volume:
$(US\$1) \times (20{,}000 - 18{,}000) = (US\$1 \times 2000) = US\$2000$

The simple calculation reveals that about 69% of the sales variance is due to the fact that the expected sales volume has not been achieved.

However, the use of aggregate figures in sales analysis presents problems. Sales may not be derived from sales territories in equal proportions, divided equally among customers, or evenly distributed across product lines. Furthermore, some product lines may contribute small unit volumes to sales but generate high levels of profit. If sales volumes derived from these various sources are disproportionate, it makes sense to disaggregate the data and to perform a micro-sales analysis. The disaggregation of sales figures can be done by using any of the information in the sales record.

Assume the above-mentioned company sells in four regions and expected sales were at 6000 units for region A, 5000 units for region B, 6500 units for region C and 2500 units for region D respectively. Compared to these figures, the actual sales volume was at 5000 units in region A, 5500 units in region B, 6500 units in region C, and 3000 units in region D respectively. The real problem seems to be region A with a 16.7% shortfall in expected sales. Sales in regions B and D exhibit a surplus in planned sales of 10% and 20% respectively, whereas in region C exactly the expected sales volume has been achieved.

This brief examination of two aspects of sales analysis provides insight into how this evaluation tool may be used. In a real business situation the analysis may have to go further, by examining sales volume by individual districts within regions and by customer groups. For instance, even if sales in region B and region D in the example were higher than expected, the excellent sales performance in those regions may cover up weaknesses in other relevant performance areas. Therefore, a study of sales volume alone is usually insufficient and maybe even misleading. Frequently, misdirected marketing efforts are not revealed, because of the effects of the so-called 'iceberg-principle' (McCarthy and Perreault, 1990). To illustrate the situation the analogy of an iceberg is used. Only a small part of an iceberg is visible above the surface of the water, the more dangerous part being submerged below the surface. Similarly to this, figures representing total sales or total costs are like the visible part of an iceberg. To reveal what is happening in the submerged segments of the market, a detailed analysis of even well-performing sales regions by sales district, sales person, product lines or customer groups is needed.

Market-share analysis

Comparing sales results with sales goals is certainly a useful form of performance evaluation. However, it does not provide any insight into how well a company is doing relative to its competitors. Therefore, in evaluating performance, market-share analysis complements sales-volume analysis. It offers a means for determining whether a company is gaining or loosing ground in comparison to its competitors. If market share is used for control purposes several questions have to be considered (Kerin and Peterson, 1993): first, how is the market defined on which the market-share percentage is based, and has the market definition changed?; second, is the market itself changing?; third, which unit of analysis is used, unit sales or dollar sales?

Considering these questions, market share can be evaluated by product, customer group, channel type, geographical region and so on. But a high market share by itself may be misleading, since overall sales in the market may be stagnating, growing or declining. Competitors may benefit from changes in economic conditions, too. Supposing a firm experiences a 10% increase in sales of a specific product. By itself, and just on the surface, this seems to be an excellent result. However, if total industry sales of the product under consideration have grown by 30% during the same time, market-share analysis would reveal that performance must have been quite poor relative to competitors. Only if market share developments are set in relation to industry-wide sales or competitors' market share, can a clear picture on performance be drawn.

Probably the major obstacle encountered in market share analysis is in obtaining valid and comparable information on total industry sales and competitors' market performance. Excellent sources for industry sales volume statistics are usually provided by government statistical offices, whereas more detailed information on specific branches may be obtained from Chambers of Commerce or other industry-related organizations.

In summary, market-share analysis should help to track a company's performance in relation to total industry and its competitors. Generally, a company is gaining relative to competitors, if the company's market share increases, whereas the firm is loosing ground on competitors, if market share decreases. According to Kotler et al. (1989), such conclusions are subject to certain qualifications: first, the assumption that external factors have the same effects on all companies within an industry is not always valid; second, the assumption that a company's performance should be judged against the average performance of all companies may not prove to be valid in any case; third, market share of all incumbent firms may decline because a new firm entered the industry; and finally, the decline in a firm's market share may be the result of a premeditated policy to improve profits. This may be the case, if a firm is streamlining product assortments.

Marketing expense-to-sales analysis

Annual plan control should also help to reveal whether a company is overspending or underspending on marketing operations to meet its sales goals. For this purpose, namely the analysis of the efficiency of marketing operations within an annual marketing plan, marketing expense-to-sales ratios are used. Usually, total marketing expenses and expenses of each strategic marketing element are evaluated in relation to sales and compared with industry standards or past company ratios.

Suppose a company's total marketing expense-to-sales ratio was at 40% for the year under evaluation. The total ratio is a compound measure of the following individual expense-to-sales ratios: an advertising-to-sales ratio of 18%, a sales promotion-to-sales ratio of 12%, a salesforce-to-sales ratio of 5%, a marketing research-to-sales ratio of 3%, and a sales administration-to-sales ratio of 2%. The monitoring of these ratios throughout the years may exhibit small period-to-period fluctuations. Frequently, the monitoring may reveal that fluctuations exceed the determined tolerance range, and imme-diate corrective action is needed. An easy way to track the period-to-period fluctuations of ratios are control charts, or control tables.

Profitability control

The control devices discussed in the previous section are quite useful in evalu-ating and controlling a company's marketing efforts according to the goals set in an annual marketing plan. Sales analysis or market-share analysis, how-ever, do not provide any information about the profitability of these efforts. Although the overall marketing of a firm may be profitable, it is very unlikely that all of its products, sales territories, advertising activities and so on, are equally profitable. Therefore, the relative profitability of various marketing factors has to be measured through marketing cost analysis. It helps to deter-mine where the firm is making or losing money in terms of the important seg-ments of its marketing operations (Hutt and Speh, 1992). These segments, or units of analysis used for control purposes, may be products, sales territories, customer groups, distribution methods, channel members, order sizes, sales persons, advertising media and so on. Overall, marketing cost analysis allows a firm to determine which segments are cost efficient or cost inefficient, and to make the appropriate adjustments.

Nevertheless, it has to be taken into account that a marketing cost ana-lysis by itself is not a performance analysis. Since marketing cost analysis is a detailed study of the operating expense section of a firm's profit and loss statement, budgetary goals have to be established for each segment to extend this analysis to a performance evaluation tool. Although this approach, to evaluate the variations between budgeted costs and actual expenses by segment seems to be logical, some major problems are appar-ent. In practice, budgeting follows accounting principles and budgets are allocated to specific expense categories (salaries, taxes, supplies or over-heads), and not to specific purposes, as for example advertising, or new product development. Therefore, profitability control has stringent informa-tion requirements. To enable management to compare marketing cost and performance figures with expected results, an accounting system is required that allocates expenses to the relevant functions and marketing segments.

Marketing cost analysis

An effective marketing programme may not inevitably be an efficient one – a marketing programme that generates large sales volumes may also create high costs. Marketing cost analysis should help to avoid inefficiencies by examining the sources of marketing costs, including evaluations on why they are incurred, how big or small they are and how they change over time (Bovée and Thill, 1992). The procedure of marketing cost analysis consists of several steps: evaluating natural account expenses, transforming natural accounts into functional accounts and allocating functional accounts to mar-keting segments.

Table 17.2. Profit and loss statement (US$).

Sales		300,000
Cost of goods sold		195,000
Gross margin		105,000
Operating expenses		
Salaries	46,900	
Rent	11,300	
Travelling	5,800	
Supplies	17,700	
Total operating expenses	81,700	
Net profit		24,300

The different steps in marketing cost analysis are illustrated with the following simplified example. A firm would like to determine the profitability of selling a specific product to three different customer groups. A profit and loss statement of this firm is shown in Table 17.2. The first step in marketing cost analysis is to determine the level of expenses for all natural accounts. They are listed in the operating expense section of the profit and loss statement, and show how the money was spent. The accounts classify costs, according to common accounting principles, by the names of expense categories (therefore the name natural accounts) and include salaries, rents, supplies, etc. Although natural accounts allow a simple cost analysis by comparing actual expenses with budgeted expense goals, or with results from previous operating periods, they do not provide sufficient information on real marketing costs (Stanton *et al.*, 1991). Effective control of marketing costs needs expenses to be allocated to the various marketing functions, such as marketing research or advertising.

In a second step, all expense entries have to be reclassified into functional accounts which provide information on why costs incurred. The critical element in the process of determining the appropriate marketing costs associated with a product or customer segment is to trace all costs to the activities for which the resources are used and then to the products or segments that consume them (Hutt and Speh, 1992). By assuming that the operating expenses listed in Table 17.2 arose from selling, advertising, pack-

Table 17.3. Reclassifying natural accounts into functional accounts (US$).

		Functional accounts			
Natural accounts	Total	Selling	Advertising	Packaging and delivery	Billing and collecting
Salaries	46,900	17,900	11,400	10,900	6,700
Rent	11,300	–	2,000	6,200	3,100
Travelling	5,800	4,200	1,000	–	600
Supplies	17,700	1,650	8,350	6,900	800
Total	81,700	23,750	22,750	24,000	11,200

Table 17.4. Basic data for allocating functional expenses to customers.

Customer groups	Selling (no. of sales calls)	Advertising (no. of advertisements)	Packing and delivery (no. of orders placed)	Billing and collecting (no. of orders placed)
Group A	600	230	190	190
Group B	450	150	130	130
Group C	200	120	80	80
Functional expense (US$)	23,750	22,750	24,000	11,200
No. of units	1,250	500	400	400
Costs per unit (US$)	19.0	45.5	60.0	28.0

aging and delivering a product, and from billing and collecting for it, four functional accounts are used to reclassify the expenses (Table 17.3). These are by no means all functional accounts a firm might create, but the selection is considered to be useful for illustrating the principles of the procedure. Occasionally, a few expenses can be assigned directly to one activity, but more often they can be prorated only after establishing a reasonable basis for allocation. Rental expenses, for instance, may be allocated according to the proportion of the total floorspace that is occupied by each functional department. Thus, packaging and delivery accounts for 55% of the total floorspace in the firm, and US$6200 in rent expense is allocated to this function. But none of the rent expense is assigned to the selling activity, since sales representatives work away from the office. According to this, Table 17.3 shows the allocation of the different expense categories to the four activities.

In the next step of marketing cost analysis, functional expenses are assigned to the relevant marketing segment, as product, distribution method, sales territory, or others. This is an intermediate step which helps to develop each classification as a profit centre. In the example, the task is to measure how much functional expense is associated with marketing the product to each of the three different customer groups. Assuming that in total 1250 sales calls were made, which amount in a total selling expense of US$23,750; the cost per call averages US$19 (Table 17.4). Similarly, the total expense for advertising is allocated according to the number of advertisements, resulting in an average cost per advertisement of US$45.5. Expenses for packing and delivery, as well as for billing and collection, are assigned according to the number of orders. Average costs are at US$60 per order for packing and delivery, and US$28 per order for billing and collecting respectively.

Finally, by combining the purchases and the cost of serving, a profit and loss statement for each customer group can be prepared. Table 17.5 shows such a statement for each group. The sum of the four major components (sales, cost of goods sold, operating expenses and net profit) is the same as on the original profit and loss statement in Table 17.2; all that has been done is to rename and rearrange the data. From Table 17.5 it can be determined that sales were highest to customer group A, but it generated the smallest

Table 17.5. Profit and loss statement for customers (US$).

	Group A	Group B	Group C	Whole company
Sales	150,000	80,000	70,000	300,000
Cost of goods sold	105,000	56,000	49,000	195,000
Gross margin	45,000	24,000	21,000	105,000
Operating expenses				
Selling (US$19/call)	11,400	8550	3800	23,750
Advertising (US$45.5/advertisement)	10,465	6825	5460	22,750
Packaging and delivery (US$60/order)	11,400	7800	4800	24,000
Billing and collecting (US$28/order)	5320	3640	2240	11,200
Total expenses	38,585	26,815	16,300	81,700
Net profit (or loss)	8665	9935	4700	24,300
Profit as per cent of sales	5.87	12.42	6.71	8.10

profit as a per cent of sales. Serving group B brought the highest total profit, and also the highest per cent of sales. However, marketing to group C generated the lowest total profit, but a more reasonable result on a per cent of sales basis than group A.

The task of allocating costs is quite difficult and expensive in time, money and manpower. Therefore, some basic points should be considered when functional expenses are allocated (Evans and Berman, 1992): first, assigning overhead costs, as for marketing administration, to different products, customers or other marketing classifications is usually somewhat arbitrary; second, the elimination of a poor performing classification would lead to overhead costs being apportioned among the remaining products, customer groups, or distribution channels. Actually, this may result in a lower overall profit. Thus, it is necessary to distinguish between those separable expenses which are directly associated with a given classification category and can be eliminated if the category is terminated and those common expenses that are shared by various categories and cannot be eliminated even if one of the categories is discontinued.

Full-cost approach and contribution margin approach

The problem of allocating costs becomes more evident, if cost totals have to be apportioned among individual categories (Stanton *et al.*, 1991). In general, operating costs can be divided into direct (separable) and indirect (common) expenses. Allocating direct expenses is quite easy, because they are directly attributable to a specific marketing function, as for instance costs-per-sales call. Such direct costs are simply allocated to the function that generated them. A real allocation problem arises with indirect costs, the common expenses which are shared by several functions. Some of these costs might be traceable (partially indirect) costs, for which the source can easily be identified and therefore allocated on a logical basis to various functions indirectly.

In contrast, some of the indirect marketing costs might be untraceable (totally indirect). These common costs are somewhat fixed and remain about the same, whether or not a marketing function is discontinued.

Table 17.6. Basic approaches in marketing cost analysis (source: adapted from Stanton *et al.*, 1991).

Full-cost approach	Contribution margin approach
Sales − Cost of goods sold	Sales − Cost of goods sold
= Gross margin	= Gross margin
− Direct expenses − Indirect expenses	− Direct expenses
= Net profit (or loss)	= Contribution margin (to cover overhead expenses plus a profit)

Typical examples for totally indirect costs are administrative expenses, interest payments and different kinds of tax payments. Since a logical basis for allocating these costs can often not be found, they have to be allocated on an arbitrary basis. Commonly used indirect cost allocation methods are to divide these costs equally among marketing categories, or in proportion to the sales volume in each category. Overall, any method used for allocating indirect expenses has some inherent weaknesses which may affect the control effort.

In a marketing cost analysis, two basic approaches in handling this allocating problem are possible: the full-cost approach and the contribution margin approach. In the full-cost approach, all functional costs (direct and indirect) are allocated to products, customers or other categories. Allocating all costs allows us to determine the profitability of any category, but, as outlined above, it requires indirect costs to be split on some basis. In contrast, the contribution margin approach focuses attention on the direct costs, and not all functional expenses are allocated in all situations (McCarthy and Perreault, 1990). The variable costs which presumably would be eliminated if the marketing function is discontinued are relevant. By deducting direct expenses from the gross margin in the profit and loss statement, the remainder equals the amount a specific function contributes to cover total indirect expenses. This approach, which considers only costs that are directly related to specific functions, may be more meaningful in situations when various alternatives are to be compared. Because the two approaches may suggest different decisions, it is important to consider the fundamental differences between full-cost and contribution margin analysis, which are summarized in Table 17.6.

According to Stanton *et al.* (1991) a real controversy exists regarding which of the two approaches is better for managerial control purposes. Advocates of the full-cost approach contend that all costs must be included if an accurate picture of the net profitability is to be drawn. Proponents of the contribution margin approach state that it is not possible accurately to allocate indirect costs among products, customer groups, or other categories. Arguments over allocation methods can be very serious, because any

Table 17.7. Some illustrative measures for strategic component control (source: adapted from Hutt and Speh, 1992).

Strategy element	Type of data required for evaluating performance
Product	Sales by market segment, sales relative to potential, sales growth rates, market share, contribution margin, percentage of total profits
Price	Price policy related to sales volume, discount structure related to sales volume and channels, margin structure related to marketing expense, price changes relative to sales volume
Distribution	Sales, expenses, and contribution margin by channel, expense-to-sales ratio by channel, sales relative to market potential by channel, logistics cost by channel
Communication	Cost per contact, sales per communication activity, advertising effectiveness by type of media, expense-to-sales ratio by type of communication

method used can make some products or customers appear less profitable than if some other allocation method is used. The method used may reflect on the performance of various managers, and may affect their salaries and bonuses (McCarthy and Perreault, 1990). To avoid these problems, firms often use the contribution margin approach which is especially useful for evaluating alternatives and to show operating managers how they are performing. It shows what they have actually contributed to covering general overhead and profit. On the other hand, full-cost analysis is useful for top-management decision-making, because it provides a clear picture if and how fixed costs are to be covered in the long-run.

Strategic component control

Strategic component control is concerned with controlling the performance of individual marketing (-mix) elements of a marketing strategy, such as product, pricing, distribution and communication strategy. Thus, data on the efficiency of resources used for each individual element has to be collected, combined and analysed on a continuous basis to provide the necessary information. Performance measures and standards utilized for controlling individual components of a marketing strategy may vary by company and situation, according to the objectives delineated in the marketing plan (Hutt and Speh, 1992). Table 17.7 provides a representative sample of the types of data required for strategic component control. As many of these measures have already been discussed in the sections on annual plan control and profitability control, respectively, they are not described in detail again. Overall, strategic component control helps management to evaluate the efficiency of each single element of the marketing mix on a continuous basis.

Strategic control

The different types of marketing control analysis discussed so far are quite helpful for planning and controlling marketing operations. However, these control processes tend only to examine the efficiency of some critical elements in marketing. They do not provide the necessary tools to evaluate the effectiveness of present and possible marketing strategies and marketing mixes (McCarthy and Perreault, 1990). Since customer needs and attitudes, as well as the behaviour of competitors can change rapidly, marketing strategies, programmes and plans may become out of date or even obsolete. Therefore, attention of management has to focus on the adjustment of marketing mixes and strategies, too. As marketing managers are usually concerned with short-term problems in planning, implementation and control, and not with evaluations of the total effectiveness of strategies, effectiveness control is a typical task for top management. To fulfil this task, a control tool is needed which allows the evaluation and monitoring of the whole marketing programme on a regular basis, and not only when crises are already evident in several operational areas. Therefore, a new concept of control – the marketing audit – has been developed during the 1950s, and it gained widespread awareness among managers.

According to the literature (Schuchmann, 1959; Oxenfeldt, 1966; McDonald, 1984; Mokwa, 1986; Kotler *et al.*, 1989) a marketing audit is defined as follows. A marketing audit is a comprehensive, systematic, unbiased, independent and periodic review and appraisal of a company's or business unit's marketing environment, objectives, strategies and activities with a view to determining problem areas and opportunities and recommending a plan of action to improve a company's marketing performance.

Therefore, the purpose of a marketing audit is to examine a company's marketing efforts so as to determine how well they are being conducted and how they can be improved. Ideally, it is conducted by an objective and experienced individual or organization relatively independent of the firm to be evaluated. By specifically analysing the market environment and the firm's internal marketing activities the audit is broader in scope than the evaluation tools already discussed. It covers all major marketing dimensions of a business and not just a few trouble spots. Kotler *et al.* (1989) contend that a marketing audit should consist of six sequential diagnostic steps including the area's marketing environment (macro- and micro-environment), marketing strategy, marketing organization, marketing systems, marketing productivity and marketing functions. The diagnostic process is followed by formulating and implementing short-term and long-term corrective action plans to improve the company's overall marketing effectiveness (Kotler *et al.*, 1989).

A well-designed marketing audit examines current conditions in the marketing environment, the effectiveness of the marketing organization, marketing's productivity, and all the functional relationships within the marketing organization (Bovée and Thill, 1992). Since it is very comprehensive and must be conducted systematically, a detailed structure for the audit has to be prepared. This structure provides the plan of operation for the auditor and includes the significant points to be examined, the questions to be asked and the type of information sought. Table 17.8 outlines such an audit structure, but it should be emphasized that not all questions are important in every situation and that sometimes additional questions have to be asked.

Obviously, a complete marketing audit is an extensive and difficult project. On the other hand, the rewards from a marketing audit can be great and may benefit the successful company as well as the company in trouble. By reviewing its strategies periodically and not just when a crisis is already apparent, a firm is likely to keep abreast of its changing marketing environment (Stanton *et al.*, 1991). Successes as well as failures can be analysed, so that the company can capitalize on its strong points or eliminate substantial weaknesses. The audit can spot deficiencies in the coordination of a marketing programme, outdated strategies, or unrealistic objectives and goals. By providing a tool for anticipating future situations and developments, too, the audit serves as an accurate tool for diagnosis and prognosis (Schuchmann, 1959).

Despite the merits of the marketing audit as an efficiency control device many firms still do not use formal marketing audits on a regular basis. According to Evans and Berman (1992) three factors account for this. First, measures of success or failure are difficult to establish. The marketing efforts may produce poor results despite the best planning if environmental factors change. On the other hand, good performance may be achieved by chance. Second, if marketing audits are completed by in-house personnel of a firm they may not be comprehensive enough or they may be biased. Third, the pressures of the day-to-day business often mean that only a small part of a firm's activities are evaluated or that audits are done on a non-regular basis. Fourth, the recommendations of an audit may be a real surprise to management. Since the auditing process is complete only after appropriate responses – such as formulating and implementing short-term and long-term corrective action plans – are taken, sometimes management may be reluctant to turn the auditor's recommendations into action.

In an ideal situation, a marketing audit should not be necessary. Good managers do their best in planning, implementing and control which goes along with a continuous evaluation of the effectiveness of the operation (McCarthy and Perreault, 1990). However, in practice managers often identify themselves with specific strategies and follow them blindly even when other strategies may be more effective. An independent view on the strategic directions followed and the plans implemented may give the needed perspective to avoid this problem and to keep a firm's marketing operations up to date.

CONCLUDING REMARKS

This chapter has examined the role of evaluation and control in the strategic marketing process. Starting out with some basic considerations on marketing control, a five-step feedback model of control is developed. Within this first section of the chapter, particularly the differences between effectiveness and efficiency and their impact on performance measurement are emphasized.

Table 17.8. Components of a marketing audit (source: based on Kotler *et al.*, 1989, and Bovée and Thill, 1992).

Marketing environment audit
The macro-environment
  Demographic
    What major demographic developments offer opportunities or pose threats to the company and what actions has the company taken to respond to these developments?
  Economic
    What major developments in income, prices, savings, and credit will affect the company and what actions has the company taken to respond to these trends?
  Natural
    What is the outlook for the cost and availability of natural resources and energy needed by the company?
    What environmental concerns have been expressed about the company and what steps has the company taken?
  Technological
    What major changes occur in product and process technology and what is the company's position in these technologies?
  Political
    What legal developments could have a serious impact on marketing strategies and tactics of the company?
  Cultural
    What is the public's attitude towards business and towards the company, its products and representatives?
    What changes in culture, values and lifestyle have an impact on the company?
The task environment
  Markets
    What is happening to market size, geographical distribution and profits?
    What are the major market segments and do they require different strategies?
  Customers
    Who are the current and potential customers and what is their frequency and quantity of purchases?
    How do customers make their buying decisions and how do they rate the company on reputation, product quality, service, salesforce and price?
  Competitors
    Who are the major competitors?
    What are their objectives and strategies, their strengths and weaknesses, as well as their sizes and market shares?
    How are competitors positioned, are there dominating firms?
    Suppliers, distributors and other supporting organizations
    What is the outlook for the availability of key resources used in production?
    What trends occur among suppliers in their pattern of selling?
    What are the main distribution channels and what are the upcoming trends?
    What are the cost and availability outlooks for financial resources, transportation services and warehousing facilities?

Marketing strategy audit
Business mission, marketing objectives and goals
  Is the business mission clearly stated in market-oriented terms and is it feasible?
  Are the corporate objectives clearly stated, and do they lead logically to the marketing objectives?

Table 17.8. *Continued*

Are the marketing objectives stated in a clear form to guide marketing planning and subsequent performance measurement?

Are the marketing objectives appropriate, given the company's competitive position resources and opportunities?

Strategy

What is the core marketing strategy for achieving the objectives and is it sound?

Are the resources budgeted to accomplish the marketing objectives sufficient?

Are the marketing resources allocated optimally to prime market segments, territories and products, as well as to the major elements of the marketing mix, i.e. product quality, service, salesforce, advertising, sales promotion and distribution?

Marketing organization audit

Formal structure

Are the marketing activities optimally structured along function, product, end-user and territorial lines?

Does the marketing manager have adequate authority and responsibility over company activities that affect the customer's satisfaction?

Functional efficiency

Are there good communication and working relations between marketing and sales?

Is the product management system working effectively?

Are the product managers able to plan profits or only sales volume?

Are there any groups in marketing that need more training, motivation, supervision, or evaluation?

Interface efficiency

Are there any problems between marketing and manufacturing, research and development, purchasing or financial management that need attention?

Marketing systems audit

Marketing information system

Is the marketing intelligence system producing accurate, sufficient and timely information about developments in the market-place?

Is marketing research being adequately used by company decision-makers?

Marketing planning system

Is the marketing planning system well conceived and effective?

Is sales forecasting and market potential measurement soundly carried out?

Are sales quotas set on a proper basis?

Marketing control system

Are the control procedures (monthly, quarterly, etc.) adequate to ensure that the annual plan objectives are being achieved?

Is provision made to periodically analyse the profitability of different products, markets, territories, and channels of distribution?

Are marketing costs being periodically examined?

New product development system

Is the company well organized to generate and screen new product ideas?

Does the company undertake adequate concept research and business analysis before investing in a new idea?

Does the company carry out adequate product and market testing before launching a new product?

Table 17.8. *Continued*

Marketing productivity audit
Profitability analysis
  What is the profitability of the company's different products, served markets, territories and channels of distribution?
  Should the company enter, expand, contract, or withdraw from any business segments, and what would be the short- and long-run profit consequences?
Cost-effectiveness analysis
  Do any marketing activities seem to have excessive costs?
  Are those costs valid and can cost-reducing steps be taken?

Marketing function audits
Products
  What are the product line objectives? Are these objectives sound and is the current product line meeting these objectives?
  Are there particular products that should be phased out and are there new products that are worth adding?
  Are any products able to benefit from quality, feature, or style improvements?
  Are there different quality levels for different markets?
Price
  What are the pricing objectives, policies, strategies and procedures? To what extent are prices set on sound cost, demand and competitive criteria?
  Do the customers see the company's prices as being in line with the perceived value of its offer?
  How does pricing compare with competition on similar levels of quality?
  Does the company use price promotion effectively?
Distribution
  What are the distribution's objectives and strategies?
  Is there adequate market coverage and service?
  Should the company consider changing its degree of reliance on distributors, sales reps and direct selling?
  Do product characteristics require specific channels of distribution?
Salesforce
  What are the organization's salesforce objectives?
  Is the salesforce large enough to accomplish the company's objectives?
  Is the salesforce organized along the proper principle of specialization (territory, market, product)?
  Does the salesforce show high morals, ability and effort? Are they sufficiently trained and motivated?
  Are the procedures adequate for setting quotas and evaluating performance?
  How is the company's salesforce perceived in relation to competitors' salesforces?
Advertising, promotion and publicity
  What are the organization's advertising and promotion objectives and are they sound?
  Is the right amount being spent on advertising? How is the budget determined?
  Are the advertising themes and copy effective? What do customers think about the advertising?
  Are the advertising media well chosen and is sales promotion used effectively?
  Is advertising integrated with promotion and sales activity?
  Do you have a clear idea of the type of company you want people to think you to be and do you have a consistent communications programme?

Taking into account that marketing performance can be examined at different levels, four different types of control are discussed in the main section of the chapter:

1. Annual plan control, which consists of monitoring the actual performance against the objectives set in the annual marketing plan. Performance standards used in annual plan control are sales volume, profit and market share. Specific control tools applied in practice include sales analysis, market-share analysis and marketing expense-to-sales analysis.

2. Profitability control, which helps to determine where a firm is making or losing money in terms of products, territories, distribution channels and so forth. Cost analysis provides the appropriate tool to evaluate whether the different marketing efforts are cost efficient or not. A major problem within marketing cost analysis is the differences between the information provided from accounting systems and the information needed for marketing control purposes. Therefore, the transformation process from natural accounts to functional accounts is described in some detail, as well as the differences between full-cost analysis and contribution margin analysis.

3. Strategic component control, which is concerned with evaluating the performance of individual elements of the marketing mix. The performance standards utilized in strategic component control vary in accordance with the specific situation of a firm, but in common they are combined measures which are also used in annual plan and profitability control.

4. Strategic control, which provides the tools necessary to evaluate the effectiveness of present and possible marketing strategies. Strategic control should help to ensure that marketing objectives, strategies and systems are optimally adapted to the current and future marketing environment. The concept of effectiveness control used in practice is the marketing audit, a comprehensive, systematic, unbiased, independent and periodic review and appraisal of a company's marketing environment, objectives, strategies and activities. A well-designed audit examines current conditions in the marketing environment, the effectiveness of the marketing organization, marketing's productivity and all the functional relationships within the marketing organization and recommends short-term and long-term actions to improve a company's overall marketing performance.

NOTES

1. This section is based on Kotler *et al.*, 1989, pp. 655–665 and Hutt and Speh, 1992, pp. 528–541.

FURTHER READING

On marketing control

Bonoma, T.V. (1984) Making your marketing strategy work. *Harvard Business Review* 62, 69–76.

Hulbert, J.M. and Toy, N.E. (1977) A strategic framework for marketing control. *Journal of Marketing* 41, 12–19.

Jaworski, B.J. (1988) Toward a theory of marketing control: Environmental context, control types, and consequences. *Journal of Marketing* 52, 23–39.

Kotler, P. (1991) From sales obsession to marketing effectiveness. In: Dolan, R.J. (ed.), *Strategic Marketing Management*. Harvard Business School Publications, Boston, MA, pp. 470–483.

Merchant, K.A. (1988) Progressing towards a theory of marketing control: A comment. *Journal of Marketing* 52, 40–44.

On marketing audits

Grashof, J.F. (1984) Conducting and using a marketing audit. In: McCarthy, E.J., Grashof, J.F. and Brogowicz, A.A. (eds), *Readings in Basic Marketing*. Irwin, Homewood, IL, pp. 318–329.

Kotler, P., Gregor, W. and Rogers, W. (1977) The marketing audit comes of age. *Sloan Management Review* 20, 25–43.

Quelch, J.A., Farris, P.W. and Olver, J. (1987) The product management audit: Design and survey findings. *Journal of Consumer Marketing* 4, 45–58.

On cost and profitability analysis

Coopera, R. and Kaplan, R.S. (1988) Measure costs right: Make the right decisions. *Harvard Business Review* 66, 96–103.

Lambert, D.M. and Sterling, J.U. (1987) What types of profitability reports do marketing managers receive? *Industrial Marketing Management* 16, 295–304.

Sandretto, M.J. (1985) What kind of cost system do you need? *Harvard Business Review* 63, 110–118.

REFERENCES

Assael, H. (1990) *Marketing Principles and Strategy*. The Dryden Press, Fort Worth, TX.

Bonoma, T.V. and Clark, B.H. (1988) *Marketing Performance Assessment*. Harvard Business School Press, Boston, MA.

Bovée, C.L. and Thill, J.V. (1992) *Marketing*. McGraw-Hill, New York.

Drucker, P. (1974) *Management: Tasks, Responsibilities, Practices*. Harper and Row, New York.

Evans, J.R. and Berman, B. (1992) *Marketing*, 5th edn. Maxwell Macmillan International, New York.

Hutt, M.D. and Speh, T.W. (1992) *Business Marketing Management – A Strategic View of Industrial and Organizational Markets*, 4th edn. The Dryden Press, Fort Worth.

Jaworski, B.J. (1988) Toward a theory of marketing control: Environmental context, control types, and consequences. *Journal of Marketing* 52, 23–39.

Kerin, R.A. and Peterson, R.A. (1993) *Strategic Marketing Problems: Cases and Comments*, 6th edn. Allyn and Bacon, Boston.

Kotler, Ph., Chandler, P., Gibbs, R. and McColl, R. (1989) *Marketing in Australia*, 2nd edn. Prentice-Hall, Sydney.

McCarthy, E.J. and Perreault, W.D. (1990) *Basic Marketing – A Managerial Approach*, 10th edn. Irwin, Homewood, IL.

McDonald, M.H.B. (1984) *Marketing Plans – How to Prepare Them, How to Use Them*. Heinemann, London.

Mokwa, M.P. (1986) The strategic marketing audit: An adoption/utilization perspective. *Journal of Business Strategy* 6, 88–95.

Oxenfeldt, A.R. (1966) *Executive Action in Marketing*. Wadsworth Publishing, Belmont.

Schuchmann, A. (1959) The marketing audit: Its nature, purpose, and problems. In: *Analyzing and Improving Marketing Performance: Marketing Audits in Theory and Practice*. Report No. 32. American Marketing Association, New York, pp. 11–19.

Stanton, W.J., Miller, K.E. and Layton, R.A. (1991) *Fundamentals of Marketing*, 2nd Australian edn. McGraw-Hill, Sydney.

Wheelan, T.L. and Hunger, J.D. (1991) *Strategic Management and Business Policy*, 4th edn. Addison-Wesley, Reading, MA.

18 International Marketing in the Midst of Competition and Partnership

Luis Miguel Albisu

Department of Agricultural Economics, Agricultural Research Service – DGA, PO Box 727, 50080 Zaragoza, Spain

INTRODUCTION

For several decades international marketing has commonly been a minor extension of national marketing management techniques applied to international markets. Up to the 1970s, its application usually took into consideration only the elements appraised in a particular country, at a firm's headquarters, plus some additions felt to be necessary to reach different countries, where the institutions and the business environment could differ from previous experiences, and with consumers of distinctive socio-economic and cultural characteristics. Concerns concentrated mainly on understanding tariffs, foreign currencies and international carriers. It was also commonly run by dispersed management structures of firms operating in different countries and no great efforts were undertaken to differentiate national from international strategies.

International marketing has attracted more attention following the international trade increase of the 1980s and 1990s. Not only agricultural commodities, but especially trade of food products has expanded at a great rate. Competition has gained strength and international marketing has become an essential activity among big firms but also a means for the expansion of small- and medium-size food firms.

Multinational firms have played a key role on international markets and they are the main organizations that have to deal with international marketing objectives, to define their corresponding corporate strategies and to implement their marketing policies. In the increasingly competitive environment of the last two decades, they have faced a serious challenge to their strategic and organizational roles. They have modified their strategies following new market requirements and their marketing approaches deserve special attention.

Nowadays, international markets reach global patterns which induce intricate marketing strategies with all sorts of alliances among firms.

Competition and partnership both intermingle and live together accomplishing previously unsuspected compromises. Continuing global proliferation of technology and managerial know-how, the reorganization of international economic boundaries and the ongoing emergence of new players in world markets promise an even more turbulent and complex competitive environment that will influence future international marketing activities.

This chapter deals with the actual approaches and elements involved in international marketing. It tries also to bring up the ideas and concepts that have been expressed in previous chapters and to relate them to international marketing. It goes over the main factors that shape international marketing and it provides insights into the reasons behind the most crucial changes. It is divided into three main sections: (i) the main issues concerning international trade in the agro-food products environment; (ii) the multinationals' strategies; and (iii) some of the marketing levers exposed. Organizational and strategic corporate aspects have been expanded to the utmost because they are considered to have paramount impacts on international markets and international marketing.

INTERNATIONAL DIMENSIONS OF AGRO-FOOD MARKETING

Agro-food trade and international marketing

The distinction which has been made throughout the book between agricultural commodities and food products fully applies to international marketing. The differences between those two concepts were explained in Chapters 1 and 2. Attention to agricultural commodity trade was concentrated in Chapter 5 and it was mentioned that agricultural commodity exports expanded at a great rate during the 1970s. The 1980s had two periods. The first half displayed no growth, but since the middle of that decade trade has followed a rising trend. Liberalization following the GATT agreements should encourage trade expansion in the future.

Trade statistics often include quantities, but not values. Although quantities are significant signals to explain trade performance they are often inadequate on their own. Most developed countries confront saturated food markets and quantities sold do not experience great changes because of the stabilization or declining of demographic figures and the already covered physiological needs of their population. However, the development of new high quality products and the addition of new services to old products can increase value added to final market prices. Statistics on the value of what has been traded are not so reliable as they are for quantities. Nevertheless, it has been estimated that total value of world trade in agricultural commodities and food products have grown more than five times in the last 20 years. The growth rate for food products trade has been greater than for agricultural commodities and food products trade now represents over 60% of total trade.

The traditional idea of trade as occurring between nation states is now misleading because national boundaries are blurred by different international agreements and the integration of nations in economic blocks. Such is the case of the European Union (EU) and NAFTA. Nations are becoming regions in this context, with a harmonization of their economic laws and a great dependence for several countries on the economic decisions that are made by the leading countries in each block.

Trade of food products is concentrated in a relatively small number of advanced countries (Henderson and Handy, 1994). Japan, Germany, the USA, France and the UK account for more than 50% of total imports and 20 countries gather around 90% of the total. Only five countries, France, The Netherlands, the USA, Germany and the UK, export around 40% of total world food product exports and 20 countries reach around 75% of that total figure. It is important to note that four developed countries (France, Germany, the UK and the USA) are among the top five, as both importers *and* exporters. Europe and the USA are the two dominant trading blocks.

The greater significance of food products trade has important influences on the kind of marketing activities to be pursued. Food products encompass all the marketing steps of agricultural commodities; but reaching, in most cases, final consumers across countries, other elements linked to advanced marketing techniques are also involved. They require more complex organizations and special knowledge about international distribution and consumer behaviour which sometimes are only attainable by large corporations.

Most of the international trade in food products is intra-industry, which means the simultaneous importing and exporting of similar foods, in comparison to inter-industry, characterized by trade involving exchange of different food products. The intra-industry trade is a consequence of consumers' tendency towards more homogeneous habits and tastes all over the world but, at the same time, it is the expression of a greater desire for differentiated food products. The intra-industry trade characterizes flows among developed countries and it provides greater scope for international marketing, since cost of production of agricultural commodities becomes then a less important share of the final consumer price. A group of competitive marketing activities separate successful from failing food products in international markets. Developing countries should be aware of those difficulties when they generate their trade expectations as a result of new trade liberalizations.

The determinant factors that influence trade among nations are of continuing analysis and controversy. Most of the actual conceptual frameworks for trade try to single out the most important components to become competitive. Porter's (1990) contribution in this matter has been ubiquitous and it is an obligatory reference for many scholars. Abbott and Bredahl (1994) have proposed the following categorization of factors determining competitiveness: (i) factor endowments and natural resources; (ii) technology; (iii) investments; (iv) human capital; (v) managerial expertise; (vi) product characteristics; (vii) firm strategy and industry structure; (viii) input supply; (ix) marketing and distribution channels; (x) infrastructure and externalities; (xi) regulatory environment; and (xii) trade policy.

Trade is associated with agricultural commodities and food products, but other trade elements are increasing their significance, such as ingredients trade for processed food products, engineering and plant construction services for food industries, food processing equipment and packaging systems, and any kind of services usually involved in high technological processes for food value-added products. They all pertain to three different sorts of exchange: (i) intellectual property transactions; (ii) technological physical structures; and (iii) capital flows.

A diverse set of foreign operations make up the food international commerce and direct physical trade represents only a small part of total sales value. Food international commerce is accomplished through different operations, such as, licensing, franchising, joint ventures, direct foreign investments (DFI), mergers and minor share-holding acquisition of firms. Recent trends show that international commerce has been promoted more through those terms of trade. However, trade is considered less risky than other commercial operations and it can be a first step to reach some countries or to sell in small markets. Later on in the chapter these concepts are expanded in relation to food company strategies.

Multinationals and the agro-food sector

Multinationals, also described as international or transnational companies, could be defined simply as enterprises which operate in at least two countries. The terms multinational and transnational are used interchangeably, in most cases, as meaning the same as international, although some authors imply subtle differences about how they operate and are organized (Dunning, 1993).

It was in the second part of the nineteenth century that multinationals found the appropriate circumstances to develop their operations. The history and evolution of agro-food multinationals follow closely the same events of multinationals linked to other sectors of the economy. There is a great deal of literature about multinationals in general, but not much research has been undertaken in the agro-food sector. Lack of secondary data, and thus the need to carry out expensive surveys, are impediments to acquiring substantial knowledge of their changing aims and internal organizational structures.

Food multinationals have not always been welcomed by the governments of developed or developing countries. National governments have often been reluctant to provide facilities for their installation in order to protect national firms and to keep control over key agro-food products. Nevertheless, multinationals have increasingly benefited from an improving institutional environment in many countries. They have been considered as important forces to boost their economy by incorporating high technology processes, the required capital and the appropriate methods to create skilled employment able to produce and market high quality products. Because of their recognized brand names and their foreign marketing contacts, they have been more successful than domestic firms and their very success has often created market entry barriers to national companies.

There is considerable evidence of the role which the multinational companies have played in providing information, advice and credit to their suppliers. They also incorporate new business environments and standards which are internationally accepted and they might have a spillover effect on other firms in the countries in which they operate. The new atmosphere of more liberal economic policies, the national economic integration into large economic blocks and the overriding evidence of their power have been important determinants for their continuous expansion.

Agro-food multinationals tend to rely more on national management for their foreign subsidiaries, and their headquarters blend professionals from different countries quite apart from the country where it is located. It proves their multicultural engagement to understand better their operations all over the world. International marketing used to be the job of a small number of executives, at headquarters firms, dedicated to understand foreign markets; but now it is usually the interaction between a greater number of professionals located in different countries trying to find out common marketing policies or at least an understanding of their differences. Probably an extreme example is Nestlé, which is run by a German; its 11 top executives are from seven different countries and among the 1300 people who work at Switzerland headquarters there are 60 different nationalities.

The more open the economies of the world become, the more national food firms turn into multinationals. There is a growing number of agro-food multinationals, many of them very small, but at the same time the largest ones are acquiring a larger share of the total agro-food market. The so-called Triad, integrated by Western Europe, the USA and Japan, are the geographical areas where the headquarters of the biggest agro-food multinationals are based. The UK stands out among European countries as having the best performing multinational food companies. However, some of the most well-known multinationals have their headquarters in one country but their capital has different countries of origin and their shareholders are more internationalized.

It is estimated that the top 100 agro-food multinationals account for around 30% of the total world agro-food sales (Rastoin and Oncuoglu, 1992). Their rate of growth is greater, at about 3–4%, than other agro-food corporations, which means that by the year 2000 they will account for around 50% of total sales. Among the top 100 multinationals there were 50 from the USA in 1978 and only 29 in 1991. In 1991 there were six American food multinationals among the top ten: Philip Morris, Cargill, Pepsico, Coca-Cola, Conagra and RJR Nabisco. The decline of the USA has been compensated by the growth of European firms which accounted for, in 1991, 43 firms among the top 100 and 16 from Japan, which in spite of its lack of raw agricultural materials is increasing its agro-food multinationals. The USA has some of the largest multinationals, but the most profitable ones are European, especially those from the UK. Most of the leading firms are packaging food manufacturers but ingredient firms and traders are also consolidating and becoming major multinational corporations.

In 1991, out of the 21 most important European multinational agro-food groups, the majority (13) had their headquarters in the UK, four in France, two in The Netherlands, one in Switzerland and one in Italy. But if we concentrate on the top ten, then six were from the UK. Unilever (The Netherlands) and Nestlé (Switzerland) are at the top with similar sales figures and the rest follow them at a distance. The ranking position, if value added was considered, was quite similar although there were some changes (Vieille, 1993). France, the first agricultural and agro-food producing country in Europe does not have the leadership when taking into consideration agro-food multinationals; it is the UK which stands out. The annual rate of growth for total sales, between 1987 and 1991, for the top 21 has been close to 10%. Overall finance performance is better for beverages than other food items and this is probably the reason that some top multinationals have directed their investments in that food group.

Economic and socio-cultural environments

It is important for food international marketing practitioners to understand how markets are reacting, the degree of homogeneity which exists among consumers in different countries and the motivations which support their decisions. It has direct implications for their marketing strategies and performance. An understanding of how cultural differences affect international marketing decisions is important to a firm's external operations.

There is a tendency for consumers in developed countries to follow similar food patterns as a consequence of common lifestyles and constraints. The labour market with a greater integration and participation of women, consumer concern about health and food security, the increase of food consumption eaten away from the home and the lack of time to prepare food at home are some of the common features which affect many households.

Diets tend to be more homogeneous all over the world. Blandford (1984) has verified that, for the period 1956–1978, the per capita apparent food consumption trends were converging among the more advanced OECD countries. Thus, despite the recognition by medical experts that the Mediterranean diet generates a balanced and healthy combination of food products, the Mediterranean countries have been moving away from those beneficial habits during the last decades and approaching typical consumption patterns of central and northern European countries (Gracia and Albisu, 1994). Total current calory intake among European countries is more similar to that applied at the beginning of the 1960s. The proportion of calories coming from animal and vegetal products shows that diets, in the EU countries, are converging because of the increase, in the majority of them, of the proportion of animal calories, with higher rates in those countries where those proportions were lower.

Connor (1994) has analysed food consumption trends convergence between the USA and Western Europe based on household income, relative prices, demographic changes, concerns by consumers about the nutritional impacts and preventative health possibilities of dietary habits. He believes that the operations of multinational agro-food companies and the international trade between the USA and Western Europe are emerging as major instruments fostering the convergence in food expenditure patterns between those areas. Globalization of food systems seems to be replacing other instruments of change.

The USA, Western Europe, and Japan, which constitute the major world markets, appear to be becoming fairly homogeneous based on macroeconomic indicators. But scholars observe that as people around the world become better educated and more affluent, their tastes actually diverge. Economic similarities favour market standardization but, rather than looking at the target market in terms of rich and poor nations, it may be possible to identify segments, in both developed and developing countries.

There are strong economic interests in Europe to know if large amounts of people could be considered as 'Euroconsumers' described by common consumption habits. In spite of a continuous convergence of diets among European countries there are still discernible national characteristics which are the result of national culinary attachments developed through history. It seems appropriate to state that there are economic factors which lead towards the convergence of food diets. However, there are other factors, such as ethnical and socio-cultural-differentiated characteristics, which still prevent a total adaptation and provide ample scope for differentiated food products.

International marketing has been favoured through policies among countries to increase trade by diminishing tariffs and abolishing or diminishing other non-tariff barriers. During the last two decades it has been common also for governments to facilitate the investment of foreign capital and to diminish bureaucratic measures which made it difficult for businesses to become established in many countries. Governments have searched for agro-food companies that were able to transform agricultural commodities into finished food products and be capable of international marketing activities with good connections in the distribution system. However, as Huber (1984) has synthesized, the features of future economic and socio-cultural environments will be dominated by three characteristics: (i) more and increasing diversity; (ii) more and increasing knowledge; and (iii) more and increasing turbulence.

Standardization versus adaptation in international marketing

The existence of global markets was pointed out by Levitt (1983) and Ohmae (1985), among other authors. It implies that consumers throughout the world have similar preferences for high technological products with quality and trust as two items high in their ranking priorities.

A complete standardization of international marketing refers to using a common product, price, distribution and promotion programme on a worldwide basis. Huge markets like the EU, NAFTA and other more or less integrated economic blocks, encourage agro-food processors to follow the idea of producing and marketing standardized food products. Lowering international tariffs across nations, not integrated in regional blocks, is another incentive to reach a greater number of countries with the same product concept. Recent advances in communication have led to a greater exchange of information across national markets and the similarities more than the differences should characterize world markets. This permits attaining high quality products at lower prices.

It has been argued that adopting a standardized approach, instead of a customized approach, to serving international markets is desirable because sales can be increased by developing a consistent image of the product across national markets. Economies of scale are achieved not only on production but also in distribution and in marketing. Costs can be lowered by pooling production activities across countries, moving production to low-cost locations without redefining the production process and capturing the economies associated with formulating and implementing a single marketing programme. Jain (1989) has distinguished five factors to define standardization of one or more parts of a marketing programme: (i) target market; (ii) market position; (iii) nature of the product; (iv) environment; and (v) organization factors. The three market conditions that influence the standardization decisions are cultural differences, economic differences and differences in customer perceptions in foreign markets. But standardization, at best, is difficult, although there are well-known examples, like Coca-Cola.

With a few exceptions, most of the literature on standardization addresses marketing programmes referring to various aspects of the marketing mix. A standardization approach does not mean that it will not be desirable to have a degree of differentiation for individual marketing mix elements across national markets, such as advertisements and promotions, use of different channels of distribution and other marketing activities. The controversial debate about standardization of marketing strategies concerns whether it is possible to combine common product concepts with the ability to customize *all* marketing variables (including the product) to meet the different international market requirements. The opposite view offers the idea that international marketing should rigorously accommodate specific consumer needs across countries. Firms have to cope with fragmented worldwide markets, and the numerous segments differ more significantly in preferences and consumption habits because of the enormous ethnical and cultural diversity. Further, segments will change and restructure continuously, spurred by the intensity of innovation, competition and information technology.

International segmentation approaches either classify countries on a single dimension or on multiple socio-economic, political and cultural criteria. An alternative approach is cross-country segmentation which derives groups of customers who are alike and encourages firms to look for niche markets in different countries. Each country might contain several clusters that cross the borders. International marketing across international segments benefits from the addition of common accumulated experiences in most developed countries. When penetrating new markets, a firm can borrow from experiences collected in similar segments pertaining to countries that were entered earlier.

In recent years the debate on the pros and cons of pursuing a strategy of total standardization across national markets versus complete adaptation to individual markets has given way to a more fruitful dialogue focusing on: (i) the desired degree of standardization (or adaptation) with respect to various competitive strategy variables such as branding, advertising, sales promotion and pricing; and (ii) the moderating effects of organizational and environmental contingencies on the desired degree of standardization (or adaptation) with respect to these variables (Szymanski *et al.*, 1993).

GLOBAL CORPORATE STRATEGY

Global versus multidomestic strategies

Hallen and Johanson (1989) have pointed out that international markets can be considered as networks of connected exchange relationships between interdependent companies that may extend beyond a specific industry or may include only parts of a specific industry. Four types of networks can be visualized: those of companies, specific technologies, products and nations. Relationships in networks have elements of competition as well as of cooperation.

A series of broad forces have led to growing international competition and widespread globalization of industry scope. These trends are: (i) growing similarity of countries in terms of available infrastructure, distribution channels and marketing approaches; (ii) fluid global capital markets, national capital markets are growing into global capital markets because of the large flow of funds between countries; (iii) technological restructuring, the reshaping of competition globally as a result of technological revolutions; (iv) the integrating role of technology, reduced cost and increased impact of products have made them accessible to more global consumers; and (v) new global competitors, a shift in competitors from traditional country competitors to emerging global competitors.

Yip (1992) has argued that the development of a global corporate strategy has three components. The first component entails developing the core strategy, which constitutes the basis for a company's strategic advantage. The second component involves internationalizing the core strategy by adapting it to the various international markets, resulting in many cases in wide strategies from country to country. The third component, not usually found among multinationals, is a globalization strategy which integrates and manages for worldwide leverage and competitive advantage. Thus, a global corporation strategy is a more advanced criteria that considers the amalgamation of production and organizational structures towards the achievement of an integrated and coordinated network of all their activities and performance.

Global diversity involves much more than just operating globally from a domestic base. Moreover, headquarter's base could become a difficult-to-handle monster with managers too far away from customers and local needs. Nowadays some multinationals are downsizing their headquarter's offices and thinking globally, but operating more locally with more freedom for their subsidiaries that are able to take responsibilities for global products. This is, for example, the case for Nestlé, which has moved the headquarters of its pasta business to Italy and similar decisions have been taken by Unilever. They have created or strengthened management units, with a high degree of specialization, that control whole areas and not only countries. There is a tendency for multinationals to rely more on their local managers and executives but also on their low-level staff as creative sources of ideas that later on can be applied worldwide through their subsidiaries.

The term multidomestic is used to refer to multinationals which operate with affiliates in different countries which take decisions quite independent from the headquarter's company and try basically to fulfil local market needs. Multidomestic strategies are the consequence of national markets which have been, in some cases, quite closed to the rest of the world and this reinforces the notion of producing and marketing for confined areas. In other cases, multinationals have acquired national enterprises which have been doing business only in the home country and, as a first step, they have let them continue with their usual business waiting for further future integration.

Multinational companies operating in foreign countries, under a multidomestic framework, traditionally have viewed international markets as either extensions of their home markets or markets with different needs than the home market. In the first situation, the multinational firms make a minimum of modifications to their business strategies when addressing the needs of international markets. In the second, the firms resort to adapt their core business strategies to the various international environments. However, under the global strategy approach a complex effort is aimed at managing multinational business on an integrated, worldwide basis and not as a loosely knit federation of subsidiaries.

To become a globally diverse company, the organization of tomorrow will evolve in a global partnership of skills and resources. Also the successful organization of the future will be organized to optimize its information-processing capacity. The global information-based organization leverages time efficiency and its reaction capabilities to market opportunity. This will mean flexibility to react to market signals with a customized and immediate response, but also marketing strategy will increasingly involve anticipating the competitor's strategic moves rather than solely analysing the market itself.

Big multinationals, once thought to be so powerful as to be infallible, have sent out a different message during the last decade. Some great failures have occurred and they have fallen into huge capital losses. This has precipitated all sorts of business combinations, from downsizing to merging, from getting into more specialized products to joining financial firms but, in any case, their strategy has been fully revised and implemented following new rules. Globalization and deregulation are two phenomena that have been running together and they have given opportunities to middle- and small-size firms, as much as to big ones, to enter new markets which they were unable to enter previously.

A food firm has a long path to reach a global strategy. The first step is to be transformed from a national to a multinational coverage. But this important decision can be accomplished through different procedures. The idea of a multinational having subsidiaries in different countries is commonly accepted but it is not so well known how this situation is achieved. In most cases direct foreign investment is a requisite but other formulae are used, especially licensing. Finally, becoming a big multinational is not sufficient to reach global markets and to have dominant positions in many activities. So, multinationals feel the need to get involved in alliances with other firms which can complement their activities and/or physical coverage. The significance of direct foreign investment, alliances and licensing is expanded in the next sections.

Direct foreign investment

Direct foreign investment (DFI) is usually considered to be all foreign activities which involve capital investment, but technological transfer, management of executives in foreign affiliates and many other sorts of resource transferred towards foreign countries may also be involved. It is not necessarily the case that firms applying direct foreign investments obtain their finance from their own resources or country of origin, because host countries are often willing to provide capital in order to attract high technological processes. There are many theories about the reasons behind direct foreign investments which are closely linked to multinationals' behaviour. Since 1983 it has been calculated, by the International Monetary Fund, that all DFI has grown five times faster than world trade and ten times faster than world output[1].

For most multinationals the goal is to establish production facilities in foreign markets through direct investment. Reasons cited include satisfying local wholesale and retail customers, understanding local market and regulatory conditions, better access to local transportation and distribution systems, reducing transportation costs, and overcoming trade barriers (Henderson and Handy, 1994). According to the same authors, data on DFI generate a number of observations. First, most DFI is horizontal rather than vertical. Second, most DFI outlays are to acquire existing facilities rather than for new construction. Third, most sales by foreign affiliates are made in the same markets as those served by other firms in the host country. Also data provide no support for the suggestion that exports and foreign production are substitute strategies; in general, as one rises, both rise. However, this might be the case only when the relationship is between developed countries and not when between a developed and a less-developed country.

As with trade in food products, both the sources and destinations of DFI are concentrated in developed countries. However, there is a growing interest in developing country DFI, in emerging fast growing economies, especially in Asia but also in Eastern Europe. There is also a tendency for foreign investment to be directed towards normal areas of geographical influence, if possible without language barriers, and where previous business contacts already existed. For example, there has always been a great flow of capital between the USA and the UK, or between some developed countries and their former colonies.

Mergers are the major vehicle for facilitating foreign direct investment. The preferred strategy of most multinationals is to establish fully controlled operations in major foreign markets, either by acquisition or investment in new facilities. However, by acquisitions some firms usually only want to get the market share and brands, marketing organization and commercial networks. So in the process of the merger many plants are closed down. Firms appear to favour the acquisitions and mergers and joint equity routes in pursuance of their strategic goals. Notable examples include Nestlé and Unilever, both of whom, since the mid-1980s, have embarked on a systematic and strategically related series of cross-border acquisitions and mergers. Nestlé in the last ten years has been growing two-thirds based on acquisitions and the other third because of internal growth. However, it intends to reverse this trend in the next ten years. So one-third will be as a consequence of acquisitions and this could amount to US$1 billion a year.

Handy and MacDonald (1989) have reported some findings about USA food manufacturers emphasizing the role of foreign investment and its relation to food trade in the following terms: (i) major USA food manufacturers do relatively little exporting in comparison to other OECD countries, probably as a consequence of the size and relative isolation of the USA market, quite opposite to the European nations and markets; (ii) USA food manufacturers nevertheless have extensive overseas interests, through direct investments; (iii) the geographic pattern of food industry foreign direct investment has changed sharply in the last 20 years and Europe is the focus of that investment; and (iv) trade with affiliates accounts for only a small share of food manufacturing exports and imports.

Henderson *et al.* (1993) have investigated determinants of international commercial behaviour for a multinational sample of about 600 food manufacturing firms. Major findings were: (i) firms tend toward foreign investment and away from product trade as a firm's dominance (e.g. market share) in its home market increases; and (ii) as the range of diversity of food products produced by a firm increases, the greater its involvement in both exports and DFI.

The largest USA food manufacturers tend to rely on direct foreign investment to gain sales in international markets (Handy and Henderson, 1991). For example, in 1989, Coca-Cola and CPC International received more than 50% of their sales from foreign affiliates. At the same time, foreign firms are gaining ground in USA markets by purchasing USA firms and building new processing plants. Nevertheless, sales from USA-owned food processing affiliates abroad are much larger than sales from foreign-owned affiliates in the USA (Henderson and Handy, 1993). Sales from foreign-owned firms in the USA, however, are growing at a faster rate. From

1982 to 1990, sales from USA-owned affiliates abroad grew from US$39 billion to US$75 billion. The average annual growth rate for this eight-year period was 8.5%. During this same period, sales from foreign-owned affiliates in the USA grew from US$15 billion to US$45 billion – an average annual growth rate of 14.6%.

Prior to the 1980s, American investment in other countries was so large that it aroused widespread fear and resentment of an American economic invasion in some countries. Inside the USA, critics argued that American firms were shifting manufacturing bases, substituting USA exports and exporting jobs. In the late 1980s, foreign investment in the USA became a contentious issue as Americans became concerned with a foreign takeover of USA assets. Nearly 62% of USA total assets of affiliates were located in Europe in 1991 and the UK accounted for 17%. In the same year more than 65% of total assets of USA foreign affiliates originated in Europe and 39% originated in the UK (Ning *et al.*, 1994).

The USA multinationals have key positions in many European markets for highly differentiated foods and beverages. Their European rivals have strategically responded with similar marketing techniques: in many cases they have turned for help to the very American advertising companies that followed USA multinationals to Europe. There is accumulative evidence that the foreign direct investment activities of major food manufacturers have been responsible for the increasing convergence of food advertising methods between the USA and Western Europe (Connor, 1994).

Strategic alliances and partnerships

Under the common umbrella of commercial strategies can be considered: (i) direct product trade; (ii) direct foreign investment; (iii) licensing; (iv) joint ventures; and (v) strategic alliances. The last three can be considered as different forms of establishing partnerships among firms. Previously in the chapter, trade and direct foreign investment have been considered. In this section an overview is offered to understand joint ventures and strategic alliances. The following section singles out the importance of licensing.

Nowadays alliances and partnerships are the focus of a great attention among scholars and businessmen. Although in this chapter the term partnership embraces licensing, joint ventures and alliances, there are no clear definitions and common understanding about the essence of partnerships and alliances. Some authors give those two terms the same meaning, others make a distinction, and also there are others who differentiate between vertical and horizontal alliances.

For example, Badaracco (1991) has defined strategic alliances, in general, as accomplishments scores of new arrangements, which join forces with competitors, customers, suppliers, government agencies, universities and labour unions. Personnel from each organization must work together and he strongly defends the idea that knowledge forces are reshaping competition. The new kind of alliances and the knowledge links make possible the transfer of the embedded knowledge which is enhanced by networks of personal relationships and not by legal contracts between allies. Contracts are not essential and the terms of any agreement might be rapidly changing so reliability is a firm requirement. Trust is the pillar to develop these relationships among firms but also among managers inside firms. In explaining the strategic intent behind a number of foreign alliances several factors can be considered, such as filling out product lines, cutting costs, reducing risks, accelerating products' speed to market, building flexibility, guiding the migration of knowledge and monitoring and/or neutralizing competitors.

A distinctive approach has been given by Hughes (1994), who has defined partnerships as some arrangements between buyer and seller entered into freely, to facilitate a mutually satisfying exchange over time, which leaves the operations and control of the two businesses substantially independent of each other, which commonly happens between successive links in the food chain; whereas, alliances occur horizontally at the same level in order mainly to increase their power in the market.

Taylor (1991) has made the distinction between horizontal and vertical alliances. A horizontal alliance is where business or organizations with complementary skills and capabilities come together to achieve a common goal. Examples where they are applied are joint ventures, the retail buying groups and agricultural ventures in developing countries. Vertical alliances are the many relationships which develop between the separate links of the food chain.

Sporleder (1993) has argued the idea, supported by some analysts, that alliances between firms are identical to joint ventures. This author understands joint ventures as operations in which two firms of different nationalities but with similar product lines develop a jointly-owned and operated facility, often in a third country market in which neither parent previously operated. In contrast, the nature of control provided by inter-firm alliances is weak and malleable compared to other arrangements. Inter-firm alliances embody learning, technology transfer or exchange of tacit information between the parties to the alliance. In the event of a mistake, exit costs are relatively low with alliances compared with other arrangements. Breach of expectations by either party results in termination of the alliance, typically without legal recourse or third-party enforcement. This contrasts to contracts or joint ventures where legal recourse is the primary means of preventing breach.

Joint ventures are a popular strategy for accessing foreign markets, particularly markets where local customs differ widely from those in the multinational's home market (Henderson and Handy, 1994). Joint ventures allow firms to tap into the production, marketing, and regulatory know-how of host-country firms without the expense of acquiring wholly-owned-subsidiaries. With a partner that understands the host market, joint ventures can provide access to locally accepted brands, distribution systems and merchandising know-how. Joint ventures have an inherent problem in maintaining a balance of power over the long-run, and often end with one partner buying out the other. Also some executives express reluctance in using joint ventures because they do not want to share in-house technology and management expertise. Generally, international joint ventures do not seem long standing and large multinationals seek smaller firms as joint venture partners. Exceptional cases among big firms has been the joint venture formed by Nestlé and General Mills to produce breakfast cereals in Western Europe; another example is the joint venture set up in 1990 between Nestlé and Coca-Cola to distribute chilled Nestlé products in Coca-Cola's vending machines.

One critical feature of any alliance is to create a sustained, competitive advantage which comes about with critical mass. No company can withstand the isolation or the decentralization of manufacturing development entities in all the major markets of the world. All firms, especially the big ones, are aware that they have to reach global markets and they have found out through alliances the way to cover such expansion. Also regulatory and political constraints are encouraging alliances. Big companies are becoming bigger and boundaries among them are blurred, in many instances, as a result of their alliances and also markets are becoming decreasingly definable. What is new is the phenomenal growth of non-equity strategic alliances between enterprises who, outside the alliances, may fiercely compete with each other.

Food retailing alliances in Europe are a topical subject, but they are not all new, especially buying alliances. Food retailing alliances have a number of common features: (i) members do not compete with each other; (ii) all profess a beneficial role; and (iii) all are designed to secure volume-related lower prices from food manufacturers whether through combined buying power, pan-European overriding discounts, or just shared information.

Licensing of branded food

Licensing the production and marketing rights for a branded product to a foreign manufacturer is called brand name licensing. According to Sheldon and Henderson (1992), a licence is a transaction by a food or beverage manufacturing firm with a brand name that is well established in one country (the licensor), contracting for a firm in another country (the licensee) to manufacture and sell the branded product in the licensee's home and/or third-country markets. In addition to the right to exclusive use of the brand name in the specified market(s), the licensor often provides some technical production assistance and a quality control regime.

Also, the licensor may provide the product formula or recipe, some critical ingredients and some financial assistance for market development. This generally requires none or little direct investment in foreign production facilities, but considerable investments are required to identify appropriate licensees, develop production and marketing procedures, and establish quality control safeguards. Henderson and Sheldon (1992) have established two general motivations for brand name licensing: (i) when changes in demand create a need for product line extension, licensing represents an alternative to in-house development of new brands; and (ii) where there are barriers to direct entry, such as high levels of advertising or vertical ties, licensing is a means for a foreign firm to establish a market presence.

Licensing can be a particular manner to be introduced in the international arena before a direct foreign investment decision is made. It is often used as a step in a sequence leading to local production. A common expanding business path for agro-food firms has been from trade, to licensing, to joint ventures, to fully owned foreign operations. Licensing is undertaken among big multinationals to capitalize on their cross experiences and many of the foreign licensees are among the world's leading food processors. Nevertheless it has been shown that American consumers tend to prefer an imported food product to a foreign branded product that is manufactured in the USA under licence (Henderson and Sheldon, 1992).

Business arrangements are of a very different nature and agreements usually involve terms for the length, the geographical area, the marketing involvement, the production control and royalties to be paid. Licences tend to be relatively long-term agreements, providing sufficient time for the licensee to recoup market development costs. Licensees prefer longer arrangements than do licensors. A compromise appears to be the inclusion of extension and cancellation clauses that clarify performance expectations. For example, contracts may allow for automatic periodic extensions based upon criteria such as obtained market share, sales volume or level of promotional expenditures.

Multinationals' executives see licensing as a relatively low risk, low investment means of entering markets; also applicable where the market is too small to warrant a wholly-owned subsidiary. Firms sometimes use licensing to avoid a specific trade barrier. It is estimated that international sales of licensed foods may exceed those through direct exports and imports. Even so, executives claim to prefer to avoid licensing, finding it less profitable than direct investment.

International brand name licensing therefore appears to be an important aspect of the globalization of the food system. Licensing of food branded products has received little attention in comparison to exports and foreign direct investments. A review of corporate annual reports from 120 of the world's largest food manufacturing firms indicates that at least half of those with international operations are engaged in product licensing (Henderson and Sheldon, 1992). Because branded product licences are privately negotiated contracts, subject to confidentially clauses, there is little in the public domain about their contribution to firm's profits. Nevertheless it is estimated that the total value of international sales of licensed food products exceeds that of direct product trade.

GLOBAL MARKETING STRATEGY IMPLEMENTATION

Product strategy

The global economy requires a global product strategy but completed with products strategically produced for regional markets to fulfil their consumers' tastes and desires. This is the philosophy adapted by big multinationals like Nestlé and Unilever which have established product groups with a global and regional market coverage. Product strategy is closely linked to brand strategy and multinationals tend to have global as well as regional brands to enhance product differentiations between global and regional products.

The firm able to match market diversity with product diversity and adjust to shifting preference structures will emerge in a dominant position. A wide variety of markets (especially mature industries) are of this kind and tend to be dominated by firms often cited as exemplars of the marketing concept-oriented approach, as it is the case in cereals for Kellogg.

New product development is a requisite to have a sound position in international markets, but is a risky operation. A great percentage of new consumer products fail when introduced in the market (further explanations about product policy can be found in Chapter 13). Because of the high hurdles in the regulatory arena there are not too many companies that take advantage of a patented product before we see the generic copy on the world market. Consumers are willing to pay a price premium for superior new products that cannot be readily duplicated by competitors. Innovative new products and improvement in existing products resulting from investments in research and development are less likely to be vulnerable to price wars and more likely to command a price premium.

New products are one of the main strengths that multinationals have to spread over a large diversity of markets. Competition in that respect is high and competitors react very rapidly. The impact of new product introduction and the competitive nature of markets is a key issue in international marketing. The pricing, advertising and distribution strategies that a firm, which is introducing a new product, considers may depend in part on when competitors are expected to respond. If they are expected to respond relatively quickly, the time it takes for a new product introduction strategy to become effective is important.

A number of alternative structures can be used to achieve new products development at international markets, ranging from traditional bureaucratic structures that coordinate interfunctional interaction through a centralized position higher up in the organizational hierarchy, usually at headquarters firm, to the increasingly popular team structures, where responsibility for coordination and decision-making are decentralized and shared among members of a development team that could be spread in different countries. The distinction is between firms that spread their resources, maintaining generalized skills applicable in a broad spectrum of the environment (thus balancing their risks), and firms that concentrate resources in a skill to maximize use of a narrow segment of the environment (thus, maximizing returns) (Lambkin and Day, 1989).

What distinguishes top food multinationals is their performance in achieving leader and second ranked brands. Leader brands are associated with big market share, usually at least 20%, to profit from large investments. In Europe, the price margin differences between the leader and the second have been estimated to be more than 10%, and between the leader and the third to be around 25%. Leader brands are thus very profitable (around 45% return on investment) and costs are covered in just over two years. The inconvenience of not being a leader is that there are about the same costs of production and promotion but with a lower rate of return. It has been recognized that the high value added of agricultural-based products often resides in the way in which they are packaged and branded. For example, Mountain coffee sells in Japan for four times the price of Brazilian coffee; Koshihikari rice is 30% more expensive than normal rice, while Kobe beef is two to three times more expensive than any kind of Japanese-produced beef (Ohmae, 1990).

Quality is a requisite to sell in any country all over the world but the perception of this term can vary among consumers from country to country. A common feature to reach high quality standards is the introduction by manufacturers of processes that incorporate technological elements to create new products or services. Product specialization is the unique way for small and medium enterprises which enter international markets to reach confined niche markets.

The product life cycle (as explained in Chapter 13) has been used by international marketing researchers in the context of product management and strategic planning. Different national markets for a given product are in different stages of development. A convenient way of explaining this phenomenon is through the product life concept. If a product's foreign market is at a different stage of market development, appropriate changes in the product design are desirable in order to make an adequate product/market match. Several authors have argued that countries belonging to the same macro-level segment should reveal comparable product life cycles. The international marketing analyst might consider diffusion patterns for past new product introductions and he should be interested in incorporating marketing mix variables in the diffusion studies.

Vernon (1966) has contributed a great deal to link product cycle patterns and multinationals' behaviour. Vernon's point was that a product follows stages since it is created until it is replaced from the market. At early stages multinationals are willing to produce new products at their country of origin to later move operations to their subsidiaries. Innovation is first accomplished at headquarters' base, where consumption is able to accept the new product. It is a consequence of technical and human resources' abilities at the firms's base but also matching with the most favourable conditions existing in nearby markets.

Recognizing the need for a broader framework pertaining to product growth, Tellis and Crawford (1981) have drawn from concepts in the field of biology to suggest the product evolutionary cycle as an alternative to the product life cycle concept, which fits better to international marketing than the previous concept. Product evolution is a function of three underlying forces: (i) market dynamics (actions of consumers and competitors); (ii) managerial activity (promotional themes and changes); and (iii) government mediation.

Research and development

Innovation is essential to explain international marketing, but developing and introducing new products can require large investment in research and development (R&D) and it is mostly only big firms which are able to have departments for R&D. Multinationals rely heavily on innovation and investments tend to have long gestation periods. The most important food multinationals dedicate around 3% of total sales to R&D. In 1990, around US$1 billion were invested by Nestlé, about the same quantity by Philip Morris and close to US$0.8 billion by Unilever.

R&D directed at both product innovation and process innovation (e.g. improving quality and/or lowering production costs) are likely to endow a business with enduring competitive advantages in international markets. R&D activities take place in countries that are major sources of innovation in the industry, where workers are highly skilled and where consumers have very high demands.

The terms 'external' versus 'internal' R&D orientation are used to refer to the extent to which companies stress their internal compared to their external research environment in finding and developing new products. The high cost of R&D have dictated the necessity of strategic alliances with shared services in R&D. Big agro-food manufacturers have to rely on outsourcing R&D activities where alliances with universities, research centres and other small companies play an important role. The return on investment in R&D by small firms is reported to be five times that of medium-size and 21 times that of large corporations.

R&D activities should react promptly and be flexible in order to integrate design and development into marketing so that diverse consumer preferences are analysed by persons close to international markets and quickly converted into products. Multinationals benefit from being introduced in different countries with consumers of different income levels. Their mature products in developed markets can be transferred to lower income countries and their short life can be extended as they move from one country to another.

Technology transfer not only is provided through the development of new products to foreign affiliates and other firms, it is also part of the linkages that exist between ingredient firms and food processors, and between engineering firms and their food clients requiring processing plants all over the world. Thus, an emerging trend among large agro-food manufacturers is a shift towards greater reliance on fewer ingredient suppliers.

Production strategy

The location dimension refers to the choice of where to locate each of the activities of the multinational firm, from research and development to post-purchase servicing. In selecting a manufacturing location, low cost of raw materials and labour, location relative to major markets, manufacturing presence of global competitors and a favourable view of the country of origin are of great significance. A specific location is also the obvious strategy for products with a unique geographic production origin, often the case for products such as wine. New corporate strategies involving outsourcing, foreign investments and strategic alliances have created numerous products with multiple country affiliations. The country label is no longer a unidimensional phenomenon. A product can be designed in one country and assembled in another with parts sourced in a third country.

If production is subject to scale economies and the market is limited, the least-cost arrangement is likely to be production at home and export to foreign countries. Most multinationals are also consolidating production into fewer but larger plants due to economies of scale. These plants are assigned mandates to serve an entire region rather than a specific country. A global strategy for activity location argues for locating each activity in one or a few countries most appropriate for the respective activities. Processors often first acquire existing plants in different countries and later on concentrate their efforts in selected locations where they try to achieve economies of scale. Economies of scale have been commonly attained by acquisitions but this is switching towards productive alliances as explained earlier.

The United Nations Conference on Trade and Development (UNCTAD) has established the difference between simple and complex production integration. In the first case, multinationals locate simple basic production in developing countries' subsidiaries and they keep their more sophisticated operations in the home country. Complex integration requires location of production facilities according to market rationale and it is accompanied by a decision-making procedure through the entire organization.

However, there are geographical locational conditions and internal strategic conditions related to each firm's organization and control. Also other external conditions to be mentioned are the economic and political environment. Transportation costs rarely account for important shares of the final price of differentiated products and production strategy has been closely linked to the product strategy. Because it may be infeasible for foreign affiliates to replicate a highly diverse product line, it is more likely that highly diverse firms use exports to 'fill in' product lines after the establishment of a foreign affiliate (Malanoski *et al.*, 1995).

Distribution and procurement

As explained in Chapter 16, retail groups are very concentrated in most developed countries with large investments in foreign countries. This concentration provides them with an enormous power to bargain with agro-food manufacturers. Agro-food manufacturers themselves encounter dealing with buying establishments that might be spread in many countries but with a unique or interconnected buying decision scheme. Agro-food manufacturers not able to match their requirements can suffer an unbearable penalty for their business because of the consequences of being out of their distribution systems.

Multinationals, especially in Europe, have a prominent role in food distribution. Large distribution firms have their own international marketing policies and they tend to be more uniform across countries. Nevertheless, retail organizations require for many purposes a country-specific approach. Regulations vary quite substantially from country to country and multinationals adapt to those circumstances. For example, the average period distributors arrange to pay for their supplies ranges from a few days to a few months. In some countries there are restrictions fixing the maximum period but in other countries it is totally free. These policies have important financial consequences for distribution and agro-food manufacturing firms. Also multinationals face quite distinctive country regulations, with respect to the establishment of large supermarkets. Although providing a more efficient distribution system, they may put out of business many local retail outlets usually located in the old town quarters.

Procurement and distribution rely heavily on logistics. Transportation costs among countries continue to decline as a result of better transportation means. Containerization and air freight are two examples, in the area of physical distribution, of common features in modern distribution. The extensive use of computer networks has facilitated such practices as 'just in time management' objectives in international markets and logistic platforms where food is procured from different countries.

The agro-food sector has the distinctive feature from other sectors that distribution is not usually integrated with production. Moreover both production and distribution firms have their own views about marketing that sometimes clash and are a source of conflicts. Nevertheless, there is a growing trend to increase alliances among distributors in different countries and also between agro-food processors and distributors, as a means of better achieving consumers' preferences across countries and establishing stable business relationships (Hughes and Ray, 1994).

Distribution brands (or 'own label'), as explained in Chapter 16, are expanding and their penetration varies from country to country, but it is also important to indicate that multinationals follow different policies across countries trying to adapt to the national economic environment. Thus, they do not necessarily apply the same distribution brand in different countries and follow the same intensity in that respect across countries, although marketing margins for their own distribution labels are greater than from agro-food producers. Differential prices in different countries of the European Union are of great significance and, for example, for biscuits the difference is between 50% and 100% among national markets. In the future, buying euro-centrals for all of Europe, with probably a unique or similar price, will force the concentration of production units in order to achieve economies of scale.

Only big food firms, with well-established international brands, are able to follow a determined marketing policy across countries. If a food product is dominated by a few suppliers or if suppliers' products are clearly differentiated, there is a tendency towards a greater emphasis of manufacturer's brands. Most middle size and small companies have to accommodate their marketing plans to the requirements of the distribution firms that might be different from country to country.

Pricing strategy

Pricing policy can be directly related to business policy, in general, and specifically export marketing policy. Pricing is only one element, but a critical one in building up the export marketing position (Chapter 14 provides a thorough description of pricing policy). It has its roots in the firm's overall strategy and should be consistent with the other elements in the export promotion plan. It is important to understand the range of complex factors, both inside and outside the firm, which will influence pricing policy. The pricing decision seldom can be made in isolation and is usually made after decisions on the other international marketing variables.

Obviously, any international firm tries to maximize its profits but this goal can be accomplished in a short-term or long-term period of time. Consequently its pricing policy will be planned accordingly. Its export prices should not necessarily follow its domestic prices policies, as risks involved in international markets are much greater than those encountered in domestic markets. It will depend on the competitive environment that faces in each country but also on the costs involved in each operation.

Transfer pricing in international marketing refers to prices placed on goods sold within the corporate network, that is, from division to division or to a foreign subsidiary. This practice can be a means by which a multi-national headquarters can either help or penalize their subsidiaries, transferring profits. There has been great concern from countries where the multinational subsidiaries are located since, with this system, multinationals might adversely affect their interests. Multinationals' internal overall organizational reasons can influence this practice, such as the taxation level in different countries, the control improvement and the coordination of cash and income flows from foreign subsidiaries. Transfer pricing can face internal and external constraints.

Leading food firms have a policy of gaining a high market share in the first years and, later, emphasizing the rate of return of investment. This approach can affect their pricing policy and their promotion budget during the early years to build an appropriate image in the market. It is also linked to the stage of the life cycle of the food product across countries, although international trade liberalization policies tend to homogenize international markets' competitive environments.

CONCLUDING REMARKS

International agro-food trade has been constantly increasing during the last few decades and international marketing has been evolving accordingly, trying to incorporate new functions and complexities. Food trade has been growing at a greater rate than agricultural commodity trade, and the main international food flows have occurred among economically developed countries with analogous business environments enabling the application of sophisticated marketing techniques.

Multinationals are key business organizations to understand how international marketing management has been applied. Their strategies in different socio-economic situations and increasing international competition has encouraged them to find out and implement an arrangement of diverse partnerships. Their international expansion paths intermingle among competitors and cooperators which, in some cases, are the same agro-food firms.

The global economy requires a global corporate strategy which moves away from the so far better known international multidomestic approach. All the common elements influencing agro-food marketing, at national level, need to be reconsidered. In this chapter some attention has been given to product strategy, research and development, production strategy, distribution and procurement, and pricing strategy.

NOTES

1.   The Economist (1995) Survey multinationals. *The Economist*, June 24–30, pp. 1–24.

FURTHER READING

Badaracco Jr., J.L. (1991) *The Knowledge Link: How Firms Compete Through Strategic Alliances*. Harvard Business School Press, Boston, MA.

Dunning, J.H. (1993) *Multinational Enterprises and the Global Economy*. Addison-Wesley Publishing Company, Wokingham.

Hughes, D. (ed.) (1994) *Breaking with Tradition. Building Partnerships & Alliances in the European Food Industry*. Wye College Press, Wye, Ashford, Kent.

Jain, S.C. (1993) *International Marketing Management*, 4th edn. PWS-KENT Publishing Company, Boston, MA.

Meulenberg, M. (ed.) (1993) *Food and Agribusiness in Europe*. Haworth Press, Binghamton, NY.

REFERENCES

Abbott, P.C. and Bredahl, M.E. (1994) Competitiveness: definitions, useful concepts, and issues. In: Bredahl, M.E., Abbott, P.C. and Reed, M.R. (eds), *Competitiveness in International Food Markets*. Westview Press, Inc., Boulder, CO, pp. 11–35.

Badaracco Jr., J.L. (1991) *The Knowledge Link: How Firms Compete Through Strategic Alliances*. Harvard Business School Press, Boston, MA.

Blandford, D. (1984) Changes in food consumption patterns in the OECD area. *European Review of Agricultural Economics* 11, 43–45.

Connor, J.M. (1994) Northern America as a precursor of changes in Western European food-purchasing patterns. *European Review of Agricultural Economics* 21(2), 155–174.

Dunning, J.H. (1993) *Multinational Enterprises and the Global Economy*. Addison-Wesley Publishing Company, Wokingham.

Gracia, A. and Albisu, L.M. (1994) Food diets in EC countries. *Medit: Rivista di Economia, Agricoltora e Ambiente* 9(1), 113–126.

Hallen, L. and Johanson, J. (ed.) (1989) *Advances in International Marketing*, Vol. 3. CT: JAI Press, Inc., Greewich.

Handy, C.R. and Henderson, D.R. (1991) Industry organization and global competitiveness in food manufacturing. In: Gorman, W.D. and Litzenberg, K.K. (eds), *Global Agribusiness for the '90s*, Proceedings of the Inaugural Symposium of the International Agribusiness Management Association. Printing and Duplicating Center of the New Mexico State University, Las Cruces, NM, pp. 88–101.

Handy, Ch.R. and MacDonald, J.M. (1989) Multinational structures and strategies of US food firms. *American Journal of Agricultural Economics* 71(5), 1246–1254.

Henderson, D.R. and Handy, C.R. (1993) Globalization of the food industry. In: Padberg, D. (ed.), *Food and Agricultural Marketing Issues for the 21st Century*. The Food and Agricultural Marketing Consortium, FAMC 93–1. Texas A&M University, College Station, TX, pp. 21–42.

Henderson, D.R. and Handy, C.R. (1994) International commerce in food: Market strategies of multinational firms. Paper presented at the XXII Congress of the International Association of Agricultural Economists, August 22–29, Harare, Zimbabwe.

Henderson, D.R. and Sheldon, I.M. (1992) International licensing of branded food products. *Agribusiness. An International Journal* 8(5), 399–412.

Henderson, D.R., Vörös, P.R. and Hirschberg, J.G. (1993) Industrial determinants of international trade and foreign investment by food and beverage manufacturing firms. Paper presented at the NC-194 Conference on Empirical Studies of Industrial Organization and Trade in the Food and Related Industries, April 7–8, Indianapolis, IN.

Huber, G. (1984) The nature and design of post-industrial organizations. *Management Science* 30, 928–951.

Hughes, D. (ed.) (1994) *Breaking with Tradition. Building Partnerships & Alliances in the European Food Industry*. Wye College Press, Wye, Ashford, Kent.

Hughes, D. and Ray, D. (1994) Types of partnerships & alliances in the European food industry. In: Hughes, D. (ed.), *Breaking with Tradition. Building Partnerships & Alliances in the European Food Industry*. Wye College Press, Wye, Ashford, Kent, pp. 33–49.

Jain, S.C. (1989) Standardization of international marketing strategy: Some research hypothesis. *Journal of Marketing* 53(1), 70–79.

Lambkin, M. and Day, G.S. (1989) Evolutionary processes in competitive markets: Beyond the product life cycle. *Journal of Marketing* 53, 4–20.

Levitt, T. (1983) The globalization of markets. *Harvard Business Review* 61, 92–102.

Malanoski, M., Handy, C. and Henderson, D. (1995) Time dependent relationships in US processed food trade and foreign direct investment. Paper presented at the NCR-182 Conference on Foreign Direct Investment and Processed Food Trade, March 9–10, Arlington, VA.

Ning, Y., Reed, M. and Marchant, M. (1994) US direct foreign investment abroad and foreign direct investment in the US in food and kindred products. Poster presented at the AAEA Annual Meeting, August 7–10, San Diego, CA.

Ohmae, K. (1985) *Triad Power*. The Free Press, New York.

Ohmae, K. (1990) *The Borderless World*. Harper Business, New York.

Porter, M. (1990) *The Competitive Advantage of Nations*. The Free Press, New York.

Rastoin, J.L. and Oncuoglu, S. (1992) Les multinationales et le système alimentaire mondial: Tendances stratégiques. *Economies et Sociétés, Série Développement Agro-alimentaire* 21, 137–174.

Sheldon, I.M. and Henderson, D.R. (1992) International licensing of branded food products. In: Gorman, W.D. (ed.), *Evolution of the Food Chain in a Changing International Environment*, Proceedings of the II Symposium of the International Agribusiness Management Association. Newman Printing Co., Inc., Bryan, TX, pp. 275–283.

Sporleder, T.L. (1993) Strategic alliances as a tactic for enhancing vertical coordination in agricultural channels. In: Gorman, W.D. (ed.), *Managing in a Global Economy*, Proceedings of the III Symposium of the International Agribusiness Management Association. Department of Agricultural Economics, Texas A&M University, College Station, TX, pp. 56–64.

Szymanski, D.M., Bharadwaj, S.G. and Varadarajan, P.R. (1993) Standardization versus adaptation of international marketing strategy. *Journal of Marketing* 57(4), 1–17.

Taylor, J.F. (1991) Developing global strategic alliances in the food and business sector. In: Gorman, W.D. and Litzenberg, K.K. (eds), *Global Agribusiness for the '90s*, Proceedings of the Inaugural Symposium of the International Agribusiness Management Association. Printing and Duplicating Center of the New Mexico State University, Las Cruces, NM, pp. 73–76.

Tellis, G.J. and Crawford, C.M. (1981) An evolutionary approach to product growth theory. *Journal of Marketing* 45, 125–132.

Vernon, R. (1966) International investment and international trade in the product cycle. *Quarterly Journal of Economics* 80, 190–207.

Vieille, J.N. (1993) *Les Leaders Européens de l'Agro-Alimentaire: Evaluation Economique et Financière des Performances*. Collection 'Stratégies Industrielles et Financières'. Eurostaf, Paris.

Yip, G.S. (1992) *Total Global Strategy: Managing for Worldwide Competitive Advantage*. Prentice-Hall, Englewood Cliffs, NJ.

19 Strategic Marketing Cases

Daniel I. Padberg

Department of Agricultural Economics, Room 308G Anthropology Building, Texas A&M University, College Station, Texas 77843–2124, USA

INTRODUCTION

Classic economic analysis explains the behaviour of a firm in response to other economic factors and conditions within a context where technology does not change. This classic behaviour is a pattern of adjustments of the flow of inputs and outputs in relation to costs and prices. When a new threshold of technology invades this environment, there is a shock to costs and prices and the firm makes the classic accommodation of these changes – reactively. Throughout the food marketing channel there are many small firms behaving in this way. They are price takers. Their behaviour is reactive.

At the other end of the firm size spectrum are giant multinational conglomerates. Their intrinsic behaviour is different. Their chosen niche is developing new technology – especially new products. They frequently distance themselves from the classic commodity type markets. In many cases, the classic markets do not provide necessary inputs they need nor a market for outputs of new products. Also, the price instability of the classic markets is an irritant to the large firms over the extended period of product development.

Unlike the classic price taker firms, innovation-oriented food manufacturing firms must find their own strategic sense of direction. They must know where they are going, because they get little direction from the market. The strategic direction for one firm may not be the same as for another. Managing the process of new product development may be quite different across equally successful firms. There is great latitude for exploring new patterns of organization and administration. Compared to firms responding to commodity markets, which may be very old, this type of competition has been in operation only a few decades. Perhaps this kind of strategic management is more of an art than a science. Yet, there seem to be patterns developing – an emerging orderliness. It may be too early to distil a full set of strategic management principles, but it is not too soon to learn from the considerable experience acquired over the last couple of decades.

Compared to other consumer products industries, the food industries have a large development of new product-oriented firms. In most years, food manufacturers dominate the list of largest consumer product advertisers in the USA. While this type of firm is most active in high income countries, its importance is not restricted to these countries. Often the large multinational manufacturers set patterns which influence the food trade worldwide.

Case studies have been chosen as a way to deal with the subject matter of strategic management. These cases give a vivid account of the kinds of decisions which must be made and the setting and context in which they occur. The following sections present two cases of large multinational food marketing firms. Both of them have dozens or hundreds of product brands. One has a well-articulated strategic direction while the other has evolved from history with not much of a sense of direction. A short analysis will follow each case. The conclusions section will compare and contrast the strategic management of both firms.

Harvard Business School   **9-391-191**
Rev. April 7, 1995

Beatrice Companies—1985

In early August 1985, the board of directors of Beatrice Cos. (1985 sales of $12.6 billion) convened an emergency meeting to consider the future of James Dutt, the company's president, chairman, and CEO. Dutt, elected chief executive of Beatrice in 1979, had attempted to significantly alter the strategic course of the sprawling conglomerate through a series of acquisitions, divestitures, and corporate reorganizations. However, as Wall Street began to question the viability of Beatrice's new strategy, Dutt was increasingly criticized. At the August board meeting, the directors had to decide not only the fate of James Dutt, but also the future of Beatrice's corporate strategy.

The Early Years

George Haskell formed the partnership of Haskell & Bosworth as a wholesale produce dealer in Beatrice, Nebraska, in 1891. After three years of modest success, Haskell entered the dairy processing industry; later, in 1897, Haskell & Bosworth incorporated as the Beatrice Creamery Company, which churned and packaged butter.

Beatrice quickly expanded into the surrounding countryside, buying smaller creameries that were close to raw material sources and shipping the final product to market. In 1905, Beatrice acquired Continental Creamery Co. of Topeka, Kansas. Continental was one of the oldest dairies and had a strong regional brand in "Meadow Gold," which became the cornerstone of Beatrice's dairy business. At that time, this was the largest merger in the U.S. dairy industry.

In 1927, Beatrice faced financial and management difficulties. William Ferguson, who had succeeded George Haskell as president, decided to sell the firm to National Dairy (later Kraft), at the time the largest dairy company. Ferguson traveled to New York to negotiate the sale price with National Dairy, but National's highest offer of $45.50 a share significantly undershot Beatrice's over-the-counter price of $60 a share. After negotiations failed to reduce the gap, Ferguson ended talks with National Dairy. The following year, Ferguson retired and the board elected Clinton Haskell, nephew of the company's founder, as president of Beatrice.

Research Associate Toby Stuart prepared this case under the supervision of Professor David Collis as the basis for class discussion rather than to illustrate either effective or ineffective handling of an administrative situation. This case is essentially a condensation of Professor George Baker's paper, "Beatrice: A Study in the Creation and Destruction of Value," with some minor additions, changes, and reinterpretations of evidence.

Haskell had two goals as president of Beatrice. First, he would expand the creamery, focusing on extending processing plants to the East Coast; second, he would diversify into additional product lines within the dairy industry, such as ice cream. Haskell expanded Beatrice mostly by acquiring creameries rather than by constructing new facilities. By the end of 1928, Beatrice had purchased 13 dairy companies, the largest of which was for $1.5 million (**Exhibit 1**). In stride with these acquisitions, Beatrice's revenues increased from $40 million to $84 million between 1928 and 1930.

In 1930 and 1931, Beatrice acquired an additional 44 dairies. Among these were eastern facilities, such as Carry Ice Cream of Washington, D.C., and Maryland Creamery of Baltimore. These purchases established Beatrice, along with National Dairy and Borden, as one of the big three dairy firms. During this time, the dairy industry was consolidating as technological innovations changed the fundamentals of dairy processing. Refrigeration technology coupled with advances in dairy-processing machinery dramatically increased the minimum efficient plant size for dairies. In addition, the federal government introduced a grading system for milk and mandated pasteurization of dairy products, a process of applying sustained heat to milk to eliminate harmful organisms, which meant that more sophisticated machinery was required to operate a dairy plant.

Beatrice typically acquired small dairy companies, and looked for competent incumbent managers who would stay on after the acquisition. The company built market share in a geographic area through selective acquisitions of regional companies. Unlike National Dairy or Borden, Beatrice generally discarded the brand names of its acquisitions. Instead, it packaged all of its dairy products under the Meadow Gold logo. As early as 1930, Beatrice advertised Meadow Gold in such nationally circulating journals as *Life* and the *Saturday Evening Post*. Another difference between Beatrice and the other large dairies was that National Dairy and Borden focused on the highly urbanized markets, while Beatrice concentrated more heavily on less-populated areas.

In 1931, Beatrice expanded by purchasing a minority stake in Chicago Cold Storage, a refrigerated-warehousing concern, which operated as a wholly owned subsidiary of Beatrice. The company diversified again in 1938 when it began to distribute frozen foods under the "Birds Eye" label.

The organizational structure that the early Beatrice assumed was decentralized with many geographic divisions and a central office. Each of Beatrice's acquired plants operated as an integrated unit that both processed and distributed its goods. The central office housed the corporate officers and handled all financial, legal, research, advertising, quality control, and general policy formulation activities. Seven district managers linked the field units to the central office: plant managers reported directly to district managers also located in the field, who in turn reported to corporate. Control laboratories located at each of the major plants monitored product quality.

Diversification: 1940-1976

On November 1, 1943, Beatrice acquired La Choy Food Products of Archbold, Ohio. A maker of Chinese specialty foods, La Choy was Beatrice's first non-dairy related acquisition. Symbolically, in the company's proxy statement filed for 1945, the board of directors recommended that the company change its name from Beatrice Creamery to Beatrice Foods Co. since "The company had long ceased to be just a creamery."

In 1952, William G. Karnes succeeded Clinton Haskell as Beatrice's President. Karnes was a Northwestern Law School graduate who had served as chief financial officer under Haskell, but had never held an operating position in the company. However, Karnes had been heavily involved in the negotiations for many of Beatrice's previous acquisitions. In the year of his election, Karnes formed a committee, comprised of himself and four other executives, to set long-term objectives for Beatrice. The committee decided to continue Beatrice's strategy of growth through acquisition. In expanding its dairy operations, the committee recommended that Beatrice follow the population shift from the Midwest to the Southeast and Southwest.

In 1953, Beatrice acquired six dairy companies, including Creameries of America, the nation's seventh largest dairy producer with 1952 revenues of $49 million. Creameries of America had operations in the western states and Hawaii; after the acquisition, Beatrice processed and sold its dairy products from coast to coast. From 1951 to 1961, Beatrice continued expanding its dairy business, participating in about 175 dairy mergers during this period and increasing its sales 136% to $539 million in 1961 (**Exhibit 2**). Beatrice established its first overseas operations in 1961 with the construction of a condensed milk plant in Malaysia. In 1962, it purchased a Belgian dairy.

In the mid 1950s, Congress altered Beatrice's growth strategy when it passed the Celler-Kefauver Act, that greatly strengthened Section 7 of the Clayton Act, the existing anti-trust law. Following the enactment of Celler-Kefauver, the Federal Trade Commission (FTC) challenged mergers made by large dairy corporations from 1950 to 1956. Among the complaints was one filed against Beatrice for five acquisitions, including the Creameries of America merger. (The FTC also filed anti-trust complaints against National Dairy and Borden.) After extensive negotiations, Beatrice agreed to divest certain plants amounting to $27 million in sales and a smaller percentage of its net earnings. In addition, Beatrice accepted a moratorium on all dairy acquisitions for a 10-year period, with the exception of purchases that the FTC approved.

Faced with declining margins in the dairy industry and pressure from the FTC, Karnes decided that Beatrice needed to expand into new industries to maintain its historical earnings growth. Margins in the dairy industry were declining as consolidation intensified competition and as many large grocery store chains backward integrated into dairy processing.

As the FTC reviewed Beatrice's dairy mergers, Karnes led the company into the confectionery business in 1955 with the acquisition of the D.L. Clark Co., a national manufacturer of candy bars. Two years later, Beatrice purchased the Bond Pickle Co. and concurrently established a grocery products division that included its non-dairy food operations.

Fueled by the steady cash flow generated by its dairy operations, Beatrice launched an aggressive acquisition campaign in the mid 1960s. Beatrice took its first step toward unrelated diversification in 1964 when it acquired Bloomfield Industries, a manufacturer of institutional food-service equipment for restaurants and hotels. Bloomfield was quickly followed with the $16.9 million acquisition of Stahl Finish and Polyvinyl Chemical in 1965, manufacturers of polymers and raw materials for polishes.

Under Karnes' direction, Beatrice executives analyzed a variety of industries to determine areas for profitable expansion. For example, in 1967, Beatrice conducted a review of the "do-it-yourself market for home consumers" and judged this to be "a very rapid growth and potential profit industry."[1] In the same year, Beatrice purchased Melnor Industries, a manufacturer of do-it-yourself gardening equipment, and followed this acquisition with the purchase of seven additional home products companies in the upcoming two years.

Throughout Karnes' tenure as CEO, Beatrice followed a similar pattern in its acquisitions: after analyzing an industry and orchestrating a first merger, the company followed the beachhead with subsequent purchases in the same business area. Among the list of industries Beatrice expanded into in a similar manner were agricultural products, bakery products, soft-drink bottling, food-service equipment, industrial forging and fabricating, specialty chemicals, recreational vehicles, and graphic arts.

Karnes maintained four guidelines that he steadfastly used to evaluate potential takeover targets. First, Beatrice would pursue only profitable companies in industries growing at a faster rate than food. Second, Karnes only considered producers of branded products; Beatrice was not interested in commodities. Third, only companies judged to possess high-quality managers who were willing to remain in their operating roles after Beatrice gained ownership of the company were acquired. This criterion was extended for overseas acquisitions, where Karnes insisted that managers maintain a 10% to 20% equity stake in the target companies. Finally, Karnes only sought small companies compared to Beatrice so that no individual acquisition posed a serious financial risk to the overall company. In making acquisitions, Beatrice preferred not to issue large amounts of debt, instead financing expansion with stock swaps, cash, and lease backs.

Employing these guidelines, Beatrice grew rapidly with 78 acquisitions between 1965 and 1970 (**Exhibit 1**). Indeed, over $750 million of Beatrice's 1970 sales of $1.83 billion was directly attributable to these purchases. In choosing which companies to acquire within an industry, Beatrice considered the quality of the company's incumbent managers to be the most important single criterion. Particularly when initially entering a new industry, Beatrice pinpointed the company that it desired, and Karnes often became personally involved in negotiating the merger. The vast majority of the companies Beatrice acquired under Karnes were privately-held, family-run businesses (of the more than 330 acquisitions made while Karnes was CEO, all but five were of privately-held firms). Part of the reason for the stress on incumbent managers in Beatrice's first acquisition in a new industry was that these individuals were often valuable sources of future acquisition leads within their industries.

After Beatrice decided on a target company, Karnes would often approach managers four or five times, patiently waiting until they were ready to negotiate. In many cases, there were unusual circumstances at the company prior to the acquisition, such as the death of the company founder. In making acquisitions, Beatrice had a reputation for not being transaction driven: Karnes viewed divestitures very negatively, and throughout his entire tenure as CEO, Beatrice divested only three companies. Beatrice's commitment to growing rather than divesting companies facilitated friendly acquisitions of family-owned businesses.

Under Karnes, Beatrice continued to function in its traditional manner as a decentrally-run (and now diversified) company because the companies it acquired usually retained their managers. Every Beatrice division had its own CEO, and division managers· possessed the

---

1. Federal Trade Commission, *FTC Decisions*, "Decisions on Beatrice Foods," Docket 8864, July 1975.

responsibility to hire and fire, promote employees, determine pay scales, purchase supplies, and advertise and promote products. Corporate headquarters, however, retained control over capital expenditures and determined inventory quotas for each division. Headquarters received information about each division through monthly financial reports containing sales and profit data.

Under Karnes, most acquisitions were made individual profit centers when they became part of Beatrice, and Karnes rarely consolidated any of the company's myriad subsidiaries. In addition, Beatrice divisions had very little interaction with one another at the operational level (for example, there were no joint materials purchases, and divisions did not internally source). In 24 annual reports with Karnes as president, not one included the word synergy. Corporate headquarters provided each of the hundreds of profit centers with nearly complete manufacturing and marketing autonomy during Karnes's tenure.

Beatrice was liberal in granting funds for capital improvements to its divisions, and target-company managers often agreed to be acquired because Beatrice was known to be generous in providing capital for expansion with few restraints from company headquarters. Corporate headquarters played a bank-like role in loaning capital to divisions. A profit center simply submitted a loan-request form along with its five-year sales, earnings, RONA, and cash-flow numbers, and Beatrice headquarters decided whether to grant the loan, which typically carried a market interest rate, based on the division's historical performance.

Beatrice also installed an incentive system for division managers under which plant managers received a base salary plus a bonus of about 2% of plant-level pretax profits. Certain profit-center managers were also entitled to stock options under an incentive program instituted in 1957. An additional incentive aspect of the Beatrice system was that every Beatrice corporate officer appointed under Karnes was chosen from within the organization. Divisions therefore served as the training grounds for future corporate officers, and Beatrice regularly held meetings for division leaders at which management problems were discussed and business school professors gave lectures.

The Interim CEO: 1976-1979

William Karnes resigned in 1976 when he reached Beatrice's mandatory retirement age of 65. During Karnes's tenure as president and CEO, Beatrice's annual revenues multiplied from $52 million in 1952 to $5.3 billion in 1976. In that year, Beatrice's return on assets was over 8% (**Exhibit 3**). At the time of his resignation, Beatrice was considered a multinational conglomerate with over 20% of its revenues generated overseas.

When Karnes retired, he had left in place what he thought was an adequate succession plan. William Mitchell, an attorney and CFO of Beatrice, was elected chairman and chief operating officer, while Wallace Rasmussen, an operating manager who had ascended to the executive suite through the food segment, was made chief executive officer. Mitchell was said to be Karnes's protege; like Karnes, he was a Northwestern Law School graduate with a background in finance. Rasmussen was 62 years old, and was expected to be a caretaker until he retired at 65 and passed the job of CEO to Mitchell. However, a conflict quickly developed between Mitchell and Rasmussen concerning different opinions over how the company should proceed. While Mitchell felt that Beatrice should focus on digesting its many recent acquisitions, including instituting more thorough reporting structures and tightening control of its roughly 400 profit centers, Rasmussen wanted to aggressively forge ahead with Beatrice's expansion program.

The controversy was resolved when Mitchell was "outflanked and outmuscled," and was forced to resign 15 months after accepting his position.[2] James L. Dutt, a 52-year-old operations manager with a background similar to Rasmussen's, replaced Mitchell. According to one ousted executive, in the following months, Rasmussen sought to "clean out anyone who had a close connection to Karnes."[3]

Rasmussen introduced important strategic changes at Beatrice. In 1978, Beatrice made its first substantial acquisition under Rasmussen when it acquired publicly-held Tropicana Products Inc. for $490 million (**Exhibit 4**). This price greatly exceeded the second highest bid for the company, a $344 million offer from Kellogg Co. In dollar amount, the Tropicana deal was more than six times the size of Beatrice's second largest acquisition—Samsonite luggage—which Beatrice had purchased for $80 million in 1972. While Tropicana possessed a strong brand name, orange juice was considered by many at Beatrice to be a commodity good, and many also opposed the acquisition because Tropicana competed against the nation's premier marketing companies. One food-company executive said of the deal, "With Tropicana, Beatrice not only has to contend with the logistics of distributing the fresh product, but it also now faces the brutal prospect of going head-on with Coca-Cola."[4] In the same year, Beatrice also acquired Culligan International, a maker of water softeners, for $50.8 million, and Harman International Industries, a $137 million manufacturer of hi-fi equipment.

While Rasmussen supported Beatrice's decentralized operating philosophy, he began to make some operating changes. For example, the fiscal 1977 annual report stated that Beatrice had created five executive vice president positions to "supervise specific sections of our operations permitting us to concentrate on corporate directions and goals." The company established a pyramid management structure, under which 54 group managers reported to 17 division managers, who in turn reported to one of the five executive vice presidents. Each group manager regularly visited his profit center managers, watching key variables and collecting monthly financial reports that were also reviewed by division managers.

Beatrice also began to emphasize corporate marketing under Rasmussen. Under Karnes, Beatrice had dedicated the vast majority of its advertising expenditures to unmeasured media (coupons, premiums, packaging, and point-of-sale promotions), and even in 1977, $54 million of Beatrice's $76.8 million in ad spending went to unmeasured media. In his letter to shareholders for fiscal 1977, Rasmussen noted, "We have taken a number of steps to strengthen our marketing resources. Special marketing groups now report to each of the five executive vice presidents. These groups give us the flexibility to bolster the marketing activities of individual operating units and to seek out and capitalize on totally new opportunities."

Controversial Leadership: 1979

Rasmussen reached mandatory retirement age in July 1979, and the board of directors elected James Dutt as his successor in a meeting that came to be known as Beatrice's "board-room brawl." At the time, the members of the board who worked outside the company purportedly supported Richard Voell, then Beatrice's deputy chairman, as Rasmussen's successor.

2. Meg Cox and Paul Ingrassia, "Discord at the Top: Beatrice Foods' Board, Officers Split Bitterly in a Battle for Control," *Wall Street Journal*, May 21, 1976.
3. "The Man Who Came to Dinner," *Forbes*, February 19, 1979, p. 86.
4. "Beatrice Foods: Adding Tropicana for a Broader Nationwide Network," *Business Week*, May 15, 1978, p. 114.

Rasmussen, however, secretly persuaded all of the inside directors and two outside board members to support Dutt. When Dutt was elected, Voell and two outside directors resigned over the incident.

Dutt began his career at Beatrice in 1947, and had held many domestic and international operating posts. By the time he became CEO in 1979, Beatrice had sales of $8.3 billion and employed 84,000 workers at operations in over 90 countries.

As CEO, one of Dutt's first actions was to move company headquarters from a one and one-half floor office to a five floor office in a glass tower in downtown Chicago. Dutt's next major step was to streamline Beatrice; he reportedly ranked all of the company's more than 500 subsidiaries according to profitability, and targeted for divestiture those with the lowest return on assets. In 1979, Beatrice began systematically divesting businesses for the first time in its 88-year history. In addition, by 1980, Dutt had come to view international expansion as crucial, complaining about "how exceedingly badly we do in the export business."[5] In 1980, Beatrice signed an agreement with the People's Republic of China that authorized the company to begin a number of joint ventures in China. Domestically, Dutt pushed to expand distribution of Beatrice's many regional products, hoping to roll them out nationally.

Dutt also set out to restructure Beatrice in 1981, forming 10 major divisions to focus the company's 400 profit centers around 10 disparate business areas (**Exhibit 5**). All subsidiaries which fell outside these areas were also slated for divestiture. By late 1981, Beatrice had divested about 50 companies with roughly $1 billion in sales. Included among the spin-offs were Dannon Yogurt, the largest of Beatrice's early divestitures, which was sold for $84 million; parts of the unprofitable Harman Kardon Inc., a maker of audio equipment; and Beatrice's Royal Crown Soft Drink division, which bottled RC Cola. Many of these companies failed to meet Beatrice's ambitious new financial performance goals. Others were divested because they fell outside the company's core areas, and Beatrice executives, most having risen through the company's food business, felt uncomfortable running some of the less-related subsidiaries. Nonetheless, some core companies were maintained despite their failure to meet Beatrice's ROA target.

In November 1981, Dutt was highly criticized for investing the proceeds of prior divestitures to acquire the Coca-Cola Bottling Division of Northwest Industries. This acquisition continued Beatrice's push into beverages, which began with the Tropicana acquisition. However, the price of Northwest, $580 million, exceeded the book value of the bottler's assets by about $450 million, and represented 22 times 1980 earnings. Many shareholders felt that Beatrice should have done a large stock buy back rather than the acquisition.

Focusing on marketing, in 1982, Dutt established a $25 million corporate marketing fund which could be assigned to operating divisions at his discretion. The fund was designed to support product introductions and geographical expansions that were too costly for individual profit centers. Dutt felt that the fund would enable him to become "closer to the operating end of the business," and would expand his authority in deciding how advertising funds were spent[6] (**Exhibit 6**). Beatrice's Swiss Miss division requested support from the marketing fund, and introduced a low-calorie hot-chocolate drink that reached $40 million in sales in its first year and increased Swiss Miss's market share by 50%.

---

5. "Beatrice Foods Chief Reins in Growth by Acquisition," *Wall Street Journal*, July 21, 1980.
6. Nancy Giges, "Beatrice has a Big Thirst for Beverages," *Advertising Age*, June 7, 1982, p. 12.

Strategic Redirection

At the 1983 annual shareholder's meeting, Dutt unveiled a completely new strategic course for Beatrice, formally abandoning the decentralized management system that the company had employed for its entire history. Justifying this move, Dutt stated, "In the past, our system of small, regional profit centers worked well. But to be competitive and to grow in the future, we need broader product lines, with national marketing and advertising and with national distribution systems." To expand distribution of its products, in 1983, Dutt replaced Beatrice's regional distributors with nationwide food brokers. The centerpiece of Beatrice's new strategy, however, was a "total commitment to marketing." To facilitate this goal, Beatrice underwent a second major realignment, consolidating operations to increase their marketing effectiveness. Dutt again altered the number of its internal divisions to six, explaining, "The result will be a much smaller number of free-standing, self-supporting businesses operating in the company's principal marketing areas. These new "businesses of Beatrice" will be larger, cohesive business units"[7] (**Exhibit 5**). As a component of the reorganization, Beatrice planned to consolidate its 350 profit centers into 50 to 100 divisions. Ultimately, Dutt succeeded in reducing the profit centers to 27 divisions.

To coordinate the division consolidations, Beatrice brought in William Reidy, a former Dart & Kraft executive who helped reorganize Kraft's food operations in the 1970s and who was named senior vice president of strategy at Beatrice. Under the new organization, profit centers that faced common competitors were grouped together. For example, Beatrice consolidated its eight confectionery companies into a single profit center. In another instance, Beatrice combined its Rosarita and Gebhardt Mexican food brands, which formerly had competed against one another, and began to sell all of the division's products under the Rosarita name because of its superior brand recognition.

In addition to the consolidations, Beatrice formed a corporate marketing department under a senior vice president of marketing, that focused on taking regional brands national (for example, in 1982, Beatrice's Midwestern cheese brand, County Line, spent $142,000 on advertising, compared to the $2.8 million that Kraft spent on its Cracker Barrel line alone). The marketing group's staff was largely recruited from other consumer products marketers, and the new department's mandate was broad, including working with Beatrice's brands on advertising, product development, and market research. Under Dutt's direction, the group reduced the number of advertising agencies that Beatrice used from about 140 to 10, and the corporate office began to set ad spending levels for the divisions. Discussing the role of the new group, one advertising executive explained, "We clustered our individual operations into a small number of groups that could effectively exercise their collective marketing muscle, generate synergies, and capitalize on economies of scale."[8] The goal of the new program was to "give us the opportunity to piggyback brands with similar programming objectives and similar demographics."[9]

In reorganizing Beatrice, Dutt hoped to follow the company's competitors, including General Mills, General Foods, and Kraft in consolidating operations and expanding marketing. These companies had many brands that possessed market share leadership (**Exhibit 7**). Speaking about Beatrice's competitors, Dutt said, "The degree of sophistication of management and the

---

7. Neil R. Gazel, *Beatrice From Buildup Through Breakup*, University of Illinois Press, 1990.
8. Laura Jereski, "Beatrice Make-Over," *Marketing & Media Decisions*, May 1984, p. 75.
9. Jereski, "Beatrice Make-Over," p. 76.

qualifications of people running the businesses were changing, and we weren't changing with it."[10] To this end, Dutt initiated the "We're Beatrice" corporate advertising campaign that hoped to unify Beatrice's brands under one logo that had a quality reputation. Beatrice's 1984 annual report noted, "We have begun a television and print campaign, linking Beatrice to our most recognized brands, and we are backing this effort with the largest point-of-sale promotion in our history. We will use this identity on our packaging, signage, trucks, brand advertising, and promotions." Like Nabisco, the company began to place the Beatrice logo on many of its products. Beatrice's umbrella campaign was launched with a $30 million budget, $11 million of which was spent during the 1984 Winter Olympics. Some Beatrice product managers, however, were skeptical of the "We're Beatrice" campaign and worried that efforts to extend the Beatrice umbrella brand to the company's nonfood products was unlikely to garner improved sales. After the campaign, unprompted consumer recognition of the Beatrice name increased threefold to 8%.

While Beatrice's corporatewide advertising spending was increased under the new program, competitors often still over-spent Beatrice on advertising at the product level. For example, Tropicana spent $12 million on marketing compared to $16 million for Coca-Cola's Minute Maid, and although Beatrice increased support for its (now national) County Line cheese to $2 million, Kraft had a $68 million budget for its cheese items.

Beatrice's continual restructurings came at a short-term price. Corporate net income plummeted from $390 million in 1982 to $43 million in 1983, and in the first quarter of 1983 Beatrice reported its first quarterly loss since before Karnes had become CEO (**Exhibit 8**). Furthermore, the rapidity and uncertainty of the restructuring left many inside the company confused about the intended direction of the changes. Said one observer, "It is a very different mindset for divisions that operated with a great deal of autonomy. At the moment, no one is quite sure where one's authority begins, and the other's ends."[11] For example, Beatrice divested its candy operations six months after they were consolidated because four of the division's six top managers had left the new profit center. Formerly autonomous chief executives, these four became the equivalent of brand managers after the formation of a single confectionery division.

In May 1984, Beatrice launched a $56-a-share bid for Chicago-based Esmark, with 1983 sales of $4.1 billion. Beatrice finally paid $2.7 billion for Esmark, topping a $2.4 billion Kohlberg Kravis & Roberts-backed leveraged buyout proposal, which had already been accepted by Esmark managers. Esmark, created in 1973, was a holding company that had originally consisted of Swift & Co.'s four major lines of business (food, chemicals, energy, and financial services). In late 1973, Donald Kelly was made president and COO of Esmark, and began to run the company as an investment portfolio, buying and selling over 60 companies before the sale to Beatrice in 1984. Among Esmark's acquisitions were International Playtex, purchased for $210 million in 1975, and Norton Simon, a conglomerate consisting of Avis rental cars and Hunt/Wesson foods, which Esmark had acquired for $990 million in September 1983 (**Exhibit 4**). Prior to the acquisition, Donald Kelly had announced that he would leave Esmark to form his own investment company along with Roger Briggs, Esmark's vice chairman. Speaking of the acquisition, one industry

---

10. Sue Shellenbarger, "Beatrice Foods Moves to Centralize Business," *Wall Street Journal*, September 27, 1983, p. 1.
11. Jereski, "Beatrice Make-Over," p. 75.

observer noted, "Dutt is trying to do an awful lot of diverse things all at one time."[12] Another added, "Beatrice is an acquisition junkie. Every five years it is a different company."[13]

In acquiring Esmark, Beatrice became the largest domestic food company, with over 150 brands in 90 product categories (**Exhibit 9**). Beatrice's primary goal was to improve its distributional efficiency and its ability to roll out regional brands nationally with the addition of Esmark's Hunt-Wesson national sales force and distribution system. Previously, Beatrice had sold many of its products through regional and national brokers; after the acquisition, all of its dry grocery products would be handled by Hunt-Wesson's 500-person direct sales force and in-house distribution system. Similarly, sales and distribution of Beatrice's Eckrich meats and Esmark's Swift brand could be consolidated. In addition to sales and distribution, Dutt hoped that Hunt-Wesson's research team could be used to hasten Beatrice's new product introduction. Finally, the addition of Esmark would raise Beatrice from the sixteenth to the third largest domestic advertising spender, establishing Beatrice as one of the world's premier marketers. Speaking of Beatrice's bid for Esmark, Dutt said, "I don't lose. For us, this puts everything into place. It's the final seal on what we've been trying to do."[14] Dutt's determination to achieve his goals for Beatrice was manifested by a cartoon that he kept behind his desk. The cartoon depicted an executive at a board table surrounded by his management team, and the caption read "All of those opposed, signify by saying 'I quit'." Following the acquisition, Beatrice again consolidated the number of internal divisions from six to four (**Exhibit 5**).

After the acquisition, Beatrice's debt ballooned to $5.1 billion from $990 million prior to the Esmark takeover. Beatrice planned to reduce its debtload by divesting many of Esmark's businesses over the next two years, amounting to $4 billion in sales and $2 billion in assets. Before the end of 1984, Beatrice had signed agreements to divest its food-service business, its agri-products and leather operations, its candy divisions, and its chemical operations. The largest of these, Beatrice chemical operations, made high performance products for many niche markets and was sold for $750 million.

As Beatrice began 1985, the company faced many difficulties. The company's Tropicana subsidiary barely broke even in 1984 as freezes in the Florida citrus belt sent fruit prices soaring and as Procter & Gamble introduced Citrus Hill orange juice with a $100 million marketing budget. In addition, while Beatrice's reported earnings reached a healthy $479 million, after subtracting gains from divestitures and restructuring losses, the company netted $259 million in fiscal 1985 compared to $334 million in fiscal 1984. Beatrice's return on assets for 1985 was about 3%, down from 8% in fiscal 1980, the year Dutt replaced Rasmussen.

Beatrice also faced a management exodus in 1984 and 1985. For example, Fred Rentschler, who had come to Beatrice with the Esmark acquisition, resigned as head of Swift/Hunt-Wesson, citing philosophical conflicts with Beatrice's hands-on management style. Another former Esmark executive commented, "[Beatrice originally] said we would continue to be decentralized. Then all of a sudden there were phone calls and letters saying, 'Jim [Dutt] does it this way.' I heard his name mentioned more in two months than I heard Kelly's name in six

---

12. "Beatrice Begins Offer For Esmark Today in an Attempt to Build Marketing Muscle," *Wall Street Journal*, May 23, 1984.
13. Jo Ellen Daily, "Beatrice: An Acquisition Junkie Gets the Shakes," *Business Week*, June 3, 1985, p. 91.
14. "Beatrice Begins Offer for Esmark Today in an Attempt to Build Marketing Muscle," *Wall Street Journal*, May 23, 1984.

years."[15] Dutt was known to be a very hands-on manager, particularly in the food businesses where he initiated management changes at practically every level of the corporation and became involved in day-to-day operations. By July 1985, 37 of Beatrice's 58 corporate officers in 1979 had left the company; many of them had been fired by Dutt.

In 1985, Beatrice suffered a 20% decline in earnings in its fiscal first quarter ended May 31. The company also increased its marketing budget by 25% (to a projected $800 million for the year), although many questioned the businesses that received the additional funding (for example, Beatrice committed to spend $70 million sponsoring a professional race car team over the next three years). It was against this backdrop, in August 1985, that James Dutt entered the board meeting to consider the future of him and his corporate strategy, best expressed in his recent letter to shareholders (**Exhibit 10**).

---

15. Daily, "Beatrice: An Acquisition Junkie," p. 92.

ANALYSIS OF BEATRICE CASE

The case is about the adventures of a new management team trying to bring this firm to the forefront of modern multinational food manufacturers. It gives an opportunity to see why a new *strategic direction* is needed, what its fundamental elements are, and how difficult it may be to change a corporate culture.

What are the most important factors in the management of a firm with over 400 profit centres? Here a first question must deal with independence versus synergism. How much relatedness should be promoted between and among different profit centres. Since the 1950s, Beatrice Foods seemed to have a strategic direction which involved acquiring *autonomous* food (as well as a few non-food) manufacturing companies. There may have been some financial synergism (savings and investments) in the collective ownership of these companies. There is no indication of marketing synergisms. No evidence of company-wide engineering, food safety, product development, advertising or physical distribution is indicated in the case materials. There was no vision of important collective capabilities which acquisitions were expected to complement. As near as we can tell, the marketing activities of these subsidiaries were locally and autonomously managed. This was not a problem during the early years of development because the competition was similarly structured local or regional firms. With time, however, the competition changed.

A new level of competition emerged. Competitors developed marketing synergisms involving several of the marketing functions across many subsidiary companies and product lines. Brand families were developed which provided an umbrella for the introduction of new products and product line extensions. Advertising services were purchased at more efficient national rates. Logistics and physical distribution were carried out in conglomerate-wide facilities. Some aspects of food science and laboratory exploration of new product possibilities were centralized. The incoming Beatrice management understood the new level of competition. They set about to respond to it in a hurried and expensive programme of acquisitions, promotions and internal realignment of functions.

But Beatrice came to a period of difficult times. Earnings fell sharply and an out migration of important leadership developed. Was the new plan wrong? Was it poorly executed? Are the necessary synergisms very delicate and difficult to achieve? Is it possible to make a quick and significant change in the 'corporate culture'? Was the leadership style out of step or inadequate?

Several factors are different in the Grand Metropolitan case. The firm seemed to understand the new level of competition early on. The company was developed to provide needed capabilities. The corporate culture and communication patterns were developed to function in the uncertainties of new product development and introduction. The case makes an effort to articulate these factors.

Harvard Business School

9-590-056
Rev. 12/13/89

GRAND METROPOLITAN . . . ADDING VALUE TO FOODS

A handful of visionary firms are making strategic moves to create a global food system. These companies are formulating their strategies to coincide with the enormous changes in demographics, technology, packaging, and consolidation, as well as those stimulated by political, social, health, and environmental concerns. A leading example of such a firm is Grand Metropolitan, the eighth largest U.K. company as well as the eighth largest food company in the world (Exhibit 1). Grand Met's chairman and group chief executive, Allen Sheppard, understands the challenge of directing such a firm into the year 2000, and he has attempted to build an organization with the people, the strategy, and the assets to succeed in a rapidly changing industry.

Grand Metropolitan began life in 1962 as a small hotel and restaurant business. In the year ending September 30, 1989--27 years later--it had become an international corporation with profits before tax of £725 million ($1.1 billion) and sales of £10 billion ($15 billion) (Exhibit 2). Since 1980, the group had been undergoing a period of change. When Allen Sheppard became group chief executive in 1986 and chairman of Grand Met in 1987, his first objective was to define a clear focus for the group's long term strategy, building on its operational experiences and areas of proven success.

Although successful in the U.K. food and retailing industries, the recognized star in the Grand Met portfolio was the development of IDV, acquired in 1972, into the largest international wines and spirits company in the world. Grand Met attributed this success to brand building, securing distribution, innovative product development and marketing, and most of all, an awareness of consumer trends.

The company identified trends that the food industry would develop in international markets along similar lines to the drinks industry--with a small number of powerful companies controlling distribution and owning key international brands. Based on its food industry experience in the U.K., where it had a number of strong national brands, and on its IDV experience, the Grand Met Board set the company strategy: to be a world leader in the food, drinks, and retailing sectors in Europe, North America, and increasingly Japan and the Far East. The company wanted to be global, the low cost producer, brand leaders, and creators of new products and services compatible with global demographic changes. Most importantly, the Board determined that Grand Met should concentrate its

Janet Shaner prepared this case under the supervision of Professor Ray A. Goldberg as the basis for class discussion rather than to illustrate either effective or ineffective handling of an administrative situation.

development on the branded international food industry, a move which
resulted in the acquisition of Pillsbury for $5.8 billion in December 1988,
the largest non-oil U.S. acquisition by a U.K. company.

Grand Met had created what was considered to be the most demanding and farsighted board of directors in the industry. The four non-
executive board members consisted of Richard Giordano, an American who as
head of BOC was considered one of Britain's best and highest paid
industrialists; Sir John Harvey-Jones who led ICI to the first billion
Pound profit of a U.K. industrial firm; Sir Colin Marshall the chief
executive of British Airways who led the company in recapturing its
leadership position in the travel industry; and David Simon, deputy
chairman elect and CFO of British Petroleum. Grand Met's executive board
members consisted of Ian Martin, head of the Foods Division, Pillsbury,
Burger King, and Grand Met Inc. (North American Operations); George Bull,
head of International Distillers and Vintners (IDV) who pioneered a focus
on branding and distribution in the drinks industry in the 1970s; Clive
Strowger, group financial director (to be replaced early in 1990 by David
Nash, finance director at Cadbury Schweppes and formerly with ICI); and
David Tagg, head of Retailing and Property, previously leading the impor-
tant personnel function at Grand Met (Exhibit 3).

Rather than being threatened by such board power, Allen Sheppard
thrived on the dynamics such a group generated. Sheppard loved debate. He
demanded that his board challenge Grand Met's strategy, testing all the
options and flaws in his thinking, pushing a problem through to its
conclusion. Under Sheppard's leadership, the Grand Met board initiated a
deliberate, detailed, methodical study of the external and internal
competitive environment in which Grand Met would operate in the year 2000.
The group examined alternative scenarios to determine where Grand Met held
sustainable competitive advantages. The company had to define its unique
corporate skills and determine how the Group could add value to its
component businesses.

Grand Metropolitan made several additional acquisitions and
divestitures to implement its strategy, positioning the company to respond
to the changing competitive environment in the food and drinks industries.
Allen Sheppard realized that they were only a beginning, however. Grand
Met would have to make many more acquisitions and retain its strong
entrepreneurial culture to achieve its goals. The company had to determine
in which markets and with which products it should expand its business and
make strategic acquisitions to achieve those objectives--both for the
company as a whole and for Pillsbury in particular. Allen Sheppard
reviewed the company's evolution, its current management group, and its
asset base in order to think through his next steps.

Demographic Trends

In developing the foundation for its strategy, Grand Met
attempted to forecast the economic, political and social trends in the year
2000, generating alternative scenarios for each assumption. The group
determined that five primary "engines of change" were driving consumer
purchases.

1. **Affluence:** Personal income was increasing both in developed and developing countries, and government policy seemed to favor supply-side economics, keeping taxes low. With the population relatively stable, increased income allowed consumers to purchase a greater number of and more expensive goods.

2. **Individualism:** Mass manufacturing emerged during the post-war period. The 1990s signalled the break-up of mass manufacturing as consumers were demanding products tailored to their specific needs and wants. These consumers were unwilling to experiment, however. They wanted "character products" which stated, "I am what I eat." Consumers desired products that were familiar and trusted, not bland but safe.

3. **Healthiness:** As technology developed and income increased, consumers desired purity and an end to sickness. Jogging, healthy foods, and the environment were popular concerns. Consumers demanded the power to control the physical elements of their lives.

4. **Dual career families:** This trend translated into greater consumer buying power and increased value of time. Families consisted of fewer children, but more affluent parents indulged their children more. The microwave was becoming the "oven of the future" because individuals preferred not to waste precious time slaving over a hot stove.

5. **Consumer globalization:** People were beginning to think and act alike. As airplanes and newspapers made the world accessible to all, world consumers were becoming increasingly similar. Ethnic food, fashion and entertainment, for example were accessible and popular in many countries, gradually eroding cultural barriers previously thought unpenetrable.

   Globalization offered multinational companies new product opportunities. Previously, market segments within one country were too small to introduce products designed for very specific segments. Expanding the target market across several countries, however, created sufficient market size for these tailored products.

Internal Strengths

Internally, Grand Met examined three key areas of the Group--its business portfolio, its skills, and its corporate culture to determine if any synergies existed between the three. The Group had to determine how it could use its skills to develop sustainable competitive advantage and add value in its chosen business sectors. After a critical self-analysis, Grand Met determined that as an organization it had strengths in three general areas: an ability to develop and sustain leading brands, a corporate culture which encouraged change and swift reaction, and experience

in building and managing business portfolios including the ability to
"decomplex" businesses.

Brand Management: Grand Met derived its brand management skills
from experience with products such as Alpo pet food in the U.S. and leading
beer brands such as Fosters and Budweiser and food brands such as Ski
yogurt in the U.K. Most importantly, Grand Met built global expertise from
its wines and spirits business. Through both acquisition and new product
development, Grand Met and IDV had become the leading drinks company in the
world with major brands including J&B Rare Scotch Whisky, Smirnoff Vodka,
and Bailey's Irish Cream. The acquisitions of Heublein, Almaden wines, and
Christian Brothers allowed IDV to establish critical mass in the United
States, as did the Pillsbury acquisition in food. With a worldwide distribution network, Grand Met could continue to purchase brands to feed
through this efficient system at reduced costs, thus adding value to the
brand and to Grand Met.

Corporate Culture: Grand Met believed its people, its management
process, and its philosophy were major contributors to its success.
(<u>Exhibits 4</u> and <u>5</u> list Grand Met's organizational philosophies.) Despite
still being a very young company, British firms respected Grand Met's
style, and the company ranked tenth on the list of Britain's most admired
companies, noted particularly for its capacity to innovate and its ability
to attract, develop and retain top talent.

Fewer than 100 people worked at Grand Met's London headquarters.
Instead, Grand Met tried to create a small-company culture by retaining
sensibly sized, independently operated business units where people were
prouder of their unit than of Grand Met as a whole. The company hired risk
takers who were given considerable authority and responsibility to operate
their businesses. Performance measurements were tied to results, and
people might be "fired for never taking a chance." Top management
encouraged informal communications, openness among employees, and debate
within the organization. "Grand Met's managers impress me as forward-
looking and decisive," commented one analyst.[2]

Decomplexing Businesses: From its successful experiences in
acquiring, managing, and divesting companies in many business segments,
Grand Met believed it had developed skills in managing business portfolios
and in "decomplexing" complicated businesses. Grand Met had adopted a
policy of "controlled delegation" to transform ailing businesses. It
preserved individual company structures to encourage entrepreneurialism and
employee loyalty while implementing information and distribution systems to
retain control and improve operational efficiency. These systems reduced
the ratio of managers to employees from 1:7 to 1:11, and Grand Met believed
they were crucial to its future success.

Vision 2000

After this critical self-examination of the competitive environment and its internal strategic advantages, Grand Met forged Vision 2000--

---

[2]"Grand Met's Recipe for Pillsbury, <u>Fortune</u> (March 13, 1989), p. 61.

its unique strategy to propel the company into the future. The company
stated its position as follows:

- Grand Metropolitan is a major world corporation
  respected internationally for its management, enterprise and growth record.

- It specializes in highly-branded consumer businesses,
  where its marketing and operational skills ensure it is
  a leading contender in every market in which it operates. These businesses, which should be few in number
  but large in size, have complementary features which
  ensure that the Group can add value to them.

- Its style is restless, innovatory, and about winning.[3]

Grand Met set its focus on Food, Drinks, and Retailing in Europe, North
America, and progressively in Japan and the Far East.

Restructuring

To implement its strategy successfully, Grand Met had to divest
its extraneous operations. Over three years it engaged in "Operation
Declutter" in which it sold companies worth £3.5 billion and acquired
others for £5.1 billion. (Exhibit 6 lists the acquisitions and divestitures. Exhibits 7 and 8 present Grand Met's current business structure.)
Among the sectors divested was Grand Met's hotel operations--the corporate
foundation. Although Grand Met saw industry consolidation and the opportunity to add value in the hotel business, the company could not afford to
play in both the hotel and food arenas. When Japan's Seibu/Saison Group
offered 52 times earnings for Grand Met's Inter-Continental Hotels, the
company capitalized on the opportunity to exit the business lucratively.
Grand Met, worried about the "taint" on betting in the U.S., decided
against globalization of this business and sold its William Hill/Mecca
Bookmakers operation to Brent Walker for $1,040 million in October 1989.
(Grand Met had purchased William Hill in December 1988 for $520 million to
get critical mass in the U.K.) From the Pillsbury purchase, Grand Met sold
Pillsbury's grain merchandising division to ConAgra for $140 million, Van
de Kamp/Bumblebee for $409 million, Steak & Ale and Bennigan's restaurants
for $434 million, and announced the intention to sell its processing and
consumer operations in Central and South America for $63 million. In
divesting these operations, senior management assured the financial community that earnings would not be diluted.

At the same time, Grand Met was making "micro-strategic" acquisitions of businesses whose leading brands could be channeled through
established distribution networks, costs reduced, and sales expanded.
Grand Met's acquisition of Christian Brothers ($150 million) in 1989 and of
British-based UB Restaurants ($288 million purchase price, $236 million
sales, $18 million profits) illustrate these types of acquisitions.
Christian Brothers' quality California wines complemented brand names
acquired with the Heublein purchase including Almaden, Inglenook, Lancers,

---

[3]Business Strategy International (August 1989).

and Beaulieu, and the company should benefit from IDV's brand experience and wines and spirits distribution network. In the same way, the 813 restaurants acquired from the UB purchase (including the Wimpy's hamburger chain and Perfect Pizza), complemented Grand Met's retailing activities and gave Burger King additional scale in U.K. operations.

Pillsbury Acquisition

        If Grand Met wanted to become a major force in the global food industry, it needed critical mass in the world's single largest market-- the United States. The company analyzed U.S. acquisition opportunities and identified The Pillsbury Company as a perfect target not only to strate- gically enter the market, but to add value to the company by implementing efficient operations as well.

        Pillsbury had the #1 or #2 brands in 85% of its food business (Exhibit 9). In the 1970s Pillsbury had developed innovative new products and established itself as the leader in segments such as refrigerated dough and microwave technology. Since 1983, however, Pillsbury's performance ratios had declined rapidly (Exhibit 10), and its leadership reputation waned. Grand Met believed Pillsbury was overstaffed and inefficient, and as such it saw tremendous opportunities to add value with its brand market- ing experience, lean operations, highly developed retail systems, and property management skills. According to its financial advisors, County NatWest WoodMac:

        It provides a quantum leap in two areas, into the world's
        most dynamic food and retail market, giving a business oppor-
        tunity with which Grand Metropolitan's management has the proven
        ability to enhance shareholder value.

        Seizing this unique opportunity, Grand Met launched a $5.3 billion ($60 per share) tender offer for Pillsbury on October 4, 1988. (At that time, Pillsbury shares were trading for $39 1/8.) The market valued Pillsbury at a price of 15 times normal earnings of the fiscal quarter ended August 31, 1988, and Grand Met offered 22 times earnings. The two parties engaged in a fierce battle with Pillsbury attempting to ward off Grand Met as a suitor. By January 1989, however, Grand Met had accomplished its goal. It had purchased Pillsbury for $66 per share or $5.8 billion.

        Grand Met wanted to make Pillsbury one of the best branded food companies in the United States, and it appointed Ian Martin to lead the transformation. Martin and his team was determined to "cut costs, build brands, and develop new products."[4] Grand Met set seven key objectives:

    1. Restoring product quality to its rightful place at the very
        top of the priorities.

    2. Reorganizing the manufacturing and distribution structure to
        deliver top quality to the end consumer at an economic cost.

---

[4]Fortune, op. cit.

3. Making major cost reductions in the total business and
    rolling back the money saved into brand building e.g.,
    targeting operations cost savings of $80 million by
    September 1990.

4. Achieving a high rate of successful new product introduc-
    tions and innovation to protect future sales and income
    streams.

5. Producing annual earnings growth for Grand Met of at least
    15% per annum to ensure a steady flow of capital and re-
    investment in new products and processes.

6. Making systems and information technology a key driver of
    competitive edge. (Pillsbury committed to a $40 million
    investment in information technology.)

7. Ensuring a steady flow of tactical acquisitions to support
    and supplement existing businesses.

Within the first 144 days of the takeover, Grand Met implemented several
significant restructuring measures. These included reorganizing and
redefining the business segments, closing plants, cutting staff, selling
the grain merchandising division to ConAgra, announcing the intention to
sell milling, pasta and flour processing facilities in South American
countries and a consumer business in Guatemala, and instilling the Grand
Met corporate culture. (Exhibits 11 and 12 outline these changes and their
affects on the business.)

     Prior to Grand Met's purchase, Pillsbury was organized as
illustrated in Exhibit 13. In the restructuring process, Grand Met had to
transform Pillsbury's bureaucratic culture from one known more for its
"fluff" than its "stuff" to Grand Met's decentralized and innovative style.
Grand Met identified people with an action orientation, the willingness to
take risks, and the courage to be accountable for achieving ambitious
goals. With these people, Grand Met encouraged open debate, minimum
formality, cultural diversity, and realistic assessment of its own and
competitor's strengths and weaknesses. However, for Grand Met collegial
debate was NOT management by CONSENSUS--a culture reminiscent of the old
Pillsbury. In Grand Met, the person to whom authority and responsibility
were delegated would weigh the information and make the ultimate decision.

Pillsbury Foods

     Grand Met perceived that Pillsbury was sitting on a veritable
gold mine--the key word being sitting. With strong brand names such as the
Pillsbury Doughboy and Green Giant, the company had built an image of
quality and trust, and the Pillsbury brand name had achieved a 96% consumer
awareness level. Declining product quality threatened that image. To
revitalize the business, Grand Met resegmented into Bakery Products
(Pillsbury), Vegetables (Green Giant), and Pizza (Jeno's, Totino's,
Pappalo's), broader definitions than the previous chilled dough or cake mix
divisions.

     After reorganization, Grand Met had to act quickly to improve
product quality and become the low cost producer. Once a strong

infrastructure was rebuilt, the company could capitalize on the strength of the "Makers Marks" by developing new products, globalizing the brands, and making strategic acquisitions in weak segments which could be channeled through strong production/distribution networks. In November 1989, Pillsbury Foods took over global responsibility for developing the Green Giant and Pillsbury brands, working closely with Foods Europe and Pillsbury International.

Foods Europe

Grand Met Foods Europe encompassed the European segments of Pillsbury, the non-dairy parts of Express, and Express Foods Group International. This unit would be the key arm for the strategic thrust into continental Europe where the division had established sales of $183 million in Germany and $210 million in France, Belgium, Holland and Spain (Exhibit 14).

Pillsbury International

This division marketed products from the Pillsbury and Green Giant brands in all countries except Europe including Canada, Latin America and Asia. (Exhibit 15 illustrates the division of sales.) Previously, Pillsbury had not focused on international development. Through April 1989, sales were 7% ahead of the 1988, and the division was targeting Canada and the Asia/Pacific regions to grow sales even more.

Haagen-Dazs

The Haagen-Dazs mission was to "be the leading worldwide supplier of premium quality, delicious tasting and personally satisfying frozen dessert products." Through 1989, the brand was the leader in the adult segment of the fragmented ice cream market and fourth in the total market (Exhibit 16). To grow the brand, Grand Met would focus on producing quality products, direct store delivery to guarantee quality to retailers, product innovation, increased advertising expenditures, and world-wide expansion of the brand through Haagen-Dazs stores and foodservice distribution. By April 1989, U.S. ice cream sales were 12% greater than 1988, and in September 1989, U.S. and international operations were consolidated into one combined operation to develop Haagen-Dazs as a global brand.

Industrial and Food Service

The $600 million industrial foods business divided into food service, flour milling, and special commodities such as feed ingredients. Sales (versus marketing) drives these commodity, low-cost businesses, and the product portfolio divided into low value-added flour (35%), medium value-added mixes (55%), and high value-added entrees (10%). Pillsbury was fourth in U.S. flour milling capacity, and the company had developed a Sweet and Sour Chicken entree which was served on Air Force One. Particularly in the food service segment (defined as any food product consumed away from home), Grand Met perceived several opportunities. By 1989, 47% of food was prepared out of the home. As operators faced increasing demands from their customers relative to food quality, nutrition labeling, reduced additives, better flavor, and more freshness, operators and consumers would increasingly seek trusted suppliers to satisfy these demands. Grand Met wanted to leverage the Pillsbury name by creating branded food

service products similar to Heinz institutional ketchup and branded
crackers. Grand Met recognized the growing importance of food service in
November 1989 with the transfer of flour milling and special commodities to
Pillsbury Foods and the establishment of the Grand Met Food Service sector
to concentrate on this fast growing segment.

Burger King

Initially, analysts perceived Burger King as the thorn in the
Pillsbury crown, but upon further examination, Grand Met believed it was
one of the jewels. Burger King's 5,790 units generated revenues from three
sources: retail income from company operated units (15% of total outlets);
royalty, rental and development income from franchises; and distribution
income from servicing Burger King outlets by Distron. Burger King offered
several opportunities. It was as large as the next two competitors combined, its sales per restaurant ratio was greater than all other competitors (excluding McDonald's) by 8%, it opened new restaurants daily, and
its products were rated superior to those of competitors. Compared to
McDonald's which owned 25% of its outlets, however, Burger King had significantly less direct control over the product--and it showed. Grand Met
identified several problems with the Burger King business including a
perceived inconsistency in service quality compared to McDonald's, a
fragmentation of management focus caused by almost 100% turnover of senior
executives, and inconsistent marketing programs. Declining unit sales since
1986 highlighted these problems.

To energize Burger King, Grand Met transferred its top retailing
executive to Miami together with executives from other Grand Met activities
in the United States and the United Kingdom. By May 1989, Grand Met had
eliminated three layers of management, cut 560 people from headquarters and
regional offices for annual cost savings of $30 million, restructured
Distron (the distribution company), and refocused management on the customer. Grand Met hoped to reduce the staff by 1,500 people when restructuring was complete. (To reassure Burger King franchisees during this
period of turmoil, Grand Met sent out the letter presented in Exhibit 17.)
Through April 1989, Burger King's sales were up 4% after more than a year
of decline, and Grand Met believed that once the changes were implemented,
Burger King would contribute tremendously to its retail businesses. In
October, Burger King launched its new marketing campaign which had been
very well received.

Pillsbury was the test of Grand Met's strategy. If Grand Met
succeeded in revitalizing this bureaucratic, declining business with its
strengths in brand marketing and efficient operations, it had confidence
that it could achieve those results in any acquisition it targeted. Grand
Met had made significant progress in the first few months to transform
Pillsbury into one of the best U.S. food companies. Once it had built this
strong U.S. base, Grand Met would leverage that base by selling existing
products in new countries, by identifying new products to sell in
established markets, and by making strategic acquisitions of companies with
products to which Pillsbury could add value. Grand Met's next task was to
identify the companies and products to meet those objectives.

The Future

Grand Met would measure the success of its strategies by three specific goals: (1) breaking the barrier of £1 billion of Profit After Interest and Before Tax, (2) maintaining 15% annual growth in earnings per share, and (3) building Grand Met to be recognized as one of the 12 leading world companies. Grand Met would achieve this position both by organic and acquisitive growth, and future acquisitions would have to fit strategically with the company's sectoral and geographical focuses. An add-on purchase such as Christian Brothers or UB Restaurants should be able to contribute to earnings per share immediately in year one. If the acquisition played a "macro-strategic" role, allowing Grand Met to penetrate a new market or sector in a major way such as Pillsbury, the company should have a neutral effect on earnings per share in the second year after acquisition and contribute positively to EPS by year three.

Reflecting on where Grand Met had been and what he wanted it to become, Allen Sheppard stated:

Grand Met has been successfully entrepreneurially driven for two decades. It has just gone through a change period analyzing what it was and what it wanted to be. It still has a flair for entrepreneurialism, but it has gone on to become one of the world's first league players. It has gone from just a big company to one of the world's well thought of institutions.

In the mid-1970s we led the charge on branding and distribution to become a leading player in drinks. We want to do the same in food. In the 1980s and 1990s we want to be in front of the charge on the growth of global brands, anticipate trends well, and ride the trends effectively.

Allen Sheppard paused . . . for just a moment.

He had listened to the debate. Grand Met had formulated a strategy to transform itself from a hotel and drinks company to a leading food, drinks, and retailing company. Pillsbury represented an in-depth opportunity to apply that strategy. Allen Sheppard, Ian Martin, and their executive colleagues headed businesses with many different product lines and brand names including Pillsbury, Green Giant, Bailey's Irish Cream and Burger King. As they looked to the future, however, they had to determine what new companies and product lines would complement their current market position and competitive strengths. In the context of major food industry changes in demography, technology, packaging, consolidation, and health and environmental concerns, this was not as easy task.

Allen Sheppard knew the opportunities for implementing Grand Met's strategy in the food, drinks, and retailing sectors in Europe, North America, and Japan . . . and he knew the risks. The Grand Met team had explored all the branches of the decision tree. Now, Allen Sheppard must weigh the elements.

Advised by his strong board, guided by Grand Met's carefully formulated vision for the 1990s, and driven by its strong determination to succeed, Allen Sheppard, as group chief executive of Grand Metropolitan, would make the final decision on the fate of the company.

ANALYSIS OF GRAND METROPOLITAN CASE

Multinational food manufacturing conglomerates have histories which grow from the combination of many food manufacturing companies performing specialized functions in limited markets. Much of this experience is with commodities and some is with limited scope consumer products. The transition from the more limited background to the new international level of competition is usually difficult and complex. Crucial in the transition is developing a sense of strategic direction and purpose for the new, larger organization. The Grand Metropolitan case gives a well-developed articulation of purpose and strategic direction.

The 'strategic direction and purpose' and the 'new level of competition' are concepts developed and expressed in the mind set and vocabulary of marketing and business management. In general, it is difficult to translate this discussion into the mind set and vocabulary of economics and commodity markets. At the same time, where that can be done, it is useful. The following discussion is motivated by that goal.

Some of the salient features of the new level of competition include: (i) a central focus on new product development and introduction; (ii) acquiring favourable access to communication media; (iii) planning for efficient distribution logistics; (iv) adapting financial management to the time horizon and risk frontier of new product development; (v) developing an organizational base for effective competition in new products; (vi) developing a communication pattern and management style to facilitate new product competition; and (vii) taking this product development process across international boundaries.

This activity and focus are not easy to classify in the economist's mind set and vocabulary. Much of the economist's frame of reference is about firms that function within markets for static products. In this setting, the market defines values for prevailing quantities of known (homogeneous and unchanging) commodities or products. This arrangement can be invaded by new technology and new products (externally produced) which triggers a new pattern of market activity and market-oriented competition involving quantity and price adjustments. The economist's wisdom comes largely from understanding adjustments within a market setting with very little attention to the transition to a new product and a new market. The 'new level of competition' gives major attention to the transition to new products and loses interest when conditions approach the classic market adjustment, which includes zero profit levels.

The economist's expertise concerning firms' behaviour as conditioned and constrained by markets brings little to the understanding of these giant firms. This is more of a firm-based concept of economic behaviour than a market-based concept. It is Galbraithian in character. Almost 30 years ago, Galbraith was explaining how and why firms like these distance themselves from markets for inputs and outputs because markets serve less well than purpose-built arrangements for input acquisition and final product distribution and sales.

COMPARISON OF BEATRICE AND GRAND METROPOLITAN

Both of these firms have experienced a transition from many small independent companies, primarily involved in commodities or local consumer products, to a large multinational conglomerate. Always, this kind of experience involves a clash of ideologies. The classic ideology of cost and price competition which is dominant in commodity markets is broadly admired. We associate consumer welfare concepts with this pattern. It is often difficult a transition from a leadership position in a commodity firm to one in a consumer products firm. Yet, the consumer products firm typically has a history including roots in and acquisitions including commodity firms.

There seems to be a large degree of correspondence between the vision of strategic direction and purpose aspired to by Beatrice's new management and that articulated by Grand Metropolitan. A major difference is that Grand Metropolitan has taken more time to develop their strategic direction so that it can be understood and supported throughout the organization. Grand Metropolitan seems to have better fitting acquired components. Much is made about a participative management process. This may assist in the extensive communication needed to develop a new strategic direction.

With Beatrice, it is only in the recent past that a strategic marketing direction has been perceived. Many of the component parts were acquired before the current management took over. It is not surprising that synergies do not develop naturally from these ill-fitting parts. The new strategic direction or the need for it is not well understood and/or supported by regional and divisional managers. There is less of a tradition of participation by these managers of previously autonomous divisions. The pace of change is very fast and expensive.

Although Beatrice management understood the reasons for their competitive inferiority in relation to other food marketing firms, it was not possible for them to respond quickly. Their management style was out of tune; their industrial components were incomplete (physical distribution); and the arrangements with the autonomous management teams in each group were barriers to synergistic cooperation. Where these units had strong and successful experience in commodities or regional consumer products, it still was not enough to compete with firms of the same size, but put together with sensitivity to company-wide competitive synergisms. The process of building these company-wide competitive synergisms is often called strategic marketing or strategic management. With a poor sense of strategic management, Beatrice experienced a decline in earnings and eventually was overcome by a hostile takeover. Beatrice was acquired by a group specialized in leveraged buyouts, Kohlberg, Kravis and Roberts (KKR), in 1986 for US$8.1 billion. After the acquisition, KKR sold off assets and subsidiaries, especially non-food components, recovering US$6.5 billion in the first year. In 1990, ConAgra, Inc., a large and successful food manufacturer, purchased the remaining food components of Beatrice for US$1.34 billion. The combination of ConAgra and the residual components of Beatrice seems at this writing to be productive and successful. Grand Metropolitan has continued to this point along its expected expansion path.

STRATEGIC MANAGEMENT AND PUBLIC WELFARE

Strategic management generally refers to a set of principles, objectives and patterns of management necessary for or at least consistent with success within large firms which are not significantly constrained by classic markets. Much attention is given to success in selling products to consumers – clearly the most important interface these firms have with society. In addition, there is another interface with society that is important for these firms. That is attracting the most technically (and artistically?) competent employees. Where competition turns on the science of new products and communications (images) appropriate to introduce them, technically competent employees are a strategic resource.

Successful large marketing firms strategically position themselves for success in both of these primary interfaces with the public – selling products and acquiring employees. It is useful for them to have a positive and attractive public image. Large firms increasingly frequently devote advertising energy to their public image – always aligning themselves with popular causes and public values. They make a special effort to find harmony between firm objectives and public values.

For the most part, the statement of strategic direction in the Grand Metropolitan case is self-serving. Yet, there is a developed sensitivity to food products' nutritional function – also to changing consumer conditions and needs. While this is self-serving, it is also a harmonizing of private and public interest. It is likely that we will see greater development of sensitivity to the 'public interest' as we see more effort by firms to state their strategic direction.

This process of linking public and private incentives is probably the most significant mechanism of public control or public responsibility concerning these large firms. Traditionally, we have expected the market to discipline firms and to assure that private actions of firms were consistent with public interests in general. Economic analysis has developed carefully and rigorously the connection between private firms and public welfare within the context of markets. For these large firms, which are not limited to the classic response to markets, little has been developed for linking public welfare to the behaviour or incentives of firms.

IMPLICATIONS FOR MONOPOLY ANALYSIS

Since the functions of these large firms are not limited by or controlled by markets for commodities, it is also more complex to relate to them in the analysis of monopoly behaviour. Yet, the size of these firms together with frequently high profit levels and their insensitivity to classic markets often leads to concerns about monopoly. While no formal analysis has developed with sensitivity to these issues, there are some observations which relate to these concerns.

First, the context of the emergence of these large multinational food manufacturers relates to monopoly concerns and the likelihood of monopoly exploitation by sellers. The context is one of abundant opportunities for the development of new products. These firms and other firms, though perhaps few in number, are drawn into competition in new product development and introduction. This occurs only in high income countries. This context seems inconsistent with monopoly which operates within a context of scarcity, output restriction and lack of product alternatives.

Second, a specialized oligopoly distribution sector has an incentive to compete with heavily marketed and advertised products by offering economy priced private label products. With this vigorous competition on the economy end of the price/quality spectrum, monopoly behaviour is infeasible.

Finally, one of the major concerns with highly concentrated markets dealing with commodities is that firms have a natural motivation to collude – and through collusion evolve toward monopolistic behaviour. The economic behaviour of these large firms is centrally focused on new product development and introduction. In this competitive thrust, there is not an incentive to collude. The opposite is observed. New product producers have an incentive to be secretive about the development, testing and preparation of new products. Consequently, the presence of large firms is not likely to be compounded by a tendency toward collusion as one would expect of large firms competing within a market context.

CONCLUDING REMARKS

The large multinational food manufacturers are often ignored or scarcely considered in analyses of the food system. In most instances, neither conceptual theories nor data sets capture their presence effectively. Theories are much more developed about firms operating within markets and getting the full scope of their behaviour from market signals. Data sets usually describe particular food industries, such as beef, grain, etc. From early times, data sets have been more sensitive to the farm producing interests than to describing the largest food marketing firms.

Understanding the strategic marketing direction is especially important for the internal management of large food manufacturers. These firms are internally guided, rather than being guided by markets. There are significant potential synergies and economies of large size. The pattern of using science and communication media to respond to the higher income consumers' preferences with a stream of new products is only about a quarter of a century old. While many leaders in the food industry are insensitive to this pattern, it is still a central theme of the food industry. These are the largest firms. They are successful, high profit firms. Their activities shape what many other parts of the food industry do as well as what consumers buy.

APPENDIX A: BEATRICE COMPANIES EXHIBITS

Beatrice Companies—1985

Exhibit 1 Number of Acquisitions and Divestitures, 1910-1985

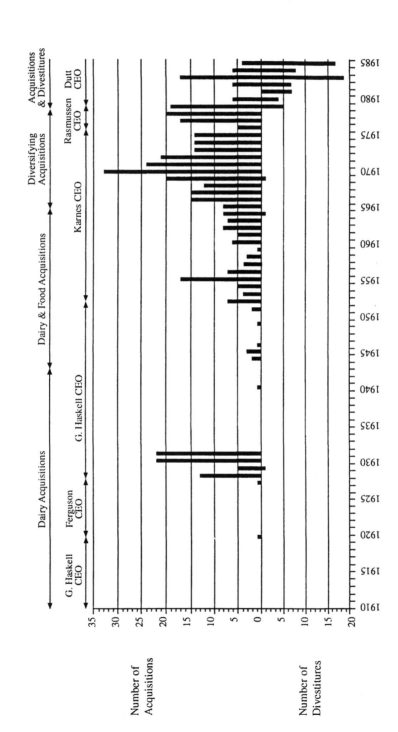

Source: George Baker, "Beatrice: A Study in the Creation and Destruction of Value".

Beatrice Companies—1985

Exhibit 2   Beatrice Cos. Financials, 1950-1979

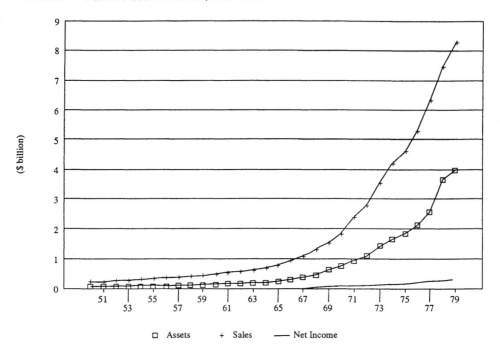

Beatrice Companies—1985

Exhibit 3   Beatrice Cos. Return on Assets, 1940-1985

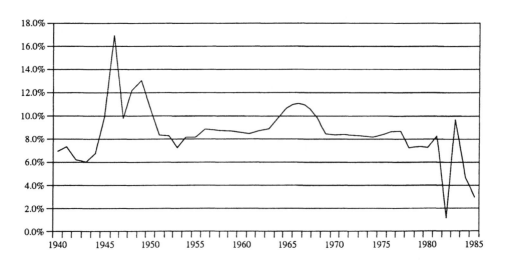

Source: George Baker, "Beatrice: A Study in the Creation and Destruction of Value".

Beatrice Companies—1985
Exhibit 4   Beatrice-Related Transactions Over $80 Million, 1973-1985

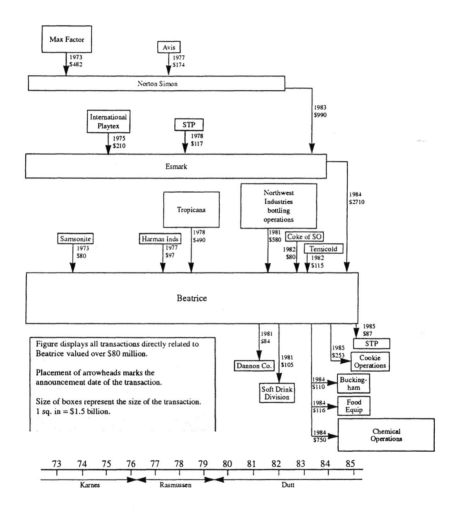

Source: George Baker, "Beatrice: A Study in the Creation and Destruction of Value".

Beatrice Companies—1985

Exhibit 5   Business Segment Reorganization, 1981-1985

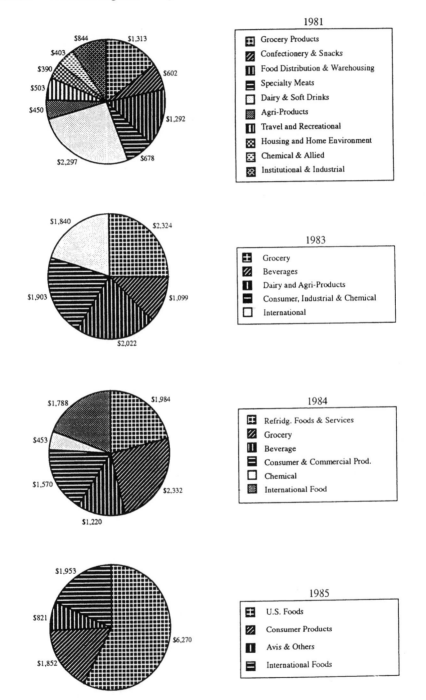

Source: George Baker, "Beatrice: A Study in the Creation and Destruction of Value".

Beatrice Companies—1985

Exhibit 6   Advertising Expenditures

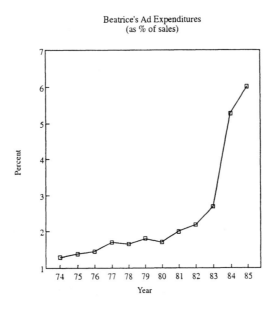

Beatrice's Ad Expenditures
(as % of sales)

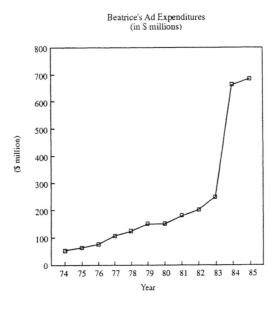

Beatrice's Ad Expenditures
(in S millions)

Source: George Baker, "Beatrice: A Study in the Creation and Destruction of Value".

Beatrice Companies – 1985

Exhibit 7   Profiles of Beatrice's Major Competitors

Dart & Kraft In 1980, Kraft Inc. and Dart Industries merged in a $2.5 billion transaction that created a diversified, consumer-oriented company. In the early 1980s, Kraft made mostly dairy- and oil-based products, and had sales of about $7 billion in 1985, and Dart Industries owned Tupperware, West Bend appliances, and Duracell batteries. Even after the formation of Dart & Kraft, the two companies operated largely unmerged.

After the merger, Kraft focused on its food businesses under leadership of Michael Miles, a former advertising executive and CEO of Kentucky Fried Chicken. In food, Kraft increased its prices while simultaneously raising ad spending to achieve brand leadership. The company also concentrated on volume growth through line extensions and marketing innovations.

General Foods   Postum Cereal Co. was incorporated in 1986 and changed its name to General Foods (GF) in 1929. The company followed a product diversification program that ultimately brought it such brands as Jell-O, Crystal Light, Entenmann's, Oscar Mayer, and Maxwell House. By 1985, GF had revenues of $9 billion and sold food products, primarily dry grocery goods, under about 60 brand names.

In 1985, GF had three major divisions (Oscar Mayer, Worldwide Food, and USA Food) and was reputed to have a swollen corporate bureaucracy that delayed its responsiveness to marketplace changes. While the company had a number of rapidly growing brands, most of these had come from recent acquisitions, and GF relied heavily on sales from mature products such as coffee and frozen vegetables. In mid-1985, Philip Morris acquired General Foods for $5.5 billion.

General Mills Beginning as a network of flour mills in 1928, General Mills moved from its base in consumer foods into a variety of industries in the 1960s and 1970s, thus stressing stability as well as growth. General Mills's "balanced diversification" strategy drove the company into toys and games, clothing, jewelry, specialty chemicals, and restaurants, and by the end of this period, the company had significant positions in 13 different industries.

By the late 1970s, General Mills's stock performance had begun to deteriorate (the company's market to book ratio fell from 3.1 in 1973 to 1.4 in 1979). Responding to its devaluation in the stock market, in the early 1980s, General Mills began to consolidate rather than diversify and to stress internal development above acquisitions, particularly in consumer foods (the company's brands included Cheerios, Total, Yoplait, and Bisquick). By 1985, General Mills had exited seven of the 13 industries that it had formerly held positions in.

Nabisco Brands   Incorporated in 1988, Nabisco had such well-known brands as Oreos, Fig Newtons, and Ritz crackers, although many considered the company unimaginative. In 1981, Nabisco became Nabisco Brands after merging with Standard Brands, a notoriously slow-moving company best known for its Chase & Sanborn coffee and Fleischmann's margarine.

Nabisco Brands shed its stodgy reputation in the early 1980s when the company emerged victorious from "the cookie war". Using its manufacturing ability, expansive distribution network, and advertising clout, Nabisco overwhelmed Procter & Gamble and Frito Lay, both recent entrants into cookie business. In 1985, RJR Reynolds, the tobacco giant, acquired Nabisco Brands for $4.9 billion.

Beatrice Companies – 1985

Exhibit 8 Beatrice Companies Financials (in $ millions)

	1985	1984	1983	1982	1981	1980
Net sales	$12,595	$9,327	$9,139	$9,021	$8,773	$8,291
Net earnings	479	433	43	390	304	290
Total assets	10,379	4,464	4,732	4,744	4,237	3,980
Long-term debt	2,587	779	772	759	691	659
Shareholders' equity	2,357	2,028	2,215	2,422	2,484	2,005

	1985		1984		1983	
Segments	Sales	Operating Earnings	Sales	Operating Earnings	Sales	Operatings Earnings
U.S. food	$6,270	$372	$4,095	$338	$3,770	$304
Consumer products	1,935	201	948	133	837	117
International food	1,852	101	1,732	105	1,758	124
Avis/other operations	821	64	28	--	33	(1)
Divested Businesses	1,699		2,524		2,741	

Beatrice Companies – 1985

Exhibit 9   Beatrice's Major Brands (1985)

Brand	Description
Tropicana	• #1 in fresh orange juice
Swift[a]/Eckrich Meats	• #5 in processed meat • #2 billion in meat sales
La Choy	• 18 frozen and 44 canned items • #1 in Oriental foods
Hunt[a]	• #1 in tomato-based products
Culligan	• Industrial/home water processing equipment
Samsonite Corp.	• Full line of hard and soft luggage • 41% of business overseas
Coca-Cola Bottling Operations	• Coca-Cola bottlers concentrated in California
Shedd Group	• Largest private-label manufacturer of margarine
Beatrice Candy Operations	• Jolly Rancher/D.L. Clarke/ Red Tulip Chocolates
Playtex International[a]	• #1 in bras with 16% market share • #2 in tampons with 32% market share • 6% market share in hosiery
Max Factor[a]	• In top 5 in U.S. market share • #1 in U.K. and Japan
Somerset[a]	• #1 premium distilled spirit importer
Avis[a]	• #2 worldwide market share

[a]Acquired as part of Esmark.

Beatrice Companies – 1985

Exhibit 10   Excerpts From James Dutt's Letter to Shareholders, Fiscal 1985 Annual Report

Fiscal 1985 Highlights:

- Successfully completed the acquisition of Esmark, Inc. for $2.7 billion.

- Received more than $1.4 billion in proceeds from divesting operations that were not critical to our focus on food and consumer products.

- Substantially increased awareness and marketing value of the Beatrice brand name.

- Simplified the organization and management structure of the company, creating four strong operating segments.

Excerpts:

"Many of the steps we've taken were dramatic, but necessary for the company to succeed in today's competitive consumer marketplace. All of our actions, both short-term and long-term, are guided by four principles: to make Beatrice the premier worldwide marketer of food and consumer products; to build strong national and international brand franchises; to gain more direct access to our trade customers and the consumers of our products; and to build and develop bigger and better people throughout the organization."

"We rapidly accelerate our market-driven strategy during fiscal 1985 with the acquisition of Esmark, Inc. The acquisition was critical to our efforts in developing stronger sales and distribution capabilities and improving our research and development efforts. While we could have developed these products and capabilities internally, we realized that it was more efficient to acquire these strengths. The integration of Esmark has proceeded smoothly, and we have already combined similar businesses into single, larger entities that have greater impact in their respective markets."

"With four strong operating segments: U.S. Food, Consumer Products, International Food and Avis/Other Operations, we are now better positioned to respond to and anticipate the needs of the marketplace. At the same time, this structure will allow more efficient management of the company. Each of these segments has the critical mass to truly lead the markets in which it competes."

"The 'New Beatrice' also has vastly increased marketing clout. As the third largest advertiser in the United States, we have the marketing muscle to establish leadership positions across a broad product range."

APPENDIX B: GRAND METROPOLITAN EXHIBITS

Exhibit 1

GRAND METROPOLITAN . . . ADDING VALUE TO FOODS

<u>Food Sector World Ranking</u>

<u>Company</u>	<u>Food Sales</u> ($BN)	<u>Position</u>
Nestle	24.6	1
Philip Morris/Kraft	19.8	2
Unilever	13.9	3
RJR Nabisco	9.4	4
Mars	7.3	5
Sara Lee	7.2	6
Con Agra	6.1	7
Grand Metropolitan/Pillsbury	5.4	8
H.J. Heinz	5.1	9
Borden	4.9	10

Excludes Japanese companies

<u>Drinks Sector World Ranking</u>

<u>Company</u>	Wines and Spirits <u>Annual Case Sales</u> (M)	<u>Position</u>
Grand Metropolitan	90	1
Seagram	78	2
Pernod	56	3
Hiram Walker	52	4
United Distillers	46	5
Martini & Rossi	33	6
Brown-Forman	28	7
Bacardi	26	8
Suntory	24	9
Jim Beam	20	10

<u>Retailing Sector World Ranking</u>

<u>Company</u>	<u>Outlets</u>[a]	<u>Position</u>
Pepsico	16,500	1
Grand Metropolitan	15,200	2
McDonald's	10,000	3

Source: Company records

[a]Includes franchised outlets

Exhibit 2 - GRAND METROPOLITAN ... ADDING VALUE TO FOODS
Financial results 1981-1989 (million pounds)

	6 Months to March 31 1989	Year End September 30							
		1988	1987	1986	1985	1984	1983	1982	1981
Sales	4,037.0	6,028.8	5,705.5	5,291.3	5,589.5	5,075.0	4,468.8	3,848.5	3,221.2
Operating income	357.0	653.6	571.6	487.4	453.2	443.9	407.0	354.8	276.6
EBIT	388.0	668.1	576.3	469.0	417.2	430.4	407.0	354.8	276.6
Interest expense	87.0	93.0	120.2	101.3	105.9	109.6	111.8	134.6	90.0
Net income	209.0	420.3	336.0	275.9	246.9	235.2	205.6	157.9	141.4
EPS (pence)[a]	23.0	46.9	38.1	31.4	28.3	28.3	24.4	19.2	18.7
Total assets	6,161.0	4,189.4	3,731.6	3,124.2	3,088.1	2,739.0	2,549.8	2,289.5	2,206.0
Total debt	4,096.0	751.3	1,358.2	932.8	1,019.4	961.0	943.9	819.5	943.3
Total equity	2,065.0	3,438.1	2,373.4	2,191.4	2,068.7	1,778.0	1,605.9	1,470.0	1,262.7
Employees		89,753	129,436	131,493	137,195	125,074			
Stock price		4.94	5.80	4.16	2.98	2.60			
Pound exchange rate	1.67	1.78	1.64	1.47	1.30	1.34			
Ratios									
Operating margin (%)	8.8	10.8	10.0	9.2	8.1	8.7	9.1	9.2	8.6
Net margin (%)	5.2	7.0	5.9	5.2	4.4	4.6	4.6	4.1	4.4
Debt/total assets (%)	66.5	17.9	36.4	29.9	33.0	35.1	37.0	35.8	42.8
EPS growth (%)	NA	23	21	11	0	16	27	3	
TIE	4.5	7.2	4.8	4.6	3.9	3.9	3.6	2.6	3.1
ROA (%)	3.4	10.0	9.0	8.8	8.0	8.6	8.1	6.9	6.4
ROE (%)	10.1	12.2	14.2	12.6	11.9	13.2	12.8	10.7	11.2
Asset turnover	1.31	1.44	1.53	1.69	1.81	1.85	1.75	1.68	1.46
Sales/employee		67,171	44,080	40,240	40,741	40,576			
Price/earnings									
Sales by Division									
Food	1,240.0	1,252.6	1,046.9	750.5	778.1	777.8			
Drinks	1,351.0	2,581.2	2,177.7	1,721.6	1,698.4	1,551.2			
Retailing	1,331.0	1,670.9	1,467.4	1,211.8	1,234.2	1,174.6			
Hotels		337.6	332.6	337.9	376.7	336.2			
Other	115.0	186.5	680.9	1,269.5	1,502.1	1,235.2			
	4,037.0	6,028.8	5,705.5	5,291.3	5,589.5	5,075.0			
Operating Income by Division									
Food	106.0	84.0	69.2	39.0	27.6	16.4			
Drinks	159.0	315.8	256.9	244.2	233.5	205.9			
Retailing	92.0	178.8	159.8	82.7	70.2	67.4			
Hotels		53.6	37.9	30.4	37.6	31.9			
Other	12.0	21.4	47.8	91.1	84.3	122.3			
	369.0	653.6	571.6	487.4	453.2	443.9			

[a] EPS adjusted for rights issues.

Exhibit 2 (continued)

	6 Months to March 31 1989	Year End September 30							
		1988	1987	1986	1985	1984	1983	1982	1981
Capital Employed									
Food		310.2	260.2	166.4	159.7	189.2			
Drinks		1,478.5	1,503.5	1,663.1	1,731.0	1,117.9			
Retailing		1,898.2	1,290.3	100.1	142.5	534.1			
Hotels		502.4	582.1	568.9	556.8	505			
Other		0	95.5	473.0	526.2	548.9			
		4,189.4	3,731.6	2,971.5	3,116.2	2,895.1			
Sales by Geography									
United Kingdom		3,835.5	3,558.6	3,281.4	3,195.0	3,016.0			
Continental Europe		220.7	214.2	236.1	267.2	249.6			
United States		1,758.3	1,720.3	1,615.8	1,892.8	1,557.5			
Rest of North America		54.1	57.5						
Africa and Middle East		127.0	120.5	125.5	165.0	186.8			
Rest of World		33.2	34.4	32.5	69.5	65.1			
		6,028.8	5,705.5	5,291.3	5,589.5	5,075.0			
Operating Income by Geography									
United Kingdom		363.9	330.9	301.0	264.2	235.0			
Continental Europe		46.2	36.1	27.3	30.8	31.4			
United States		217.6	185.4	149.4	136.1	164.0			
Rest of North America		13.8	12.8						
Africa and Middle East		6.5	5.9	7.5	9.4	4.9			
Rest of World		5.6	0.5	2.2	12.6	8.6			
		653.6	571.6	487.4	453.2	443.9			
Capital Employed by Geography									
United Kingdom		2,699.6	1,944.6	1,793.5	1,815.1	1,571.1			
Continental Europe		384.1	385.8	290.3	375.6	308.1			
United States		1,033.6	1,316.1	829.1	861.9	951.0			
Rest of North America		50.0	60.8						
Africa and Middle East		17.9	18.5	43.1	26.9	27.8			
Rest of World		4.2	5.8	15.5	36.7	37.1			
		4,189.4	3,731.6	2,971.5	3,116.2	2,895.1			

Source: Annual Report

Exhibit 2 (continued)

Combined Profit and Loss--Forecast Year to September
(£M)

	1989	1990	1991
Turnover			
Grand Metropolitan	6,840	7,690	8,450
Pillsbury	2,380	3,520	3,850
Total	9,220	11,210	12,300
Trading Profits			
Core Grand Metropolitan			
Drinks	366	413	460
Food	120	142	158
Retail	196	235	270
Discount	18	0	0
Grand Met	**700**	**790**	**888**
Pillsbury & Burger King			
Food	153	250	290
Retail	90	150	200
Pillsbury	**243**	**400**	**490**
Total	943	1,190	1,378
Interest	**255**	**300**	**278**
Exceptionals	**8**	**10**	**0**
Pretax profit	**680**	**880**	**1,100**
Property	30	30	30
Reported PBT	**710**	**910**	**1,130**
Tax	**204**	**264**	**365**
Minorities/Prefs.	**10**	**10**	**10**
Earnings	**496**	**636**	**755**
Ex. props.	**466**	**606**	**725**
Average shares[a]	930	990	1,010
EPS	54.7	64.2	75.0
Ex. props.	**51.4**	**61.0**	**72.0**
DPS	18.0	21.0	25.0

Source: James Capel & Co Investment Report

[a]The average number of shares figures fully dilutes for the rights issue on a pro-rata basis with payments. EPS also reflect the nominal £12M saving on interest coupons on the convertible loan stock.

Exhibit 2 (continued)

Pro Forma Balance Sheets
(At September 30--£M)

	1989F	1990F	1991
Fixed assets	3,600	3,775	4,00
Investments	145	145	145
Brands	2,350	2,350	2,35
Working capital	30	20	0
Net debt	3,190	3,045	2,71
Net assets	2,935	3,245	3,78
Shareholders' funds	2,900	3,210	3,75
Minorities	35	35	35
NAV/share	293p	323p	375p

Assumptions

- Consistent exchange rates over the next two to three years.
- No further acquisitions with the exception of add-on purchases of the Christian Brothers type.
- Capital expenditure gradually increasing from 1988 levels.
- Working capital will be reduced over the next two years.
- Provisions utilized steadily over the next 2-3 years.
- No factoring of a revaluation of tangible assets over the forecast period, but Grand Met should be due to revalue its property portfolio in 1990 or 1991.

Source: James Capel & Co. Investment Report

Exhibit 3

GRAND METROPOLITAN . . . ADDING VALUE TO FOODS

Grand Metropolitan Profile of Key Executives

Allen J.G. Sheppard, Chairman and Group Chief Executive, Grand Metropolitan PLC. (Born 1932): Joined Grand Metropolitan in 1975 as chief executive of Watney Mann & Truman Brewers and has served as chief executive of the Brewing and Retailing Division and the Food Division. Sheppard spent 18 years with Ford, Chrysler, and British Leyland, both in finance and marketing. He is a qualified accountant and graduated from the London School of Economics in business administration. He became Grant Met's CEO in November 1986 and Chairman/CEO in July 1987.

George J. Bull, Chairman and Chief Executive, International Distillers and Vintners, Chief Executive, Drinks Sector. (Born 1936): Joined IDV in 1961, has extensive experience in marketing and exports and was closely involved in the establishment of IDV's successful strategy and in the Heublein acquisition. Bull also serves as chairman of the Far East Advisory Committee.

Ian A. Martin, Chairman, Chief Executive Officer, and President, The Pillsbury Company, chairman and Chief Executive, Grand Metropolitan Food Sector. (Born 1935): Joined Grand Metropolitan in 1979 in the company's Watney Mann & Truman Brewers and rose to head the Brewing and Retailing Divisions. In 1985, Martin became a director of Grand Metropolitan PLC and was appointed chairman of the U.S. Consumer Products Division in 1986. Prior to Grand Metropolitan, Martin worked with Timex, Mine Safety Appliances Ltd., and ITT Europe. Martin is a chartered accountant and received a Master of Arts degree from St. Andrews University, Scotland, where he was a prize winner in Economics and a class medalist in psychology.

David Nash, Finance Director Elect. Joined Grand Metropolitan in December 1989. Previously, Nash was Finance Director of Cadbury Schweppes and before that held senior finance and planning roles during his 20 years with ICI. Nash is a chartered accountant.

Clive Strowger, Group Finance Director and previously Chief Executive, Retailing and Property. (Born 1941): Joined Grand Metropolitan in 1978 after serving as a chartered accountant with Ernst & Whinney, Ford, and British Leyland. With Grand Metropolitan, Strowger has served as commercial director and joint managing director of Watney Mann Truman Brewers, head of the Consumer Services Division, chairman of the Express Foods Group, and is a member of the board of directors. Strowger is chairman of the European Advisory Group.

David Tagg, Chief executive, Retailing and Property and previously Chief Executive of Group Services. (Born 1940): Joined Grand Metropolitan as personnel director of the Brewing Division in 1980. Tagg has had responsibility for establishing a worldwide management development program for the Grand Metropolitan Group, the Group Community Services program, the Group's Personnel Legal and Company Secretarial functions, and the Retail Betting business. Prior to joining Grand Metropolitan, Tagg worked for Massey Ferguson and Cadbury Schweppes. He obtained a Master of Arts degree in classical languages, ancient history, and philosophy from Oxford and a diploma in public and social administration.

Exhibit 4

GRAND METROPOLITAN . . . ADDING VALUE TO FOODS

Sheppard's Management Maxims

- With operating costs you never accept the word **impossible.**
  People here are continually probing the perimeter of what's possible.

- If you're fired, it's not because you haven't protected your left buttock, but because you're not entrepreneurial.

- I wouldn't recommend working at Grand Met for a health cure.

- The worst problem for any business is complacency and smugness.

- We welcome professional competitors.

- Business is not complicated, though people can make it so. We decomplicate everything we do.

- We delegate the capacity to succeed.

- A lot of management skills have to do with common sense.

- It's all about earning growth as that's how we're measured at the end of the day.

- You don't have to be in a go-go business.

- Introducing new products you have to accept a fairly high failure rate. If you expect that by planting 10 bulbs you will end up with 10 blooms, you're in for a nasty surprise.

Exhibit 5

GRAND METROPOLITAN . . . ADDING VALUE TO FOODS

The Organizational Philosophy

- Decentralized authority

- Open communications

- Challenge culture

- Small corporate center adding value

- Avoidance of bureaucracy

- Strong performance orientation

- No compromise on high management standards

- Developing people

- Tolerance of different personalities and styles

- Action orientation

- Major role in trade and industry affairs

Exhibit 6 - GRAND METROPOLITAN . . . ADDING VALUE TO FOODS

Major Acquisitions/Disposals Since 1980

Acquisitions

1980	The Paddington Corporation; Carillon Importers; ALPO Pet Foods; Atlantic Soft Drink Company (sold 1988); Pepsi Cola San Joaquin Bottling Company (sold 1988); Liggett Group Inc. (sold 1986); L&M do Brasil (sold 1985); The Pinkerton Tobacco Company (sold 1985); Diversified Products Corporation (sold 1987).	$450m(net)/£190M (net)[a]
1981	Inter-Continental Hotels Corp (sold 1988)	$500M/£265M[a]
1982	Rumple Minza--liqueur	£4M
1983	Children's World (U.S.)--early childhood education services (sold 1987)	$35M/£25M[a]
1985	Quality Care (U.S.)--home health care (sold 1987)	$125M/£110M[a]
	25% interest in Cinzano	Not disclosed
	Minority interest in Ritz and Casanova Casinos	£16M
	Pearle Optical (U.S.)--eyecare	$385M/£280M[a]
	Sambuca Romana--liqueur	Not disclosed
	G. Ruddle & Company--brewing	£15M
	Hamard Catering--contract catering (sold with Compass 1987)	£5M
	Strathleven Bonded Warehouse	£10M
1987	Heublein Inc.--wines and spirits (Smirnoff)	$1.2 billion/£800M
	Almaden Vineyards Inc.	$128M/£85M
	S. Reece	£2M
	Dairy Produce Packers Ltd.--dairy manufacturing	£20M
	Sacconne & Speed and Roberts & Cooper--wines and spirits and off licences	£50M
	Martell Cognac (10% interest)	£30M
	MacCormac Products, Connacht Foods--specialty milk powder manufacturing	IR£16M/£15M[a]
	Jim Dandy Company--dry pet food (U.S.)	$25M/£15M[a]
	Fleur de Lys--frozen gateaux and pastries	£12M
	The Hervin Company--Blue Mountain pet food (U.S.)	$10M/£7M[a]
1988	Vision Express (U.S.)--optical superstores	$40M/£21M[a]
	Healthworks--health food snacks	Not disclosed
	Eye + Tech (U.S.)--optical superstores	$32M/£17M[a]
	J. Thayer & Sons--premium ice cream	Not disclosed
	Kaysens--frozen gateaux	£21.5M

Exhibit 6 (continued)

Acquisitions (continued)

1988 (cont.)	Peter's Savoury Products--meat & pastry products	£75 million
	The William Hill Organization--retail betting	£331 million
	Wienerwald/Spaghetti Factory--German & Swiss restaurants	£20 million
1989	The Pillsbury Company (including Burger King)	$5.8 billion/£3.3 billion[a]
	Sileno--Portuguese wine & spirit distribution	Not disclosed
	Christian Brothers	$150 million/£94 million
	Metaxa-Greek wine & spirits distribution	Not disclosed
	UB Restaurants	$288 million/£180 million

Disposals

1984	CC Soft Drinks	£30 million
1985	Express (Northern milk activity)	£50 million
	Pinkerton--chewing tobacco	$140M/£100 million[a]
	L&M do Brasil--tobacco leaf	$30M/£20 million[a]
	Mecca Leisure	£95 million
1986	Stern Brauerei	£15 million
	Brouwerij Maes	£30 million
	Watney Mann and Truman Maltings	£5 million
	Liggett Group Inc.--cigarettes	$140M/£100 million[a]
	Drybrough & Co., Ltd.	£50 million
1987	Express Foods USA	$7M/£5 million[a]
	North Sea oil interests	$4M/£3 million[a]
	Compass Services, Compass Vending, GIS, GM Health Care	£160 million
	Quality Care--in-home health care	$100M/£65 million[a]
	Diversified Products--fitness products	$35M/£25 million[a]
	Children's World--child care	$115M/£70 million[a]
	MacGuinness Distillers--Canadian spirits	$45M/£20 million[a]
1988	Meurice Hotel Paris	£35 million
	Atlantic Soft Drinks/Pepsi Cola San Joaquin (U.S.)--soft drink bottling	$705M/£400 million[a]
	Inter-Continental Hotels	$2 billion (net)/£1.2 billion (net)[a]
1989	William Hill/Mecca Bookmakers (including overseas operations, etc.)	£775 million
	Steak & Ale, Bennigans Restaurants	$434 million
	Pillsbury South America operations	$63 million
	Pillsbury Grain Merchandising	$140 million
	Pillsbury Seafood Operations	$409 million

Source: Company report

[a]Notional sterling equivalent at date of transaction

Exhibit 7

GRAND METROPOLITAN . . . ADDING VALUE TO FOODS

Grand Met Business Sectors

Division	Number Companies	Number of Employees	Outlets	Products	Major Brands
Food					
Pillsbury	1	62,800		Dough, pizza, vegetables, ice cream	Pillsbury, Green Giant, Häagen Dazs, Totinos
Other	12	12,820		Yogurt, cheese, milk powder, pastry and savoury products, pet food	Eden Vale, Ski, Express, Fleur de Lys, Kaysen's, Thayer's Alpo
Drinks					
IDV and Grand Met Brewing	29	29,000		Wines and spirits, beer, nonalcoholic drinks	J&B Scotch Whisky, Bailey's Irish Cream, Smirnoff, Aqua Libra, Grand Marnier, Piat d'Or, Watneys, Ruddles
Retailing (excluding betting)					
Burger King	1	39,000	5,790	Retail food	"The Whopper"
Pearle Inc.	1	4,750	1,250	Eyecare retailer	Pearle Vision/Pearle Express
European Retailing and Property	3	37,750	9,280	Restaurants, alcohol retailing, property management	Berni, Chef & Brewer, Peter Dominic, Pastificio, Wenerwald

Exhibit 8

GRAND METROPOLITAN . . . ADDING VALUE TO FOODS

	United Kingdom and Ireland	United States	Continental Europe	Japan	Other North America	Other
Food						
Dairy	X					
Yogurt	X					
Milk	X					
Cheese			X			X
Butter						
Bakery						
Bread	X	X	X			
Cereal						
Pasta						
Biscuits						
Snack Foods						
Chocolate	X	X				
Sugar Confections	X	X				
Ice Cream			X	X		
Snacks						
Miscellaneous						
Frozen Foods		X				
Canned Foods		X				
Baby Foods						
Pet Foods	X	X	X	X		
Soups			X			
Drinks						
Carbonated	X	X				
Mineral Water						
Fruit Juices						
Beer	X	X	X	X	X	X
Wine	X	X	X	X	X	X
Spirits	X	X				
Retailing						
Restaurants	X	X	X	X	X	X
Optical		X	X	X	X	X
Licensed Retailing	X					

Exhibit 9

GRAND METROPOLITAN . . . ADDING VALUE TO FOODS

Pillsbury: Leading Branded Food Products

Main Brand	Category	Company Brand % Share	Company Brand Position		Estimated Share of Pillsbury's Consumer Food Sales %
Pillsbury	Chilled dough	80	1		21
Totino's	Frozen pizza	41	1		16
Green Giant	Canned vegetables	22	1		14
Häagen-Dazs	Luxury ice cream	NA	1		8
Green Giant	Frozen vegetables	21	1		12
Pillsbury	Baking mixes	21	2		14
Proportion of total consumer sales					85%
Alpo	United States	Premium canned dog food		Alpo	1
	United States	Dry dog food		Alpo	2
Express	United Kingdom	Yogurt		Ski	1
	United Kingdom	Yogurt		Munch Bunch	3
	United Kingdom	Drinking yogurt		Ski Cool	1

Source: County NatWest Wood Mac, research report, November 18, 1988.

Exhibit 10

GRAND METROPOLITAN ... ADDING VALUE TO FOODS

Pillsbury--Recent Performance ($ million)

Food Division: Sales and Profit--1983-1988

(Year to May)	1983	1984	1985	1986	1987	1988
Net sales	2,192	2,566	2,790	3,185	3,363	3,560
Operating profit	155.8	181.3	196.6	206.9	246.1	214.8
Unusual items				12.6	23.8	65.7
Operating profits ex-unusuals	155.8	181.3	196.6	219.5	269.9	280.5
Operating margin	7.1	7.1	7.1	6.5	7.3	6.0
Ex-unusuals	7.1	7.1	7.1	6.9	8.0	7.9
Identifiable assets	1,212	1,335	1,346	1,660	1,822	1,884
Return of assets	12.9	13.6	14.6	12.5	13.5	11.4
RonA ex-unusuals	12.9	13.6	14.6	13.2	14.8	14.9
Cash flow (including retailing)		313	361	437	475	434

Retail Division: Sales and Profit Record--1983-1988

(Year to May)	1983	1984	1985	1986	1987	1988
Net sales	1,495	1,769	2,053	2,663	2,764	2,630
Operating profit	135.3	184.7	220.0	304.7	219.2	72.0
Unusual items				8.8	17.0	144.2
Operating ex-unusuals	135.3	184.7	220.0	315.5	236.2	217.2
Operating margin (%)	9.1	10.4	10.7	11.4	7.9	2.8
Ex-unusuals (%)	9.1	10.4	10.7	11.8	8.5	8.3
Identifiable assets	1,026	1,191	1,316	1,841.1	1,875.3	1,819.9
Return on assets (%)	13.2	15.8	16.7	16.5	11.7	4.0
Return ex-unusuals (%)	13.2	15.5	16.7	17.0	12.6	11.9
Stock price ($)	18.19	22.25	30.69	33.88	35.13	65.88
-Close, May 31	19.37	19.00	27.37	40.12	38.75	

Source: County NatWest Wood Mac, research report, November 18, 1988.

Exhibit 11

GRAND METROPOLITAN . . . ADDING VALUE TO FOODS

	Action Steps:	Cost Savings/Income
1.	Cut 550 people from headquarters	$ 30 million annually
2.	Sold five production plants to Cargill	$ 13 million annually
	(Goal to reduce operating cost by $80 million by September 1990)	
3.	Sold Steak & Ale/Bennigan's	$434 million
4.	Divested Latin American Food investments	$ 63 million
5.	Sold grain merchandising division to ConAgra	$140 million
6.	Cut 560 people. Reduced management layers from 13 to 10	$ 30 million annually
7.	Sold Van de Kamps/Bumble Bee	$409 million

Results

1. Prepared dough volumes up 2%.

2. Ice cream sales up 12%.

3. International sales up 7%.

4. Overhead cost reductions of $60 million-- twice forecast.

5. Burger King headcount reduced by 1,500 for 7% cost reduction.

6. Burger King profit per nonstore employee 25% better.

Exhibit 12

GRAND METROPOLITAN ... ADDING VALUE TO FOODS

Pillsbury Foods Annual Sales
(FY 1990 estimated--$ millions)

Prepared dough	$ 510
Green Giant vegetables	715
Pizza	295
Desserts/sweet snacks	440
Europe	500
Total	$2,460

Express Foods Annual Sales

Expres Dairy (liquid milk)	$ 400
Express Foods (cheese and other dairy)	640
Eden Vale	310
Ireland	180
Peter's Savoury Products	110
Other	340
Total	$1,980

£ at $1.55

All Others Annual Sales

Ice cream	$ 260
Industrial	600
Pillsbury International	340
Alpo	560
Burger King	2,040

Source: Company records

Exhibit 13

GRAN METROPOLITAN . . . ADDING VALUE TO FOODS

Pre-Takeover

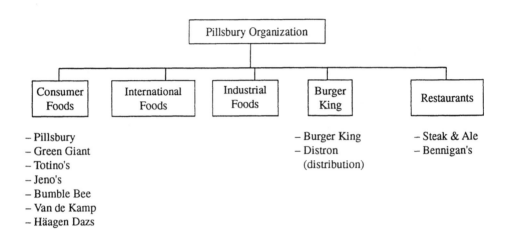

- Pillsbury
- Green Giant
- Totino's
- Jeno's
- Bumble Bee
- Van de Kamp
- Häagen Dazs

- Burger King
- Distron
  (distribution)

- Steak & Ale
- Bennigan's

Post-Takeover
(June 1, 1989)

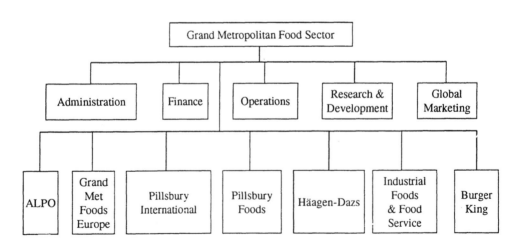

Source: Company records

Exhibit 14

GRAN METROPOLITAN . . . ADDING VALUE TO FOODS

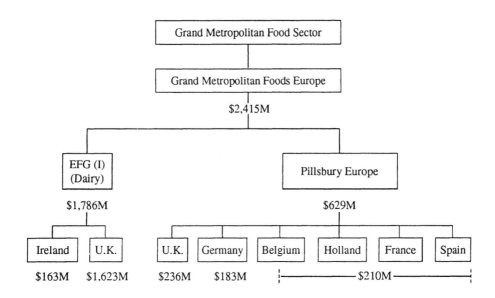

Source: Company records

Exhibit 15

GRAN METROPOLITAN . . . ADDING VALUE TO FOODS

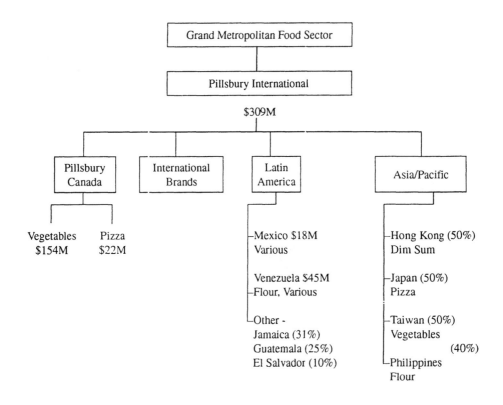

Source: Company records

Exhibit 16

GRAND METROPOLITAN . . . ADDING VALUE TO FOODS

Häagen-Dazs Ice Cream Ranking and Share

Adult Segment		All Ice Cream	
Häagen-Dazs	4.9%	Breyer's	13.4
Ben & Jerry's	1.7	Dreyer's/Edy's	6.1
Frusen Gladje	1.4	Blue Bell	5.2
Steve's	.8	Häagen-Dazs	4.9
		Sealtest	3.7
		Ben & Jerry's	1.7

Häagen-Dazs Overall Novelty Ranking and Share

Adult Segment		Total Novelties	
Häagen-Dazs	3.8%	Popsicle	7.8%
Rondos	2.4	Weight Watchers	7.1
Dove	2.4	Jello	6.2
Nestlé Premium	.5	Klondike	4.9
Ben & Jerry's	.2	Betty Crocker	4.4
Forever Yours	.1	Häagen-Dazs	3.8
		Dole	3.6
		Dove	2.5
		Rondos	2.4

Source: Company records

Exhibit 17

GRAND METROPOLITAN . . . ADDING VALUE TO FOODS

February 2, 1989

TO: All Burger King Franchises

FROM: Barry J. Gibbons

Dear Colleague:

As you may have noticed, I have now arrived in Miami and I have completed about ten working days. Let me tell you what we've done so far; otherwise, I'll fall into the trap of "communicating" through the press.

I'm meeting as many people, and seeing as many things, as quickly as I can--but you will appreciate the sheer scale of the business means it will be some time before I can meet even all the key players.

For those who haven't met me, I have two eyes, two ears, and one mouth--and, for a period, I'm using them in that proportion. I'm looking and listening--and not doing too much talking because there is much for me to see and understand before I start signaling changes.

I've exchanged a lot of letters and telephone calls with individual franchisees--and have been very gratified by many good wishes and much support. I've met some individual franchisees and some groups--and plan to meet more, again, on an informal basis to start with.

In the first ten days, I've made some decisions about restructuring the senior management responsibilities of Burger King, and I've identified six or seven short-term priority areas that we need to concentrate on (that doesn't mean we do nothing else; it means just what it says: they are our priorities). I've put task teams in place on all of them and told them to press the "fast forward" button. Attached is an internal announcement which explains the last bit in more detail.

I'll keep you in touch with progress, and look forward to meeting with you all as soon as possible.

Good trading!

Regards,

Barry J. Gibbons
Chief Executive Officer

BJG:SB
M:BJG47

Index